Songbirds of Turkey

Cover: Crimson-winged Finch *Rhodopechys sanguinea*.
Original illustration by F.J. Maas
Maps by Peter de Vries, Amsterdam, based on originals by C.S. Roselaar
Printed in the Netherlands

CIP - KONINKLIJKE BIBLIOTHEEK, DEN HAAG

Roselaar, C.S.

Taxonomy, morphology, and distribution of the Songbirds
of Turkey : an atlas of biodiversity of Turkish passerine
birds / C.S. Roselaar - Haarlem : GMB
Met lit. opg.
ISBN 90-74345-07-7
Trefw.: vogels ; Turkije

Published in the Netherlands by:
GMB uitgeverij
Vrijheidsweg 86
2033 CE Haarlem
ISBN 90-74345-07-7

Published in the U.K. by:
Pica Press
The Banks, Mountfield
Robertsbridge
East Sussex TN 32 5JY
U.K.
ISBN 1-873403-44-5

© 1995 C.S. Roselaar, Zoölogisch Museum, Universiteit van Amsterdam.
All rights reserved. No part of this publication may be reproduced in any form or by any means - graphic, electronic or mechanical, including photocopying, recording, taping or information storage and retrieval systems - without the prior permission in writing of the publishers.

CONTENTS

I	Introduction		5
II	Outline of this book		6
III	The distribution maps		9
IV	Geographical variation		6
V	New subspecies of birds from Turkey and the surrounding countries		22
VI	Endemic subspecies of Turkey		26
VII	The non-passerine birds of Turkey		27
VIII	Acknowledgements		28
IX	Abbreviations used in the text and symbols used on the maps		28
X	Species accounts and maps of Turkish passerine birds		30
		map	
	Alaudidae Larks *Tarlakusugiller*	1-9	30
	Hirundinidae Swallows and Martins *Kirlangiçgiller*	10-14	45
	Motacillidae Pipits and Wagtails *Kuyruksallayangiller*	15-21	49
	Pycnonotidae Bulbuls *Gribülbülgiller*	22	57
	Cinclidae Dippers *Su Karatavuklari*	23	58
	Troglodytidae Wrens *Çitkuslari*	24	60
	Prunellidae Accentors *Bosbogazgiller*	25-27	62
	Turdidae Thrushes, *Ardiç Kusugiller*	28-48	65
	Sylviidae Warblers *Ötlegengiller*	49-79	95
	Muscicapidae Flycatchers *Sinekkapangiller*	80-82	129
	Timaliidae Babblers *Biyikli Bastankaralar*	83	133
	Aegithalidae Long-tailed Tits *Uzunkuyuruk Bastankaralar*	84	135
	Paridae Tits *Bastankaragiller*	85-89	136
	Sittidae Nuthatches *Sivacikusugiller*	90-93	144
	Tichodromadidae Wallcreepers *Duvar Tirmasiklari*	94	150
	Certhiidae Treecreepers *Agaç Tirmasiklari*	95-96	151
	Remizidae Penduline Tits *Çulhakuslari*	97	153
	Oriolidae Orioles *Sariasmagiller*	98	155
	Laniidae Shrikes *Çekirgekuslari*	99-102	156
	Corvidae Crows and Jays *Kargasgiller*	103-110	160
	Sturnidae Starlings *Sigircikgiller*	111-112	171
	Passeridae Sparrows and Snowfinches *Serçegiller*	113-120	174
	Fringillidae Finches *Ispinozgiller*	121-135	184
	Emberizidae Buntings *Kirazkusugiller*	136-145	204
XI	List of non-passerine birds breeding in Turkey		216
XII	Bibliography		226
XIII	Index of Turkish names		235
XIV	Index of English and Scientific names		237

I INTRODUCTION

This book arose out of disappointment. While working on a study of geographical variation in Palearctic passerine birds (Roselaar in prep.), as a follow up to my earlier work on the taxonomy of Western Palearctic birds (Roselaar, in Cramp 1988 and 1992, and in Cramp & Perrins 1993 and 1994), I was in need of detailed maps of breeding areas to check whether certain peculiarities in the distribution of bird species and their subspecies were related to the occurrence of natural barriers in the breeding ranges of these forms, such as mountains, deserts, or other features unsuitable for breeding. For Turkey, the detailed maps which I required were not available. Turkey is a key area for the study of geographical variation in Western Palearctic birds. Species and subspecies occurring in European deciduous forests, Mediterranean scrub and wetlands, Arabian semi-desert, Caucasian mountains, and central Asian steppe all meet each other here. It is therefore important to have precise information on the distribution of the birds of Turkey. Although small maps of all Turkish passerines have been published, like those in Heinzel et al. (1972), Hollom et al. (1988), Kiziroglu (1989), and Jonsson (1992), these are not detailed enough, while others, such as those as appearing in Kumerloeve (1961a), Stresemann & Portenko (1960 and later) and Kasparek (1992) do not cover all passerine species. Though much information concerning the Turkish avifauna is available, especially through the efforts of Dr. H. Kumerloeve, hardly any of the data have been summarized, exept perhaps in Kasparek (1992), which gives brief details on the status of all the birds of Turkey. The aim of this book is to present all the information available on taxonomy, morphology, and distribution of Turkish songbirds, and to compare it with similar data from other countries in the Middle East and in south-east Europe. Characteristics and measurements of all subspecies of passerine birds of the region are given, and for this purpose a great deal of research both in the field and in museums has taken place. Much of the data presented here was collected in the course of the twenty years' research programme for Birds of the Western Palearctic, but lack of space in this handbook prevented its publication there.

Through the publication of this book the author hopes that people will become more aware that a bird seen in Turkey or elsewhere is not just a member of a widespread species, but may form part of a discrete population with a restricted range, characterised by its own morphological features. It could form a separate unit distinguishable from other populations which, because of human interference, could become irreversibly lost. The maintenance of the world's biodiversity is not just a matter of preserving species on a continental scale, but also conserving subspecies and populations at a local level.

II OUTLINE OF THIS BOOK

The sequence of species in this book is that published by Voous (1977). In a few cases, departures were made from Voous' sequence or from the species limits as recognized by Voous and the reasons for these are explained in the text of the species involved. English names follow British Birds (1993) or Beaman (1994); the Turkish names are those used by Kiziroglu (1989). When Kiziroglu lists several alternative names for a species, the one given on the plate of his book is generally followed. Unfortunately, not all Turkish characters could be properly reproduced, and hence some Turkish names and localities may look a bit strange for Turkish readers. All localities used in the texts can be found in The Times Atlas of the World (1968); those which are not in the Atlas are explained thus: 'Yesilce: north-west of Gaziantep'.

The distribution of a species is largely dependent on the occurrence of the habitat which the species favours. Therefore, each species account starts with details of the habitat in which it has been recorded in Turkey, including its altitudinal range. For a map of Turkey's main habitats, see fig. 1. Not all authors give details on habitat or altitude, and for a few species no data were available other than a general outline of habitat thoughout the entire species' range as given by Hartert (1903-10) or Vaurie (1959). On the other hand, numerous data on habitat are known for quite a number of Turkish passerines, and these data were summarized for convenience, perhaps resulting in some loss of detail may occur. For instance, combining altitudinal ranges for large areas may imply an extensive range, much larger than the ranges actually found at a local level. A statement of, for instance, 'at 1000-2000m in the Taurus mountains' has to be read in conjuntion with the habitat involved, for instance, pine forest, which may actually mean that a species in one part of the Taurus inhabits pine forest at 1000-1300m, while elsewhere in the Taurus, where similar pine forest is found only at 1800-2000m, the species accordingly lives at a much higher altitude.

The habitat section is followed by a section on distribution. As the details of distribution of each species can be found on the accompanying map (or can be ascertained by combining data of

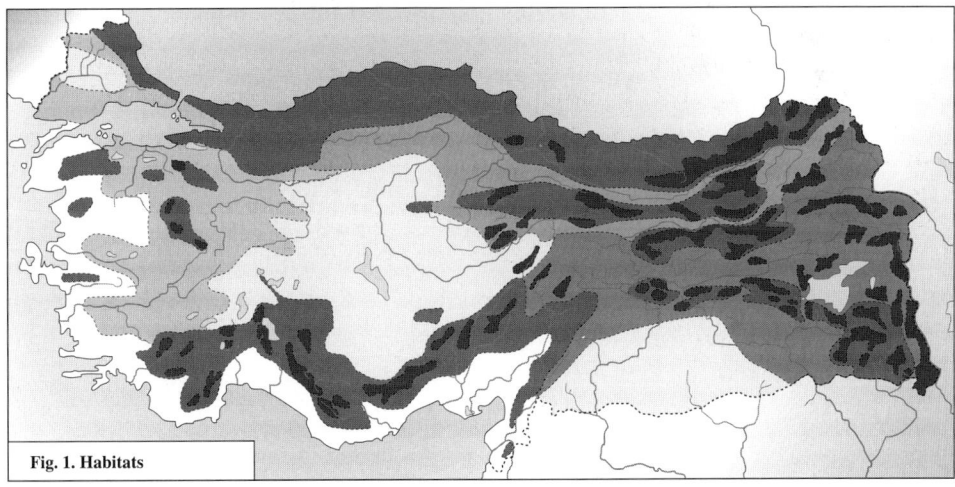

Fig. 1. Habitats

- Mediterranean vegetation
- steppe and semi-desert
- highly degraded forest
- transition from forest to steppe or lightly wooded country
- highland steppe and grassland
- remaining natural forest
- alpine vegetation

the species' preferred habitat with the habitat distribution shown in fig. 1), the texts only add information on general status in various regions and give details of certain records which are apparently out-of-range. In contrast to the atlas of Stresemann & Portenko (1960 and later), no attempt has been made to give the name of the author responsible for each dot, as this would lead to endless lists of references for each species. The subdivision of Turkey referred to in the distribution paragraph for each species is the one used by the Ornithological Society of the Middle East (OSME), repeated here as fig. 2; this subdivision has been further refined by Erol (1982-83), but the detail given in this valuable study is beyond the scope of this book. Frequently, I have avoided from mentioning these regions, as for a number of species it is more convenient to give self-explanatory terms such as 'northern half of Turkey' or 'eastern half of the country'. Sometimes, use has been made of the terms 'west', 'central', and 'eastern Turkey'; here meaning roughly the area west of 32°E, between 32°E and 38°E, and east of 38°E, respectively. 'Western Asia Minor' means Turkey west of *c.* 32°E, excluding Thrace (the European part of Turkey). Details on the compilation of the maps can be found on pp. 9-16.

The accounts of geographical variation in the next section are virtually entirely based on specimens personally examined in zoological collections. Both the characteristics given for subspecies occurring in the region and the variation described in Turkey are personal conclusions based

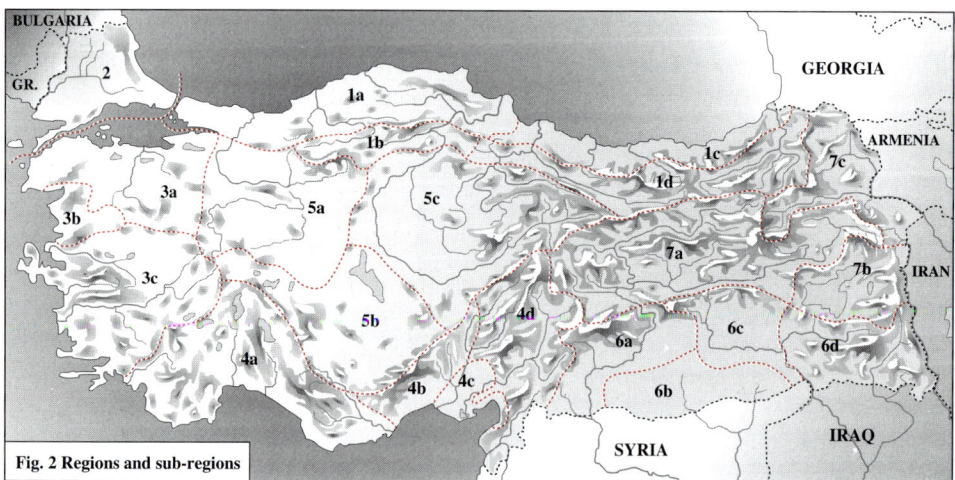

Figure 2. Regions and sub-regions of Turkey, as used by the Ornithological Society of the Middle East.

(1) - Black Sea Coastlands - a. western coastal; b. western inland; c. eastern coastal; d. eastern inland.
(2) - Thrace (European Turkey).
(3) - Western Anatolia - a. Marmara sub-region; b. Bergama sub-region; c - Izmir sub-region.
(4) - Southern Coastlands - a. Western Taurus; b. Main Taurus; c. Seyhan lowlands (or Cukurova deltas); d. Anti-Taurus (southern end: Hatay area, comprising Amanus mountains in west and Amik Gölü basin in east).
(5) - Central Plateau - a. Sakarya basin; b. Enclosed Basins; c. Kizilirmak basin.
(6) - South-East - a. Middle Firat (Euphrates) basin; b. Mesopotamian rise; c. Upper Dicle (Tigris basin; d. Kurdish Alps (eastern part: Hakkari area).
(7) - East - a. Upper Firat (Euphrates) basin; b. Van Gölü basin; c. Upper Aras basin.
Of the surrounding countries, Greece is abbreviated to 'GR.'.

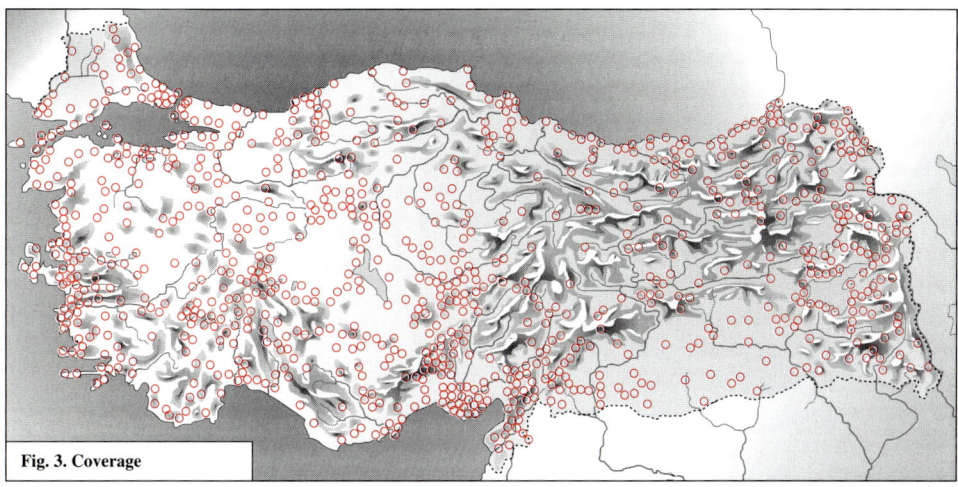

Figure 3. Position of localities in the ornithological literature and used for compiling maps of song-bird distribution. Only sites (towns, villages, lakes, etc.) for which coordinates where obtained are given, not those like 'east of...' or 'between ... and ... '.

on study skins. All Western Palearctic specimens available in the collections of the Zoological Museum of the University of Amsterdam (ZMA), the Netherlands, were measured, as were those in the Nationaal Natuurhistorisch Museum (RMNH) in Leiden, the Netherlands; the same almost applies for Western Palearctic specimens in the Zoologisches Museum und Forschungsinstitut Alexander Koenig (ZFMK) in Bonn, Germany, where only part of the extensive colection of birds from central Europe was not examined. In the Natural History Museum (BMNH) in Tring, England, and in the Naturhistorisches Museum in Wien, Austria, attention was mainly spent on birds from the Balkan countries, Greece, Turkey, and the Caucasus area, but due to the large number of specimens present and the limited time available a selection had to be made in favour of rarer species and species which are known to have some geographical variation within the region; commoner species which are supposed to show little variation within the area were neglected. A few additional species were examined in the museums of Dresden, Berlin, and Moscow. For description of colours, use was made of Smithe (1974, 1981). The measuring methods are those explained in Cramp (1988), Kiziroglu (1989), and Svensson (1992), with, for instance, wing measured while fully stretched and flattened against a ruler. All birds were measured personally, except when other authors are mentioned. As not all authors use the same method of wing measuring, data supplied by different authors may not always be fully comparable. From samples measured by both Kumerloeve and the author, it appeared that the wing data of the former were *c.* 1(-2) % lower, and the same applies to data of, for instance, Stresemann. The data of Rössner are sometimes as much as 1 - 3 % lower. Note also that the wing measurement of live or freshly dead birds if compared with the data of dry bird skins as used in this study is *c.* 2% longer (Engelmoer *et al.* 1983). For details of geographical variation, see chapter IV.

III THE DISTRIBUTION MAPS

The maps in this book are not comparable in quality with the atlases of breeding bird distribution which have appeared in various European countries during the last few decades. Turkey is an enormously large country, where local people interested in bird-watching are scarce (though increasing in number), and visiting ornithologists often keep to a relatively small number of well-known sites where rarities are certain to be found within short time, ignoring other areas which may be equally interesting but for which more time would be needed to explore. Atlas work in Turkey has been tried, using half-degree squares as a base (Porter & Beaman 1977, Albrecht 1979), but the results have not been published. Kasparek (1991) gives a survey of the few species for which maps have appeared. With a limited number of ornithologists available, a prime interest is first to survey areas which are of importance for birds in Turkey, in order to conserve those areas which are threatened. The work of Ertan et al. (1989) and Grimmet & Jones (1989) forms the basis for this survey of important bird areas, and further work on this area is co-ordinated by the Turkish Society for the Protection of Nature (Dogal Hayati Koruma Dernegi DHKD), to which all new information should be sent (address of the main office: PK 18, 80810 Bebek, Istanbul, Turkey).

The distribution maps are based on localities of specimens examined and on data obtained from the literature listed below. Though details from a surprisingly large number of localities were extracted, covering a large part of Turkey (fig. 3), and the basis for an atlas seems firm, there are some disadvantages to the mapping method used in this book:

(1) The reports span a period of more than a century. Data gathered more than 100 years ago may not be valid now. However, for passerines this is thought to be less serious than for non-passerines, for which changes in the status (especially due to human impact) of some species are known to have been large. The reader should realize that the study of geographical variation in this book also relies on specimens sometimes collected as long ago as the last century.

(2) Although, according to fig. 3, many locations have been visited, complete species' lists for most of them have not been published. Many authors mention details of rare species, but not those of widespread taxa. Accordingly, maps of rare species are thought to be more accurate than those of commoner ones. A remark like 'common everywhere' by an author is perhaps a useful statement for a birdwatcher, but of no use to someone plotting maps. Thus, the reader should realize that the maps for *Hirundo rustica* and *Motacilla alba* are less complete than those for *Irania gutturalis* and *Bucanetes mongolicus*, for example.

(3) Some observers give villages or towns as observation sites, though the true spot may have been somewhere in the surrounding country, perhaps up to 10 km away. As the red dots on the maps are c. 20 km across, this is perhaps not a serious problem, but readers should be aware that a dot may be a few mm away from the correct site. Many other observers only mention road-stretches as sites, for instance, 'seen between Ankara and Gölbasi'. When these sites are less than 50 km apart, a single dot has been placed midway; when the sites are up to 100 km apart and the species was recorded as common, 2-3 dots are placed arbitrarily in between, but none when only one bird was seen. Observations of birdwatchers travelling large distances each day and recording only the towns where night-stops had been made (for instance, 'between Ankara and Erzurum') were ignored, being too inaccurate. Likewise, records of birds obtained at smaller lakes (less than 50 km long) arbitrarily were given a dot along one of its shores (unless it was clear from the

Introduction

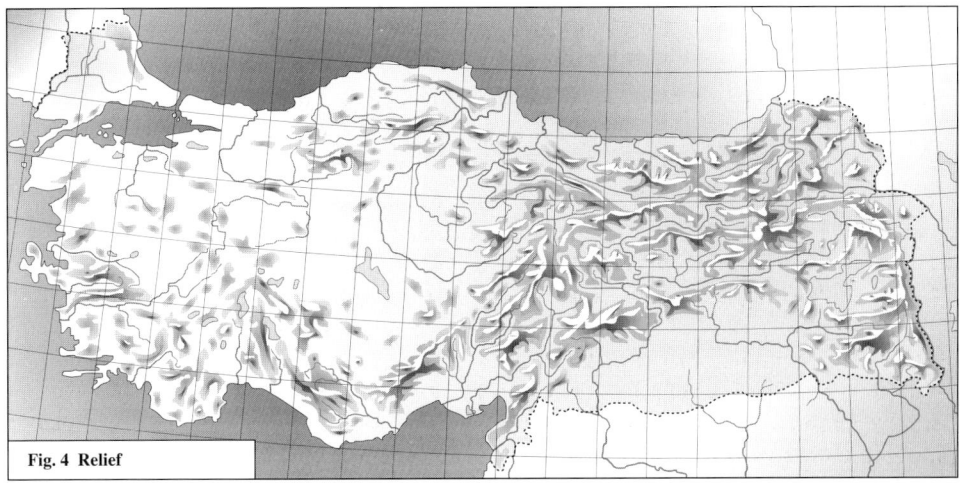

Figure 4. Base-map with relief of Turkey used for the distribution maps of the species.

itinerary what site was visited), but an observation without more detail than 'Van Gölü' did not qualify for inclusion.

(4) The maps give only 2 categories of dots, open ones for possible breeding, and black ones for probable and certain breeding. Actual breeding records are scarce (or do not reach reports), and repeated observations in the breeding season of one or more singing or displaying males in good habitat were considered to be sufficient to assume breeding status. However, many black dots are based on an author's statement of 'breeding', though it is not at all certain what evidence he or she had. Evidently, some species are reported to be breeding while nothing more of them was seen than a glimpse from a travelling car. Therefore, the presence of dots and their colour is also a matter of personal judgment by the author of this book, and a single record in the breeding season in a suitable locality within an area full of black dots is likely to be registered as a black dot, while a pertinent statement of 'breeding' far away from other black dots without further evidence generally results in an open dot.

The base-map used for distribution of the species shows the relief of the country rather than real altitudes. The average height above sea level in Turkey gradually increases from west to east, and mapping mountains only above, say, 2000m would mean that much of western Turkey would remain white and much of the east would be black. Therefore, the map shown in fig. 4 was constructed, showing those ranges which reach to approximately 500m and more above the local plateau or valley level.

For compilation of the maps the following sources were used (in sequence of year of publication):

Danford, C. G. (1877-78) A contribution to the ornithology of Asia Minor. *Ibis* (4)1, 261-274; (4)2, 1-35.
Danford, C. G. (1880) A further contribution to the ornithology of Asia Minor. *Ibis* (4)4, 81-99.
Selous, F. C. (1900) A fortnight's egg-collecting in Asia Minor. *Ibis* (7)6, 405-424.
Witherby, H. F. (1907) On a collection of birds from western Persia and Armenia, with field-notes by R. B. Woosnam. *Ibis* (9)1, 74-111.
Braun, F. (1908) Unsere Kenntnis der Ornis der kleinasiatischen Westküste. *J. Orn.* 56, 539-626.
Weigold, H. (1912-13) Ein Monat Ornithologie in den Wüsten und Kulturoasen NW-Mesopotamiens und Innersyriens. *J. Orn.* 60, 249-297; 61, 1-40.
Weigold, H. (1913-14) Zwischen Zug und Brut am Mäander. *J. Orn.* 61, 562-597; 62, 57-93.
Ramsay, L. N. G. (1914) Observations on the bird-life of the Anatolian plateau during the summer of 1907. *Ibis* (10)2, 365-387.
McGregor, P. J. C. (1917) Notes on birds observed at Erzerum. *Ibis* (10)5, 1-30.
Stenhouse, J. H. (1920) Some observations on the birds of the islands of Milos, Lemnos, and Imbros, Aegean Sea. *Ibis* (11)2, 671-678.
Kummerlöwe [=Kumerloeve], H., & G. Niethammer (1934a) Contribution à la connaissance de l'avifaune de la Turquie d'Europe (Thrace). *Alauda* 6, 298-307.
Kummerlöwe, H., & G. Niethammer (1934b) Observations sur la vie des oiseaux en Anatolie (Anatolie nord-occidentale entre la mer de Marmara et Angora). *Alauda* 6, 452-468.
Kummerlöwe, H., & G. Niethammer (1934-35) Beiträge zur Kenntnis der Avifauna Kleinasiens (Paphlagonien-Galatien). *J. Orn.* 82, 505-552; 83, 25-75.
Meinertzhagen, R. (1935) Ornithological results of a trip to Syria and adjacent countries in 1933. *Ibis*(13)5, 110-151.
Rössner, H (1935) Die Vogelsammlung de österreichischen Kleinasienexpedition 1934. *Sitzungsber. Akad. Wiss. Wien, math.-naturwiss. Kl. , Abt. I.* 144, 299-312.
Bird, C. G. (1937) The birds of southern Asia Minor from Mersin to the Euphrates. *Ibis* (14)1, 65-85.
Jordans, A. von, & J. Steinbacher (1948) Zur Avifauna Kleinasiens. *Senckenbergiana* 28, 159-186.
Wadley, N. J. P. (1951) Notes on the birds of central Anatolia. *Ibis* 93, 63-89.
Ogilvie, I. M. (1954) Bird notes from northern Asia Minor, 1946-1948. *Ibis* 96, 81-90.
Hollom, P. A. D. (1955) A fortnight in South Turkey. *Ibis* 97, 1-17.
Kumerloeve, H (1957a) Brutvogelbeobachtungen bei Savastepe und Bergama (NW-Anatolien). *Anz. orn. Ges. Bayern* 4, 712-720.
Schüz, E. (1957) Vogelkunde am Manyas-See (Türkei). *Vogelwarte* 91, 41-44.
Balance, D. K. (1958) Summer observations on the birds of the Anatolian Plateau and north-western Cilicia. *Ibis* 100, 617-620.
Etchécopar, R. D. (1959) Quelques observations en Turquie. *Oiseau* 29, 96-98.
Hollom, P. A. D. (1959) Notes from Jordan, Lebanon, Syria and Antioch. *Ibis* 101, 183-200.
Maas Geesteranus, H. P. (1959) Ornithological report on a biological excursion to Asia Minor. *Ardea* 47, 111-157.
Smith, M. Q. (1960) Notes on the birds of the Trebizond area of Turkey. *Ibis* 102, 576-583.
Hennipman, E., P. Nijhoff, C. Swennen, W. J. M. Vader, W. J. T. O. de Wilde, & A. Tulp (1961) Verslag van de Nederlandse biologische expeditie Turkije 1959. *Levende Natuur* 65 (5), bijlage, 3-27.

Kumerloeve, H. (1961a) Zur Kenntnis der Avifauna Kleinasiens. *Bonner zool. Beitr.* 12, spec. vol., 1-318.

Kumerloeve, H. ('1962'=1964a) Weitere Untersuchungen über die türkische Vogelwelt (ausgenommen Sumpf- und Wasservögel). *Istanbul Üniv. Fen Fak. Mecmuasi* B27, 165-228.

Nijhoff, P., & C. Swennen (1963) Ornithologische reisindrukken van Turkije. *Vogeljaar* 11, 8-16, 30-34.

Kumerloeve, H. (1963a) L'Avifaune du lac d'Antioche (Amik Gölü-Gölbasi) et de ses alentours. *Alauda* 30, 110-136, 161-211.

Bezzel, E. (1964) Ornithologische Sommerbeobachtungen aus Kleinasien. *Anz. orn. Ges. Bayern* 7, 106-120.

Eggers, J., & W. Lemke (1964) Ornithologische Beobachtungen in der Türkei. *Orn. Mitt.* 16, 185-188.

Kumerloeve, H. (1964b) Zur Sumpf- und Wasservogelfauna der Türkei. *J. Orn.* 105, 307-325.

Warncke, K. (1964-65) Beitrag zur Vogelwelt der Türkei. *Vogelwelt* 85, 161-174; 86, 1-19.

Schweiger, H (1965) Ornithologische Beobachtungen in Anatolien während die Jahre 1959-1965. *Istanbul Üniv. Fen Fak. Mecmuasi* B30, 177-189.

Vader, W. J. M. (1965) Bird observations by the Dutch biological Expedition Turkey 1959. *Ardea* 53, 172-204.

Herrn, C. P. (1966) Neue Sommerbeobachtungen in Anatolien, in Kilikien und im Hatay. *Vogelwarte* 23, 305-308.

Kumerloeve, H. (1966a) Ergänzungen zur Avifauna Kleinasiens. *Bonner zool. Beitr.* 17, 257-259.

Warncke, K. (1966) Ergänzungen zu meinen vogelkundlichen Beobachtungen in der Türkei. *Vogelwelt* 87, 188-189.

Kumerloeve, H. (1966-67) Migration et hivernage sur le Lac d'Antioche (Amik Gölü, Hatay, Turquie). Coup d'oeil sur son avifaune nidificatrice actuel. *Alauda* 34, 299-308; 35, 1-19.

Ganso, M., & G. Spitzer (1967) Weitere Beiträge zur Avifauna Kleinasiens. *Egretta* 10, 9-25.

Kumerloeve, H. ('1967'=1968) Neue Beiträge zur Kenntnis der Avifauna von Nordost- und Ost-Kleinasien. *Istanbul Üniv. Fen Fak. Mecmuasi* B32, 79-213.

Érard, C., & R.-D. Etchécopar (1968) Observations de printemps en Turquie. *Oiseau* 38, 87-102.

Gaston, A. J. (1968) The birds of the Ala Dagh mountains, southern Turkey. *Ibis* 110, 17-26.

Groh, G. (1968) Ornithologische Reiseeindrücke aus Griechenland und der Türkei. *Mitt. Pollichia* (3)5, 163-170.

Vielliard, J (1968) Résultats ornithologiques d'une mission à travers la Turquie. *Istanbul Üniv. Fen Fak. Mecmuasi* B33, 67-170.

Kumerloeve, H (1969a) Zur Avifauna des Van Gölü- und Hakkari-Gebietes (E/SE-Kleinasien). *Istanbul Üniv. Fen Fak. Mecmuasi* B34, 245-312.

Porter, R. F. (ed) (1969) Systematic list 1966/1967. *Bird Rep. Orn. Soc. Turkey* 1, 1-170.

Kumerloeve, H (1970a) Zur Vogelwelt im Raume Ceylanpinar (türkisch-syrisches Grenzgebiet). *Beitr. Vogelkunde* 16, 239-249.

Kumerloeve, H (1970b) Zur Kenntnis der Avifauna Kleinasiens und der europäischen Türkei (Ergänzungen - Hinweise - Fragestellungen). *Istanbul Üniv. Fen Fak. Mecmuasi* B35, 85-160.

Warncke, K (1970) Beitrag zur Vogelwelt des Zentralanatolischen Beckens. *Vogelwelt* 91, 176-184.

Keve, A. (1971) Aus den Notizen der Forscherfahrt Dr. N. Vasváris in Kleinasien. *Vertebr. Hung.* 12, 51-67.

Mertens, R. (1971) Ein weiterer Beitrag zur Vogelwelt Zentralanatoliens. *Vogelwelt* 92, 189-191.

Rokitansky, G., & H. Schifter (1971) Ornithologische Ergebnisse zweier Sammelreisen in die Türkei. *Ann. naturhist. Mus. Wien* 75, 495-538.

Renkhoff, M (1972) Bemerkenswerte Beobachtungen 1971 im Balik-Gölü-Gebiet (Nordtürkei). *Orn. Mitt.* 24, 63-73.

Sutton, R. W. W., & J. R. Gray (1972) Summer birds on Karanfil Dag. *Bird Rep. Orn. Soc. Turkey* 2, 186-205.

Vittery, A (ed) (1972) Systematic List 1968/1969. *Bird Rep. Orn. Soc. Turkey* 2, 1-184.

Warncke, K. (1972) Beitrag zur Vogelwelt der Türkei im Bereich der Südgrenze. *Vogelwelt* 93, 23-26.

Renkhoff, M (1973) Weitere Beobachtungen 1972 im Balik-Gölü-Gebiet (Nordtürkei). *Orn. Mitt.* 25, 122-124.

Vauk, G. (1973) Ergebnisse einer ornithologischen Arbeitsreise in das Gebiet des Beysehir-Gölü im April-May 1964 (Kleinasien). *Beitr. Vogelkunde* 19, 225-260.

Beaman, M (ed) (1975) Systematic list for 1970-1973, with check-list and appendix. *Bird Rep. Orn. Soc. Turkey* 3, 18-265.

Beaudoin, J.-C. (1976) Excursion ornithologique dans le Moyen-Taurus (Turquie). *Alauda* 44, 77-90.

Gallner, J.-C. (1976) Observations ornithologiques nouvelles dans la Région de Van (Turquie). *Alauda* 44, 111-117.

Louette, M., M. Becuwe, & R. Eyckermann (1977) Observations ornithologiques en Anatolie et en Thrace. *Gerfaut* 67, 427-436.

Meininger, P. L., & B. Dielissen (1977) Ornithologisch verslag van een bezoek aan Turkije 6 juli t/m 21 juli 1977. Mimeogr. rep., 15 pp.

Beaman, M. (ed) (1978) Systematic list for 1974-1975. *Bird Rep. Orn. Soc. Turkey* 4, 12-209.

Schubert, W. (1979a) Bemerkenswerte Brutnachweise und Brutzeit-feststellungen in Anatolien/Türkei. *Vogelwelt* 100, 151-155.

Petretti, A., & F. Petretti (1980) Observations ornithologiques dans les milieux desertiques et semi-desertiques de la Turquie centrale et sud-orientale. *Gerfaut* 70, 273-278.

Schmidtke, K., & H. Utschick (1980) Ornithologische Ergebnisse einer Türkeifahrt. *Anz. orn. Ges. Bayern* 19, 57-74.

Beers, M. Van (1982) Ornithologische paradijzen in Turkije. *Wielewaal* 48, 134-143.

Colin, D. (1982) Ornithologische notities bij de Wielewaal-reis naar Turkije. *Wielewaal* 48, 147-152.

Franckx, H. (1982) Verslag Wielewaal-reis Turkije 1981. *Wielewaal* 48, 144-146.

Hüni, M. (1982) Exkursion der Ala in die Südosttürkei, 3. -17. April 1982. *Ornith. Beob.* 79, 221-223.

Kasparek, M., & J. van der Ven (1983) The birds of Erçek Gölü. *Birds of Turkey* 1, 1-24. Landshut.

Baris, S., R. Akçakaya, & C. Bilgin (1984) The birds of Kizilcahamam. *Birds of Turkey* 3, 1-36. Heidelberg.

Helbig, A. (1984) Bemerkenswerte ornithologische Beobachtungen in der Türkei im Sommer 1981. *Bonner zool. Beitr.* 35, 57-69.

Husband, C., & M. Kasparek (1984) The birds of Lake Seyfe. *Birds of Turkey* 2, 1-32. Heidelberg.

Dijksen, L. J., & M. Kasparek (1985) The birds of the Kizilirmakdelta. *Birds of Turkey* 4, 1-47. Heidelberg.

Harrap, S. (1985) Birding in Turkey 1984. *Dutch Birding Travel Rep. Serv.* TR 8, 1-34. Leersum.

Kasparek, M. (1985) Die Sultanssümpfe. Naturgeschichte eines Vogelparadieses in Anatolien. Heidelberg.
Albrecht, J. S. M. (1986) Notes on the birds of Eregli, Black Sea coastlands, Turkey 1976-1978. *Sandgrouse* 8, 74-92.
Amcoff, M., S. Nilsson, B. Svensson, & M. Ullman (1986) Report of an expedition to Turkey, 27 April - 16 May 1986. Lund.
Beaman, M. (1986) Turkey: bird report 1976-81. *Sandgrouse* 8, 1-41.
Bijlsma, R., & F. de Roder (1986) Notes on the birds of some wetlands in Turkey, 17 August - 10 September 1986. *WIWO-Rep.* 12, 1-32. Zeist.
Schilperoort, L., & M. Schilperoort-Huisman (1986) Observations of waterbirds in some wetlands in Turkey July-August 1986. *WIWO-Rep.* 14, 1-41. Zeist.
Kasparek, M. (1987) The birds of Lake Kulu. *Birds of Turkey* 5, 1-42. Heidelberg.
Kiliç, A., & M. Kasparek (1987) The birds of Yeniçaga Gölü. *Birds of Turkey* 6, 1-32. Heidelberg.
Vermeulen, J. [1987] Central and eastern Turkey 14th July - 4th August 1987. *Dutch Birding Travel Rep. Serv.* TR 30, 1-28. Leersum.
Kiziroglu, I., & F. Kiziroglu ('1987'=1988) Die Vogelarten im Vogelparadies des 'Kus Cenneti/Bandirma'-Nationalparks und seiner Umgebung. *Verh. orn. Ges. Bayern* 24, 515-532.
Berk, V. M. van den, & J. P. W. Letschert (1988), in: Have, T. M. van den, V. M. van den Berk, J. P. Cronau, & M. J. Langeveld (eds) South Turkey Project: a survey of the waders and waterfowl in the Çukurova deltas, spring 1987. *WIWO-Rep.* 22, 129-163. Zeist.
Bruin, S. de [1988] Turkey 1988, July-August. *Dutch Birding Travel Rep. Serv.* TR 33, 1-32. Leersum.
Dijksen, L. J., & M. Kasparek (1988) The birds of Lake Açi. *Birds of Turkey* 7, 1-36. Heidelberg.
Gooders, J. (1988) Where to watch birds in Britain and Europe. London.
Kasparek, M. (1988) Der Bafasee: Natur und Geschichte in der Türkischen Ägäis. Heidelberg.
Snow, P. [1988] Dalyan Delta, S.W. Turkey, 12-19th Sept.'88. *Dutch Birding Travel Rep. Serv.* TR 28c, 1-2. Leersum.
Brock, G. J. [1989] Turkey - Bodrum based, 12/5/89-18/5/89. *Dutch Birding Travel Rep. Serv.* TR 28d, 1-5. Leersum.
Broome, T. [1989] Turkey incl. Cyprus, June 5th-19th 1989. *Dutch Birding Travel Rep. Serv.* TR 36, 1-36. Leersum.
Colin, D. [1989] Turkije 1989. *Dutch Birding Travel Rep. Serv.* TR 23, 1-25. Leersum.
Crosland, A. [1989] Visit to S.W. Turkey, 4/5-18/5/89. *Dutch Birding Travel Rep. Serv.* TR 28e, 1. Leersum.
Kiliç, A., & M. Kasparek (1989) The birds of the Köycegiz-Dalyan area. *Birds of Turkey* 8, 1-32. Heidelberg.
Martins, R. P. (1989) Turkey: bird report 1982-6. *Sandgrouse* 11, 1-41.
Moyle, R. [1989] Notes useful for bird watchers based in coastal western Anatolia. *Dutch Birding Travel Rep. Serv.* TR 28a, 1-8. Leersum.
Titcombe, C., & G. Hatch [1989] The Izmir area of western Turkey, 2nd-9th May 1989. *Dutch Birding Travel Rep. Serv.* TR 28b, 1-8. Leersum.
Winden, A. van, K. Mostert, P. Ruiters, M. Siki, & H. de Waard (1989) Waders and waterfowl in spring 1988 at Eber Gölü, Turkey. *WIWO-Rep.* 28, 1-60. Zeist.
Ayvaz, Y. (1990) The birds of the lake of Pinarbasi, Malatya. *Doga Turk Zool. Derg.* 14, 139-143.
Bakker, T., & W. Steenge [1990] Turkey, 9 June-8 July 1990. *Dutch Birding Travel Rep. Serv.* TR 31, 1-34. Leersum.

Colin, D. [1990] Verslag ornithologische reizen naar Turkije in juli 1989 en 1990. *Dutch Birding Travel Rep. Serv.* TR 26, 1-33. Leersum.
Dufourny, H. [1990] An ornithological report about Turkey from 1st to 23rd July 1990. *Dutch Birding Travel Rep. Serv.* TR 29, 1-52. Leersum.
Forster, A., & T. Numminen [1990] Turkey, 8.-22.6 1990. *Dutch Birding Travel Rep. Serv.* TR 32, 1-46. Leersum.
Kasparek, M. (1990a) Zur Vorkommen einiger in der Türkei seltener Vogelarten. *Bonner zool. Beitr.* 41, 181-202.
Kirwan, G. (1990) The wetlands of Eregli, Turkey. *Bull. Orn. Soc. Middle East* 25, 42-46.
Ayvaz, Y. (1991) The birds of Çildir Lake. *Doga Turk Zool. Derg.* 15, 53-58.
Dorèl, F. [1991] Turkije 1991. *Dutch Birding Travel Rep. Serv.* TR 35, 1-40. Leersum.
Mascara, R. (1991) Osservazioni ornitologiche in Grecia e Turchia. *Picus* 17, 91-94.
Schepers, F., & J. Stuart (1991) Turkse Göksu-delta, het bezoeken waard. *Vogeljaar* 39, 198-207.
Pasquali, R. ('1990'=1992) Osservazioni ornitologiche nella Turchia asiatica. *Gli Uccelli Ital.* 15, 94-98.
Eyckerman, R., M. Louette, & M. Becuwe (1992) On the biometrics of birds ringed in Turkey. *Zool. Middle East* 6, 29-37.
Frost, R., & J. Hornbuckle [1992] Eastern Turkey, July 25th - August 11th, 1992. *Dutch Birding Travel Rep. Serv.* TR 38, 1-23. Leersum.
Versluys, M. [1992] Bird observations in SW-Turkey, 29 June-25 July 1992. *Dutch Birding Travel Rep. Serv.* TR 37, 1-13. Leersum.
Kiraç, S. K. (1993) The birds of Çöl Lake, Uyuz Lake and Yagliören Pool. *Birds of Turkey* 10, 1-28. Heidelberg.
Kirwan, G. (1993) The birds of the Hotamis Marshes. *Birds of Turkey* 9, 1-44. Heidelberg.
Berk, V. M. van den, J. P. Cronau, & T. M. van der Have (1993) Waterbirds in the Van province, eastern Turkey, May 1989. *WIWO-Rep.* 34, 1-50. Zeist.
Schekkerman, H., & M. W. J. van Roomen (eds) (1993) Migration of waterbirds through wetlands in central Anatolia, spring 1988. *WIWO-Rep.* 32, 1-136. Zeist.
Hustings, F., & K. van Dijk (1994) Bird census in the Kizilirmak delta, Turkey, in spring 1992. *WIWO-Rep.* 45, 1-168. Zeist.

To this literature, all data summarized in the reports published in *Sandgrouse,* the Bulletins of the Ornithological Society of Turkey and its successor the Ornithological Society of the Middle East, and the 'European News' of the journal *British Birds* were incorporated, up to March 1994. The reference list above is not a complete survey of the existing literature on Turkish birds. For this, see the survey of the history of bird study in Turkey in Kumerloeve (1961a or 1975a) or the bibliography published by Kumerloeve (1986). As this book deals with breeding passerines, reports on wintering birds or non-passerines are not included above. In addition, some of the older literature (especially before 1900) has not been used.

IV GEOGRAPHICAL VARIATION

The species is the basic unit of our division of the animal world. In general, each individual animal or bird can immediately be recognized by us as belonging to a certain species by a combination of morphology, voice, and behaviour. These same characters are also used by individual birds themselves to find conspecifics. Individual birds which are morphologically, vocally, and behaviourally similar (apart from differences due to sex or age) and which freely interbreed with birds of the opposite sex form a population. According to Mayr (1963), a species is a group of actually or potentially interbreeding populations which are reproductively isolated from other such groups. Not all individuals of a species are necessarily similar to each other, however. Many species show individual variation, for instance, the differences between sexes or age-groups cited above, but, more importantly, many also show geographical variation: one population of the species may differ slightly from another population of the same species. When both these populations are not spatially segregated, the differences will be small because some mixing of individuals will occur. Often, two populations of a species are segregated by barriers like mountains or seas, however, and the mixing will be limited or absent. Then, the differences between both populations become more pronounced with time, especially if the species is sedentary and the environment each side of the barrier is somewhat different. They may still potentially interbreed, though this is difficult to prove, and in the course of time the morphological differences may become so large that we may arbitrarily call the populations separate species. In nature, proof that they really are separate species may occur when the barrier between both populations ceases to exist and both start to overlap in breeding range without recognizing each other as conspecific, and thus without interbreeding. In our area, this can be seen in, for instance, the larks *Calandrella rufescens* and *Calandrella cheleensis*, which are undoubtedly closely related; some forms of each are hard to separate on morphology, but some populations overlap without interbreeding.

Recognizable geographical units of a species are called subspecies, or, alternatively, a (geographical) race. A species is scientifically denoted by a binominal name, consisting of a genus name (for example, *Passer* for the sparrows) and a species name (for example, *domesticus* for the House Sparrow). Subspecies are named trinominally, for example, *Passer domesticus mayaudi* for the distinct large and white-cheeked population of the House Sparrow of eastern Turkey. Species without recognizable variation between their populations are called monotypic, and species in which at least two subspecies can be recognized are polytypic. Monotypic species are especially those with small geographical ranges, or are migratory ones which mix freely in their winter quarters and which are not inclined to breed in the area where they were reared. Sedentary species occupying a large geographical range comprising many habitats are often strongly polytypic.

The characteristics of a species or subspecies are those which are to be found in the type specimen or the type series, the bird or series of birds from a certain locality (the type locality) on which the name was based. The type specimen or type series is thus of utmost importance and should preferably be examined when assessing the characteristics of a species or subspecies. Alas, the name is sometimes based on a short description or on a plate in a rare book, especially in those names coined before about 1820. The type specimen on which such a description or plate was based may no longer be available, because scientists were not aware of the importance of a type specimen in those years and a satisfactory method of preserving skins had not yet been developed. In birds described after about 1820, the available type specimen is sometimes faded and not suitable for comparison, or it is housed in a museum which is too far away to be visited, or the type specimen may even be lost. In these cases, examination of a series of specimens from the type locality is a good alternative, as these are likely to show the same characters as the original type (as long as the latter was not a migrant or straggler). In this study, some type specimens were examined, but most descriptions and measurements were based on birds collected in later years at or near the type locality.

Variation in Turkish birds cannot be determined without a study of the variation in the surrounding countries. Although a number of species and subspecies had a type locality on Turkish territory, many others were based on specimens from elsewhere. The characters of a subspecies with a type locality in, for instance, Bulgaria, may also be present in birds from Turkey, and Turkish birds thus belong to the same subspecies. Sometimes, birds from western Turkey are similar to a subspecies originally described from, for instance, Romania, and birds of the same species in eastern Turkey are similar to a subspecies originally described from Iran, while birds of central Turkey are intermediate in their characters. When these western and eastern subspecies show a gradual transition in characters from one into the other, the variation is called clinal, with birds from Turkey forming part of a cline extending from Romania to Iran.

Turkish birds may show the characteristics of a subspecies described from any of the neighbouring countries. In all species accounts, the subspecies of importance for the assessment of geographical variation in Turkey are described in the following sequence:

(W) A subspecies described from an area **west of Turkey**, or one whose characters may influence Turkish birds from the west, in particular through the Balkan countries (former Yugoslavia, Albania, or Bulgaria), Romania, or Greece (the latter including Crete and other islands in the Aegean Sea).

(T) Refers to a subspecies originally described from **Turkey itself** (for a listing of these, see also Kumerloeve 1984).

(S) A subspecies described from an area **south of Turkey**, or possibly influencing Turkey from the south, through Cyprus, the Levant (western Syria, Lebanon, and Israel), inland Syria, or the plains of Iraq.

(E) A subspecies described from an area **east of Turkey**, or possibly influencing Turkey from the east, through the mountains and valleys of eastern Iraq, Iran, Transcaucasia, or the Caucasus.

When several subspecies have been named from, say, an area west of Turkey, only the nearest one is described, unless Turkish birds show the characteristics of more than one, for example from both the north-west (through Bulgaria or north-east Greece) and the south-west (through Crete, Karpathos, and Rodhos). The names of most subspecies described from the region are mentioned, even if they are not now considered to be separable from others. When a subspecies has always been considered to be a synonym since Hartert's time (about 1903-1922), it is only mentioned under the heading of the subspecies to which it belongs; when a subspecies has been considered valid by Hartert or later authors, it receives a full entry, even though it may not be considered valid here. This book is thus also a revision of all subspecies described for Turkey and the surrounding countries.

In the description of subspecies, no distribution is given when this is clear from the type locality and from the localities mentioned in the measurements. Details on the distribution of subspecies in Turkey are given in the sub-section on 'subspecies recognized in Turkey' and on the maps.

Although most of the characteristics and measurements given for various subspecies are based on personal research, the existing literature was not ignored. For Turkey, all literature dealing with geographical variation has already been mentioned in the list on pages 11-16; the most important authors on this subject in the list are Danford (1877-78, 1880), Witherby (1907), Weigold (1912-13, 1913-14), Kummerlöwe & Niethammer (1934-35), Rössner (1935), Bird (1937), Jordans & Steinbacher (1948), Maas Geesteranus (1959), Kumerloeve (1961a, 1963a, 1964a, 1964b, 1966-67, 1968, 1969a, 1970a, 1970b), and Rokitansky & Schifter (1971). Most of the specimens collected or commented upon by Danford, Witherby, Bird, Rössner, Jordans & Steinbacher, Maas Geesteranus, Kumerloeve, and Rokitansky & Schifter have been examined, as well as a few from Weigold and Kummerlöwe & Niethammer (these are mainly duplicates sent to Bonn), but not the bulk of the collection of the latter three authors. Apart from these, a number of birds collected long ago by Krüper and Schrader were examined (especially those housed in the BMNH and RMNH), as well as more recent ones collected by Epping (in the ZFMK and RMNH; the latter were apparently never recorded before). In the species accounts, authors like Kummerlöwe & Niethammer (1934-35) are often mentioned, not because their contributions to Turkish ornithology are more important than those of, for instance, Danford and the later work of Kumerloeve, but simply because I did not personally examine the specimens they collected and thus had to rely on the data they presented.

For judgment of geographical variation and/or measurements of birds outside Turkey, where specimens were not examined by the author, use was made of the following literature, which is listed by region:

General works

Hartert, E. (1903-10) Die Vögel der Paläarktischen Fauna 1. Berlin.

Hartert, E. (1921-22) Die Vögel der Paläaktischen Fauna 3. Berlin.

Molineux, H. G. K. (1930-31) A catalogue of birds, giving their distribution in the western portion of the Palearctic Region. Eastbourne.

Hartert, E., & F. Steinbacher (1932-38) Die Vögel der Paläarktischen Fauna, Ergänzungsband. Berlin.

Vaurie, C. (various years from 1949 to 1952) Notes from the Walter Koelz collections. *Am. Mus. Novitates*; 8 issues between nrs. 1406 and 1549 [revisions of passerine birds especially from Iran and Afghanistan].

Vaurie, C. (various years from 1953-1958) Systematic notes on Palearctic birds. *Am. Mus. Novitates*; 33 issues between nrs. 1640 and 1898.

Vaurie, C. (1959) The birds of the Palearctic Fauna, Order Passeriformes. London.

Stresemann, E., & L. A. Portenko (eds) (1960 to present) Atlas der Verbreitung palaearktischer Vögel. Berlin. [Up to 1993, 18 issues have appeared, each dealing with 10-20 bird species; after the death of Stresemann and Portenko, editors have been G. Mauersberger, I. A. Neufeldt, H. Dathe, and W. M. Loskot].

Afghanistan

Paludan, K. (1959) On the birds of Afghanistan. *Vidensk. Medd. Dansk naturhist. Foren.* 122, 1-332.

Albania

Ticehurst, C. B., & H. Whistler (1932) On the ornithology of Albania. *Ibis* (13)2, 40-93.

Bulgaria

Jordans, A. von (1940) Ein Beitrag zur Kenntnis der Vogelwelt Bulgariens. *Mitt. Königl. Wissensch. Inst. Sofia* 13, 49-152.

Harrison, J. M., & P. Pateff (1937) An ornithological survey of Thrace, the islands of Samothraki, Thasos, and Thasopulo in the north Aegean, and observations in the Struma valley and the Rhodope mountains, Bulgaria. *Ibis* (14)1, 582-625.

Niethammer, G. (1950) Zur Vogelwelt Bulgariens, insbesondere seiner nordwestlichen Landesteile. *Syllegomena Biologica*, 267-286. Leipzig.

Cyprus

Flint, P. R., & P. F. Stewart (1992) The birds of Cyprus. B. O. U. Check-list No. 6, London.

Greece

Niethammer, G. (1942) Über die Vogelwelt Kretas. *Ann. naturhist. Mus. Wien* 53 (2), 5-50.

Niethammer, G. (1943) Beiträge zur Kenntnis der Brutvögel des Peloponnes. *J. Orn.* 91, 167-238.

Makatsch, W. (1950) Die Vogelwelt Macedoniens. Leipzig.

Bauer, W., O. von Helversen, M. Hodge, & J. Martens (1969) Catalogus Faunae Graeciae 2, Aves. Thessaloniki.

Iran

Witherby, H. F. (1903) An ornithological journey in Fars, South-West Persia. *Ibis* (8)3, 501-571.

Witherby, H. F. (1910) On a collection of birds from the south coast of the Caspian Sea and the Elburz Mountains. *Ibis* (9)4, 491-517.

Stresemann, E. (1928) Die Vögel der Elburs-Expedition 1927. *J. Orn.* 76, 313-411.

Paludan, K. (1938) Zur Ornis des Zagrossgebietes, W.-Iran. *J. Orn.* 86, 562-638.

Paludan, K. (1940) Contributions to the ornithology of Iran. *Danish Sci. Invest. Iran* 2, 11-54.

Schüz, E. (1959) Die Vogelwelt des Südkaspischen Tieflandes. Stuttgart.

Diesselhorst, G (1962) Anmerkungen zu zwei kleinen Vogelsammlungen aus Iran. *Stuttgarter Beitr. Naturkde.* 86, 1-29.

Érard, C., & R.-D. Etchécopar (1970) Contribution à l'étude des oiseaux d'Iran (Résultats de la mission Etchécopar 1967). *Mém. Mus. natl. Hist. nat.* (A) 66, 5-146.

Desfayes, M., & J.-C. Praz (1978) Notes on habitat and distribution of montane birds in southern Iran. *Bonner zool. Beitr.* 29, 18-37.

Scott, D. A., H. M. Hamadani, & A. A. Mirhosseyni (1975) [The birds of Iran]. Tehran.

Vuilleumier, F. (1977) Suggestions pour recherches sur la spéciation des oiseaux en Iran. *Terre et Vie* 31, 459-488.

Iraq

Ticehurst, C. B., P. A. Buxton, & R. E. Cheesman (1921-22) The birds of Mesopotamia. *J. Bombay Nat. Hist. Soc.* 28, 210-250, 381-427, 650-674, 937-956.

Ticehurst, C. B., P. Z. Cox, & R. E. Cheesman (1926) Additional notes on the avifauna of Iraq. *J. Bombay Nat. Hist. Soc.* 31, 91-119.

Harrison, J. M. (1959) Notes on a collection of birds made in Iraq by Flight-Lieutenant David L. Harrison. *Bull. Brit. Orn. Club* 79, 9-13, 31-36, 49-50.

Ctyroky, P. (1987) Ornithological observations in Iraq. *Beitr. Vogelkde.* 33, 141-204.

Lebanon

Kumerloeve, H. (1962a) Notes on the birds of the Lebanese Republic. *Iraq Nat. Hist. Mus. Publ.* 20, 1-36.

Syria

Kumerloeve, H. (1967-69) Recherches sur l'avifaune de la Republique Arabe Syrienne, essai d'un aperçu. *Alauda* 35, 243-266; 36, 1-26, 190-207; 37, 43-58, 114-134, 188-205.

Baumgart, W., & B. Stephan (1987) Ergebnisse ornithologischer Beobachtungen in der Syrischen Arabischen Republik 2, Passeriformes. *Mitt. Zool. Mus. Berlin* 63, suppl. *Ann. Orn.* 11, 57-95.

USSR (former)

Buturlin, S. A. (1906) On the birds collected in Transcaucasia by Mr A. M. Kobylin. *Ibis* (8)6, 407-427.

Johansen, H. (1944-1957) Die Vogelfauna Westsiberiens. *J. Orn.* 91, 1-110; 92, 1-105, 145-204; 95, 64-110, 319-342; 96, 59-91, 382-410; 97, 206-219; 98, 155-171, 261-278, 397-415.

Dementiev, G. P., & N. A. Gladkov (eds) (1951-54) Ptitsy Sovetskogo Soyuza, vol. 1-6, Moskva.

Nicht, M. (1961) Beiträge zur Avifauna Armeniens. *Zool. Abh. Ber. Staatl. Mus. Tierkde. Dresden* 26, 79-99.

Stepanyan, L. S. (1990) Konspekt ornitologicheskoy fauny SSSR. Moskva.

Yugoslavia (former)

Stresemann, E. (1920) Avifauna Macedonica. München.

Matvejev, S. D., & V. F. Vasic (1973) Catalogus Faunae Jugoslaviae 4 (3), Aves. Ljubljana.

Matvejev, S. D. (1976) Preglad faune ptitsa Balkanskog Poluostrva: Conspectus Avifaunae Balcanicae 1. Beograd.

These reference lists are not complete surveys of the existing literature of each country, but only point to literature used in the species accounts. When a paper was produced in issues, only the parts referring to passerine birds are generally listed.

V   New subspecies of birds from Turkey and the surrounding countries

Inevitably, the study of many specimens of birds from Turkey and neighbouring countries led to the dismissal of some subspecies formerly considered to be valid. On the other hand, some easily distinguishable forms were found for which no name had yet been given. Rather than describing these forms in the species accounts, where they may be overlooked, I consider them below. I also take the opportunity to pay attention to a non-passerine bird.

Ceryle rudis

Ceryle rudis syriaca **nov. subspec**. Type: RMNH cat. 17, adult female, collected 'Syria', received from Maison Verreaux in 1863. Wing 151, tail 81, bill to feathers 60.8, bill to nostril 48.6 mm. Plumage freshly moulted.

This new subspecies is restricted to the Middle East. Apart from the type and another bird from 'Syria', specimens were examined from Turkey (Izmir, Antalya, and 'Harpara'), Cyprus (Larnaka; an irregular visitor only, according to Flint & Stewart 1992), Lebanon (Tyr), Israel ('plains of Genezareth' = round Lake Tiberias; also, 'near Jericho'), Jordan (Amman), Iraq (Al Faw, Al Qurnah, Qalat Salih, Baghdad, and Baqubah), and Iran (Shiraz, Mishun, Ahvaz, Mohammera, Shush, Telespid, and Lorestan). Specimens were examined in the museums of Amsterdam, Bonn, Leiden, Moscow, and Tring.

Diagnosis: similar to *C. rudis rudis* from Egypt in colour, but smaller in all measurements, except bill length (see table 1). Wing of adult *syriaca* is generally 145 mm and over, wing of adult *rudis* from Egypt (from the delta and Suez south to Manfalut) is generally less than 144 mm. In *syriaca*, only 4 of 29 adults have a wing of 144 mm or slightly less (all from Iran), in Egyptian *rudis*, only 3 out of 43 adults have a wing of 145 or over, and thus only 10% are wrongly classified when identifying subspecies on wing length; as the tail of Middle East birds is also longer than in birds from Egypt and the bill is thicker, especially in the middle, the combination of all measurements leads to an almost complete separation between the two taxa. Sexes do not differ in size and are combined in the samples but juveniles, as well as birds which have the wing-tip still formed by the juvenile primaries have an average wing length *c*. 2-4 mm less than that of adults and are excluded from the table; also, the bill is not fully grown during the first few months after fledging. In southern Egypt (south from Sohag), the wing is similar to that of birds from further north in Egypt, but the bill is shorter, apparently due to intergradation with the short-billed birds which occur in Sudan and Ethiopia. The difference in size is also supported by data from the literature, for instance, wing of birds from Syria and Jordan is 146.9 (145.5-148) (n=4, Sd 1.03) (Glutz von Blotzheim & Bauer 1980) and wing in northern Egypt (Cairo, Gizeh, and El Faiyum) is 139.8 (135-142) (n=9, Sd 1.99) (birds from the Gizeh Zoological Museum, per P. L. Meininger & S. M. Goodman).

Etymology: the new subspecies is named after the type locality.

loc./no	wing			bill		
	mean	sd	range	mean	sd	range
(1) n=14	146.0	1.94	143-149	61.73	2.77	57-66
(2) n=17	147.2	1.64	145-151	62.18	2.40	58-67
(3) n=24	140.6	2.73	136-145	60.80	2.75	57-65
(4) n=10	140.0	2.09	137-143	59.17	2.26	57-64
(5) n=11	141.4	2.80	137-146	61.73	2.80	57-66
(6) n= 7	136.9	4.15	137-140	57.10	1.32	56-60

Table 1. Length of wing and bill of adult *Ceryle rudis*, in sequence of locality, number examined, mean, standard deviation, and observed range. *Ceryle rudis syriaca*: (1) Iraq and Iran, (2) Turkey, Levant, and Cyprus. *Ceryle rudis rudis*: (3) northern Egypt (Suez Canal area, Nile Delta, and El Faiyum), (4) southern Egypt (El Minya to Aswan), (5) 'Egypt', without further details, (6) Sudan and Ethiopia. All data by CSR.

Eremophila alpestris

Eremophila alpestris kumerloevei **nov. subspec.** Type: BMNH, cat. nr. 1890.1.29.56, ♂, collected 30 April 1876 at Bereketlü (near Çamardi, eastern Taurus mountains, Turkey) by C. G. Danford. Wing of the type 118.5 mm.

Restricted to the mountains fringing the Central Plateau of Turkey. Apart from the type, another bird from Bereketlü was examined (BMNH 1879.4.5.58), as well as birds from Eregli (Taurus mountains), the Ulu Dag (Western Anatolia), 'Galatia' (where collected by Danford on 16 April 1879; according to the itinerary, in Danford 1880, the locality is on or near the Ulu Dag), and Elma Dag (near Ankara). These birds were compared with *balcanica* from the Balkan countries and Greece, with *penicillata* from eastern Turkey (Anbar, Hiniskale on eastern Bingöl Daglari, Hakkari, Baskale, Van, Erçek, and Verçenik), the Caucasus, and Transcaucasia, and with *bicornis* from the Lebanon mountains and Mount Hermon, all in the collections of the museums of Bonn, Leiden, and Tring.

Diagnosis: in moderately worn plumage (during about April-May), the following differences can be observed:

(1) the hindneck of the male of *kumerloevei* is vinous-pink, and the mantle and scapulars are greyish-pink, virtually without a buff tinge (in *bicornis*, all the upperparts are uniform vinous-buff, without a contrastingly coloured hindneck; both *balcanica* and *penicillata* also have a contrasting vinous-pink hindneck, but the mantle and scapulars of *balcanica* are purer ash-grey and those of *penicillata* are darker, dull pinkish grey-buff).

(2) the mantle and scapulars of the male of *kumerloevei* show faint dark shaft-streaks, or these streaks are virtually absent (in *penicillata* and *balcanica*, the streaks are narrow but distinct, more so than in *kumerloevei*; in *bicornis*, the streaks are absent).

(3) the upper tail-coverts and lesser upper wing-coverts of the male of *kumerloevei* are pink-grey, like those of *balcanica* and *penicillata*, but those of *bicornis* are buff or slightly rufous.

(4) the yellow of the face and throat of both sexes of *kumerloevei* is rather pale (paler in the female than in the male); the yellow of *balcanica* and *penicillata* is deeper lemon-yellow, that of *bicornis* is paler or almost white.

(5) the upperparts of the female of *kumerloevei* are pale pinkish sandy-buff with fine dull

(6) the average size of *kumerloevei* is intermediate between *penicillata* and *bicornis* (see table 2).

shaft-streaks (in the female of *balcanica*, duller sandy-grey with heavier black shaft-streaks, in *penicillata* darker grey-buff with distinct streaking, in *bicornis* sandy-buff with a pink tinge and much fainter dark streaks).

This subspecies is easily distinguished from *penicillata*, *bicornis*, and *balcanica* on colour. Although intermediate in certain aspects between *bicornis* and *penicillata* (especially the female), it is clearly different from both. In the past, the birds from the mountains surrounding the Central Plateau were either assumed to be *bicornis* (Stresemann 1928, Hartert & Steinbacher 1932-38, Vaurie 1959, Kumerloeve 1961a) or included in *penicillata* (e.g., in Vaurie 1951), apparently because insufficient series were compared with both *bicornis* and *penicillata*. In fresh plumage, the subspecies also probably differ from each other, but too few were examined to be certain of the characters. Bleaching and abrasion has a great influence on the plumage, and the autumn plumage of all subspecies is markedly different. In heavily worn plumage (June-July), the upperparts of all subspecies tends to become grey and the streaks become more distinct (least so in *bicornis*).

Etymology: this new subspecies is named in honour of H. Kumerloeve, as an esteem for his many contributions to ornithology and nature conservation in Turkey and the Middle East.

locality	wing ♂				wing ♀			
	mean	sd	n	range	mean	sd	n	range
(1)	121.3	2.74	12	118-126	109.5	2.00	5	107-112
(2)	120.4	2.42	10	117-124	109.4	1.03	4	108-111
(3)	118.4	2.73	8	114-121	107.6	2.14	4	106-111
(4)	115.8	2.26	23	112-120	107.6	1.60	10	105-110

Table 2. Length of wing of adult ♂ and ♀ of *Eremophila alpestris*. *E.a. penicillata*: (1) Caucasus and Transcaucasia, (2) eastern Turkey. *E.a. kumerloevei*: (3) western and central Anatolia, Turkey. *E.a. bicornis*: (4) Levant. All data by CSR.

Pyrrhula pyrrhula

Pyrrhula pyrrhula paphlagoniae **nov. subspec.** Type: ZFMK 39177, a male collected on 25 September 1934 at Karadere near Bolu (western Black Sea Coastlands, Turkey) by H. Rössner. Wing 87, bill to skull 11.4, bill to nostril 8.3, bill depth at base 9.1, width of lower mandible at base 9.0mm.

Restricted to the forests of the western Black Sea Coastlands; perhaps this race also occurs in the north of Western Anatolia, e.g., on the Ulu Dag. Replaced by the larger subspecies *rossikowi* in the eastern part of the Black Sea Coastlands, in the Caucasus, in Transcaucasia, and (probably *rossikowi*) in north-west Iran.

Diagnosis: belongs to the *rossikowi*-group of subspecies, characterized by a slightly more swollen base of the bill, with bill depth equal to bill width instead of width larger than depth, and with the cutting edges of the upper mandible, as seen from above, convex instead of straight. *P. p. paphlagoniae* differs from *rossikowi* in smaller measurements: wing of ♂ of *paphlagoniae* from Karadere (including the type), Bolu Dagh, Abant Gölü, and Ilgaz Daglari 88.8 (84-91.5) (n=6), of ♀ 86.8 (85-88) (n=3) (CSR, including a single ♂ measured by Jordans & Steinbacher 1948), against wing of ♂ of *rossikowi* from north-east Turkey 91.2 (90.5-92) (n=3) and of ♀ 89, 90 (n=2) (CSR), and wing of ♂ of *rossikowi* from the Caucasus and Transcaucasia 91.4 (90-95) (n=11), of ♀ 88.9 (87-91) (n=8) (Buturlin 1906, Vaurie 1949, A. J. van Loon, CSR). All but 2 of the 9 birds of *paphlagoniae* have wing of 88 or less, all but 2 of 11 birds examined of *rossikowi* have wing of 90 or more. In general size, *paphlagoniae* is close to *germanica* from west-central Europe (see the species account of *Pyrrhula pyrrhula*), but the bill structure is different, the mantle and scapulars of the male are paler ash-grey, less dark blue-grey, and the sides of the head and underparts of the male are slightly deeper flame-red, less cerise-red or pink-red; the upperparts of the female are grey with a slight amount of drab-grey suffusion, less extensively washed drab-brown than in the female of *germanica*. Both size and structure of *paphlagoniae* are close to *caspica*, a little-known subspecies inhabiting the northern foothills of the Elburz mountains in northern Iran, but the colour of the male of the latter is darker grey above and deeper red below, as far as can be judged from the two specimens examined.

Etymology: named after Paphlagonia, the ancient name of the region in which the subspecies occurs.

VI ENDEMIC SUBSPECIES OF TURKEY

The following passerine subspecies have a breeding range restricted to Turkey, and their survival is entirely in the hands of the people of Turkey:

Calandrella brachydactyla woltersi
Calandrella cheleensis niethammeri
Galerida cristata subtaurica
Eremophila alpestris kumerloevei
Prunella modularis euxina
Monticola saxatilis coloratus
Prinia gracilis akyildizi
Phylloscopus collybita brevirostris
Panurus biarmicus kosswigi
Aegithalos caudatus tephronotus
Sitta europaea levantina
Sitta neumayer zarudnyi
Certhia brachydactyla stresemanni
Garrulus glandarius hansguentheri

Passer domesticus mayaudi
Montifringilla nivalis leucura
Pyrrhula pyrrhula paphlagoniae
Emberiza cineracea cineracea

To these, the following non-passerines can be added:

Francolinus francolinus billypayni
Dendrocopos medius anatoliae
Dendrocopos major paphlagoniae

Two subspecies in this list are apparently already extinct: *Francolinus francolinus billypayni* and *Panurus biarmicus kosswigi*, both from the area round Amik Gölü. A few endemics in this list are poorly characterized and perhaps may not deserve recognition, such as *Monticola saxatilis coloratus* (probably to be included in a monotypic *M. saxatilis*), *Phylloscopus collybita brevirostris* (probably to be included in *P. c. collybita*), and *Garrulus glandarius hansguentheri* (a variable intermediate between *G. g. anatoliae* and *G. g. ferdinandi* or *graecus*). Others may eventually occur marginally outside Turkish territory (e.g., in north-west Syria, Armenia, or on the Aegean islands) and thus are not true endemics, but the length of the list of endemics is still impressive. Some subspecies which are known to occur just outside Turkey, but whose main distribution (in range or in breeding numbers) is largely confined to Turkey, can be added to this list:

Melanocorypha rufescens rufescens
Calandrella brachydactyla artemisiana
Cinclus cinclus olympicus
Oenanthe finschii finschii
Oenanthe xanthoprymna xanthoprymna

Parus lugubris anatoliae
Garrulus glandarius anatoliae
Pyrrhocorax graculus digitatus
Carduelis carduelis niediecki
Emberiza schoeniclus caspia

Moreover, a single full species can also be included in the last category, *Sitta krueperi*, and probably *Larus armenicus*.
Of all these endemics or near-endemics, the majority occur on the Central Plateau or in the forests or scrubland on the slopes of the mountains surrounding the Central Plateau, with a few each in the area of Amik Gölü and in the Van basin, all of which can be considered to be areas of endemism.

VII THE NON-PASSERINE BIRDS OF TURKEY

Readers of this book will of course be interested to know whether a companion volume will appear on non-passerine birds. This will not be the case, for three reasons:

(1) The present book on passerine birds was written because of interest in geographical variation. In most non-passerines, variation within a country like Turkey is much more limited. Many of the larger birds among the non-passerines are monotypic (thus, do not show geographical variation), or all the Turkish populations of a species belong to a single subspecies, which has a much wider breeding range than Turkey proper, e.g., extending throughout the Middle East or even throughout Eurasia. Of the 167 non-passerine birds supposed to breed in Turkey (see pp. 174-183), 47 (28%) are monotypic, while another 93 (56%) belong to widely distributed subspecies with only a single form in Turkey. Only 17 (10%) of the remaining species are known to show geographical variation within Turkey, resulting in recognition of 2 or more subspecies on Turkish territory, while 10 species (6%) are considered to be represented by a single subspecies in Turkey, though some influence of another subspecies from outside Turkey is apparent in one of the extreme corners of the country. In contrast, of the 147 Turkish passerine bird species listed in this book, 18% are monotypic, 29% are represented by a single more widespread subspecies, 9% are represented by a single subspecies but either show influence of a second or probably have a second present, while 44% are clearly represented by 2 or more subspecies. Thus, 53% of Turkish passerines show at least some geographical variation within the country, compared to only 17% of the non-passerines.

(2) When studying geographical variation, a series of specimens from each corner of a country should preferably be examined. However, a much smaller number of specimens of each non-passerine species is available than of passerine species, probably because non-passerines are generally larger than passerines and more difficult to preserve or more difficult to collect during the breeding season. For instance, of the c. 1850 bird skins mentioned in Jordans & Steinbacher (1948), Kumerloeve (1961a, 1968, 1969a, 1970b), and Rokitansky & Schifter (1971), only 18% (c. 325 skins) are of non-passerines, the remainder (c. 1525 skins) of passerines. With 167 species of non-passerines breeding in Turkey, an average of only 2 skins per species has been collected by these authors, against an average of 10 for each passerine species. It is often difficult to judge which subspecies occur in Turkey when only a few skins can be examined from widely separated localities, especially as these few skins are frequently collected in other seasons or belong to different sexes and age groups, and are thus difficult to compare. Also, when more subspecies are supposed to occur in Turkey, it is virtually impossible to map the boundaries between them withonly a limited number of specimens available.

(3) Plotting data on maps obtained from literature which is often over 100 years old would mean that for many non-passerine species one is mapping historical distribution rather than a recent picture. Breeding ranges of many non-passerines are shrinking, in contrast to those of most passerines. Either mapping should be done using only recent data, resulting in incomplete maps, or a more complicated system of symbols should be used, which not only denote possible, probable, and certain breeding but also the years in which breeding occurred.

As a starting point for research on geographical variation of Turkish non-passerine birds, a list of all non-passerine subspecies is included in an appendix. In contrast to the passerine species in the main list, no characteristics of subspecies or measurements are given in this list; for many species, these can be found in the Handbook of the Birds of Europe, the Middle East, and North Africa (BWP).

VIII ACKNOWLEDGEMENTS

Examination of specimens would not have been possible without the kind permission and hospitality of the staff of the museums I visited during the last 20 years: Dr. R. van den Elzen, Dr. G. Rheinwald, and Frau M. Krowinnus in Bonn, Dr. R. Prys-Jones, Mr. M. Walters and Mr. P. Colston in Tring, Dr. G. F. Mees and Dr. R. W. R. J. Dekker in Leiden, and Dr. J. Wattel and Ms. T. G. Prins in Amsterdam. No less valuable was the help received from Dr. P. Tomkovich in Moscow, Mr. S. Eck in Dresden, Dr. G. Mauersberger in Berlin, and from Dr und Frau H. Schifter and Dr E. Bauernfeind in Wien during an occasional visit in more recent years. Prof. Dr. K. H. Voous gave permission to consult his reprint collection, which contained many rare articles on the ornithology of the Middle East, which are otherwise difficult to obtain. Other literature was readily supplied by Mr. J. Paul, Ms. E. Zwart, and Drs. F. Pieters of the Plantage Library of the University of Amsterdam. Earlier versions of the text were read by Prof. Dr. F. Schram and Dr. J. Wattel. Drs. A. J. van Loon measured a number of flycatchers, treecreepers, and finches and Drs. W. R. R. de Batz, Drs. G. O. Keyl, and Drs. M. Platteeuw supplied data for a number of buntings. Various information was supplied by Arnoud B. van den Berg, Adriaan Dijksen, René Dekker, Euan Dunn, Steve M. Goodman, Peter L. Meininger, Hadoram Shirihai, Dorothy Vincent, Michael Walters and Mike Wilson. Last but not least I express my gratitude to my wife Irene as well as to Saskia and Jan for showing so much patience when I was supposed to behave socially in the evenings and weekends, and when plotting maps while being on a family holiday. Without all this, this book could not have been written, and I extend my sincere thanks to all who helped or showed interest in the project.

IX   Abbreviations and symbols

Abbreviations and symbols in the text

♂   male
♀   female
W   described from west of Turkey
T   described from Turkey itself
S   described from south of Turkey
E   described from east of Turkey
CSR   initials of the author, C. S. Roselaar

Symbols used on the maps

●   probable or certain breeding
○   possible breeding
<>   intermediate between the two subspecies named
>   intermediate between both subspecies, but nearer to the first-named
ssp? subspecies unknown
?   possibly this subspecies
(?)   probably this subspecies
x/y ? either subspecies x, subspecies y, partly x and partly y, or intermediates between x and y

X SPECIES ACCOUNTS AND MAPS OF TURKISH PASSERINE BIRDS

AMMOMANES CINCTURUS

Bar-tailed Desert Lark   Çöl Toygarı

Habitat Sparsely vegetated sandy desert, usually flat or gently sloping.

Distribution No map. 'Seen at 4 places between Birecik and Diyarbakir, showing breeding behaviour' (*Bull. Orn. Soc. Turkey* 13, 2-5, 1976). This record is from a period before the breeding of *A. deserti* in Turkey was proved, and the species was perhaps wrongly identified. Note that *A. cincturus* is highly erratic in breeding, settling anywhere in the Syrian-Arabian desert where rains may have fallen, and occurrence or even breeding in Turkey is not entirely unlikely.

Geographical Variation

Subspecies described or recorded in the region:

(S)   ***A. c. arenicolor*** (Sundevall), lower Egypt. [A pale sandy-pink subspecies, breeding from north-west Africa east to Arabia and the Middle East; not as dark rufous as *cincturus* from the Cape Verde Islands, nor as dusky brown-grey as *zarudnyi* which breeds from central Iran east to Pakistan. Wing of ♂ from north-west Africa 95.4 (91-97) (n=16), of ♀ 89.6 (86-93) (n=19) (Roselaar in Cramp 1988).]

Subspecies recognized in Turkey: never collected. If the species really occurs, it is unlikely to belong to anything else than *arenicolor*.

AMMOMANES DESERTI

Desert Lark

Habitat Sparsely vegetated stony plateaux and slopes with scattered rocks and boulders.

Distribution See map 1. Restricted to the Birecik area, where first found in 1983; breeding proved (*Bull. Orn. Soc. Middle East* 13, 8-12, 1984; *ibid.* 20, 11-20, 1988). A record of an '*Ammomanes*' near Igdir in the East on 6 August 1992 (Frost & Hornbuckle 1992) is in need of verification; perhaps a juvenile of another species was involved (e.g., *Eremophila alpestris*).

Geographical Variation.

Subspecies described or recorded in the region:

(S)   ***A. d. coxi*** Meinertzhagen, 1923, El Qaryatein (Syria). [A pale sandy-grey or brown-grey subspecies of fairly large size. Wing of ♂ from Syria and western Iraq 105 (102-108) (n=7), of ♀ 97.5 (95-100) (n=3) (Hartert & Steinbacher 1932-8; Roselaar in Cramp 1988). Much paler than *annae* from the black lava plateaux of Jordan, larger than the equally pale *isabellinus*, which occurs from southern Jordan and southern Iraq south-west to Egypt.]

Subspecies recognized in Turkey: none collected and thus no certainty about subspecies, but only *coxi* of Syria occurs nearby and this subspecies is likely to be involved.

RAMPHOCORIS CLOTBEY

Thick-billed Lark

Habitat Sparsely vegetated stony desert.

Distribution See map 1. Seen 31 May 1975 [thus, in breeding season] near Amik Gölü (*Bull. Orn. Soc. Turkey* 13, 2-5, 1976). No proof of breeding, and confirmation is required of the identification. The species is an erratic breeder, like *Ammomanes cincturus* (above) and, e.g., Dunn's Lark *Eremalauda dunni*, though to a less extent than these. It has been found breeding as nearby as the 'Syrian desert' (probably the Qaryatein-Palmyra area) by Aharoni (1931, 1932).

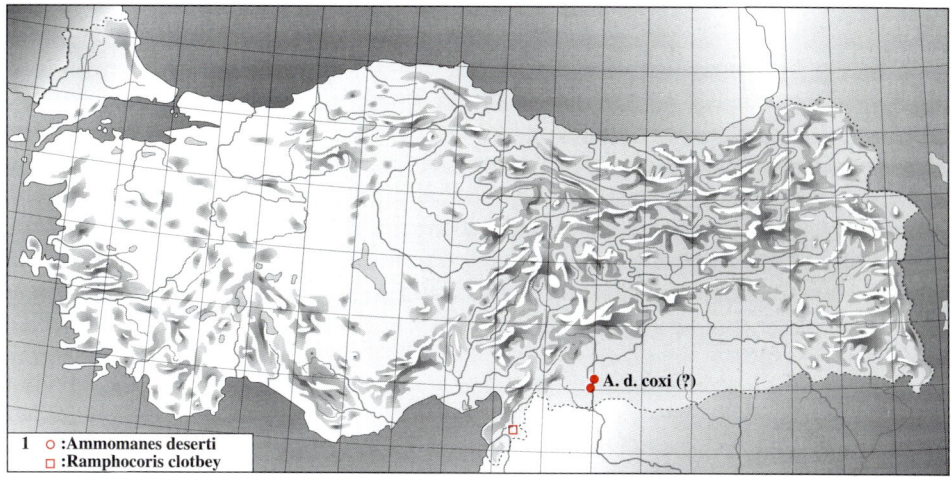

Geographical Variation None. *R. clotbey* (Bonaparte), 1850, originally described from the Egyptian desert, and occurring along the northern fringe of the Sahara from Morocco to Egypt as well as along the western fringe of the Syrian-Arabian desert, is considered to be monotypic. Only a few birds from the Syrian desert were examined, and these did not differ in size or plumage from birds of North Africa (Roselaar in Cramp 1988). Never collected in Turkey.

References Aharoni, J. (1931) Brutbiologisches aus der Syrische Wüste und dem Libanon. *Beitr. Fortpfl. Biol. Vögel* 7, 161-166, 222-226. Aharoni, J. (1932) Bemerkungen und Ergänzungen zu R. Meinertzhagens Werk 'Nicoll's Birds of Egypt'. *J. Orn.* 80, 416-424.

MELANOCORYPHA CALANDRA
Calandra Lark   Bogmakli Tarlakusu

Habitat Arid open steppe or arable land on flat or undulating ground, covered with tufts of low grasses, herbs, or crops, generally devoid of larger stones, mainly at lower altitudes (below 1200m) but locally up to 1800-2000m in the east.

Distribution See map 2. Common on plains and fields of west and central Turkey, but scarce or absent at higher altitudes in the east, where restricted to some wide valleys and open plains. Absent from the Black Sea Coastlands.

Geographical Variation

Subspecies described or recorded in the region:

(W/E) ***M. c. calandra*** (Linnaeus), 1766, Pyrénées (France/Spain). [Upperparts brown-grey or medium olive-grey with little or no cinnamon tinge and with rather large dark marks on centres of feathers; breast cream or greyish-white. Wing of ♂ from southern Europe 130.7 (126-134) (n=15), of ♀ 118.5 (116-121) (n=15), of ♂ in the Caucasus area and northern Iran 131.8 (128-137) (n=8), of ♀ 117 (n=1) (CSR). *M. c. hollomi* Kumerloeve, 1969, described from Yüksekova (extreme south-east Turkey) is a synonym. Upperparts of the

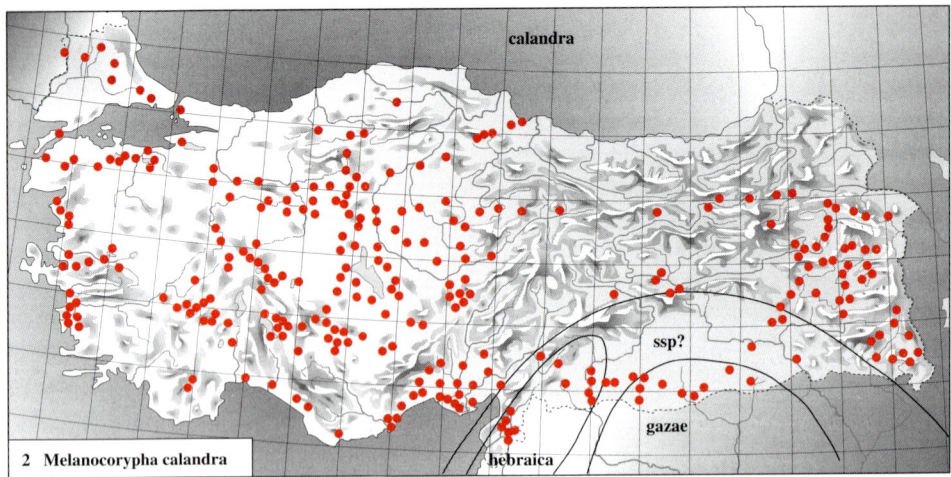

2 *Melanocorypha calandra*

latter were described as pale brown-grey with little or no cinnamon tinge and with narrow dark marks, paler than *calandra*, less cinnamon and larger in size than other subspecies (Kumerloeve 1969a). The type of *'hollomi'* and other specimens examined from the type locality show slightly larger dark feather centres than *calandra* and slightly more pronounced cinnamon feather fringes (*contra* Kumerloeve 1969a), somewhat tending to *hebraica*, but the difference from *calandra* is not marked enough to recognize *hollomi* and the difference in size from other populations of *calandra* from Turkey is slight (see below).]

(S) **M. c. hebraica** Meinertzhagen, 1920, Jenin (northern Israel). [Feathers of upperparts narrowly fringed rufous-cinnamon (when fresh) to dull sandy-grey (when worn), dark feather centres on upperparts and spots on cheeks and breast rather large, black, often more contrasting with remainder of feathering than in *calandra*; breast cinnamon with distinct black streaks. Wing of ♂ from Israel, Lebanon, western Jordan, and western Syria 134.4 (131-142) (n=17), of ♀ 120 (n=1) (CSR).]

(S) **M. c. gazae** Meinertzhagen, 1919, Shellal near Gaza (Sinai). [A pale grey race with narrower olive-brown or dull black streaks on upperparts than previous subspecies and with broader and paler pink-cinnamon or sandy-cinnamon feather fringes; breast cream-cinnamon with fairly pronounced dark streaking. Wing of ♂ from north-east Sinai, Gaza, and Amman (Jordan) 133.0 (129-136) (n=6), of ♀ 119.2 (115-124) (n=8) (CSR). *M. c. dathei* Kumerloeve, 1970, described from Ceylanpinar (south-east Turkey) is a synonym (CSR).]

(E) **M. c. psammochroa** Hartert, 1904, Dur Badom (Eastern Iran). [Feathers of upperparts broadly fringed pink-cinnamon (when fresh) to sandy-grey or sandy-yellow (when worn), dark feather centres on upperparts as well as dark spots on cheeks and breast small and inconspicuous; breast cream or whitish; the palest subspecies. Wing of ♂ from southern, central, and eastern Iran 133.4 (125-138) (n=17), of ♀ 119.6 (117-125) (n=7) (Paludan 1940, Diesselhorst 1962, Érard & Etchécopar 1970, CSR).]

Subspecies recognized in Turkey: most birds examined are *calandra*, but those from Birecik are best considered to belong to *hebraica* (though with some influence of *calandra*), while those of Ceylanpinar (*'dathei'*) are *gazae*. Wing of ♂ of *calandra* from the west coast (Aydin, Priene) 129.6 (125-135) (n=9), of ♀ 116.0 (113-119) (n=6) (Weigold 1913-14, CSR), of ♂ of *calandra* from the Central Plateau (Eskisehir and Burdur to Çubuk, Tuz

Gölü, and Eregli) 133.5 (127-138.5) (n=10), of ♀ 121.1 (118-123) (n=4), of ♂ of *calandra* from the east of the Southern Coastlands (Mersin, Haruniye, Maras, and Amik area) 132.2 (128-135) (n=6), of ♀ 123.5 (119-128) (n=4); wing of ♂ of *'hollomi'* from Yüksekova, Van area, and Erzurum 135.2 (131-139) (n=9) (CSR). Wing of ♂ of *hebraica* from Birecik 134.3 (132-136) (n=3) (CSR). Wing of ♂ of *gazae* (*'dathei'*) from Ceylanpinar (including the type) 136.1 (133-140) (n=4), of ♀ 121.5 (n=1) (CSR). No *psammochroa* breeds in Turkey but as this subspecies winters nearby in Iraq, Jordan, and Israel, it may occur in winter in Turkey.

MELANOCORYPHA BIMACULATA
Bimaculated Lark   **Küçük Bogmakli Tarlakusu**

Habitat Mainly on stony or gravelly uncultivated and sparsely vegetated ground at lower altitudes of the Central Plateau and the South-East (in contrast, *M. calandra* is often on cultivated fields here); in the Taurus and the East, in all open cultivated or uncultivated habitats, also at higher altitudes (*M. calandra* here largely absent); occurs to over 2000m in the Taurus and up to *c.* 2400m in the east.

Distribution See map 3. Restricted to the Central Plateau, the Taurus, the South-East, and the East, avoiding all western Turkey and the coastlands.

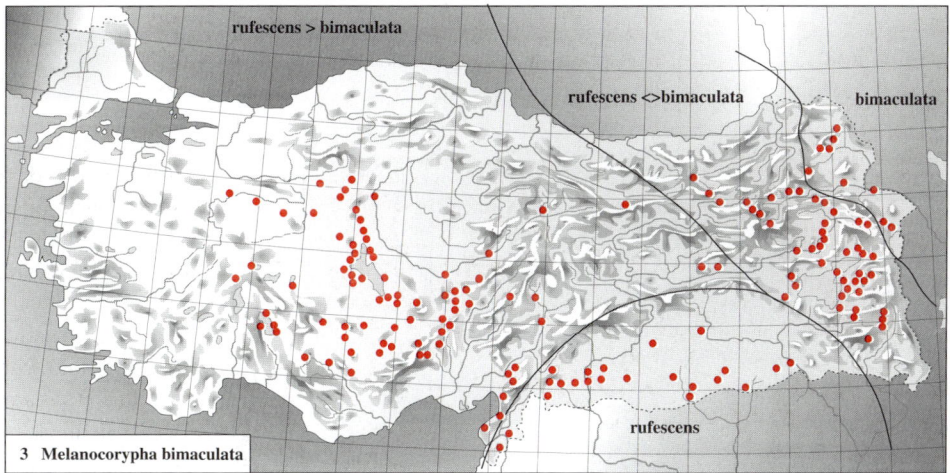

3 *Melanocorypha bimaculata*

Geographical Variation
Subspecies described or recorded in the region:

(S)   ***M. b. rufescens*** C. L. Brehm, 1855, Blue Nile (Sudan, in winter). [Upperparts brown- or olive-grey, feathers with ill-defined blackish centres and deep pink to tawny fringes, paler and more uniform sandy-grey or buff-grey when worn; supercilium tawny-pink, inconspicuous when plumage fresh; ear-coverts, breast, and fringes of flight feathers tinged tawny. Wing of ♂ from Israel, Jordan, and southern Syria 121.3 (120-123) (n=5), of ♀ 116 (n=1) (CSR).]

(E)   ***M. b. bimaculata*** (Ménétries), 1823, Talysh mountains (Azerbaijan, south-east Transcaucasia). [Upperparts paler and more uniform grey than in *rufescens*, slightly suffused with paler cinnamon only when plumage quite fresh, and with dark streaks slightly heavier; supercilium whiter and more contrasting; ear-coverts, breast, and fringes of flight feathers tinged grey-buff. Wing of ♂ in north-west and south-west Iran 124.0 (119-128) (n=13), of ♀ 112.6 (110-116) (n=8) (CSR). Occurs in Transcaucasia and northern Iran. East from north-east Iran and Transcaspia, it is replaced by *torquata* Blyth, 1847, described from Afghanistan, which is still paler and greyer, showing less pronounced and narrower dark streaks on upperparts, and with wing of ♂ of 120.1 (116-125) (n=11), of ♀ 114.7 (114-117) (n=5) (Érard & Etchécopar 1970, CSR.]

Subspecies recognized in Turkey: *rufescens* runs very smoothly into *bimaculata*, and the boundaries between both are hard to define. Birds from the mountains and hills of the Levant are similar to wintering birds from the type locality of *rufescens* in Sudan (CSR), and birds from Ceylanpinar are inseparable, too. Birds from Igdir and Kars in north-east Turkey are indistinguishable from typical *bimaculata*. All other birds from Turkey are more or less intermediate between both races. Populations breeding on the Central Plateau and in the Taurus, as well as those of the Amanus, Gaziantep, and Birecik areas are best included in *rufescens*, being close to typical specimens of this subspecies; birds from Van and Erçek areas are fully intermediate. Wing of ♂ *rufescens* from south-central Turkey (Haruniye, Gaziantep, Birecik, Ceylanpinar) 124.0 (119-129) (n=11), of ♀ 113.5, 114 (n=2), of ♂ of *rufescens* from the Central Plateau and the Taurus (Ankara, Beynam, Beysehir, Tuz Gölü, Anasha) 124.0 (122-126) (n=6), of ♀ 113.9 (111-115) (n=4), of ♂ of intermediates from Van and Erçek areas 122.5 (118.5-125) (n=14), of ♀ 115.8 (113.5-117.5) (n=3) (CSR), of *bimaculata* from extreme north-east Turkey 125.7 (125-127) (n=5), of ♀ 114.5 (n=1) (Érard & Etchécopar 1970, CSR). No *torquata* has been recorded in Turkey.

CALANDRELLA BRACHYDACTYLA
Short-toed Lark   Bozkir Toygari

Habitat Dry flat or hilly ground, barren, fallow, or cultivated with low crops, open or covered with scrub alternated with open spaces; generally restricted to areas below 1000m in the west and below 1500m in the east.

Distribution See map 4; also Kumerloeve (1967a). Widespread and locally common in the west, scarce in the east, avoiding higher altitudes everywhere. Virtually absent from the Black Sea Coastlands and the Kurdish Alps.

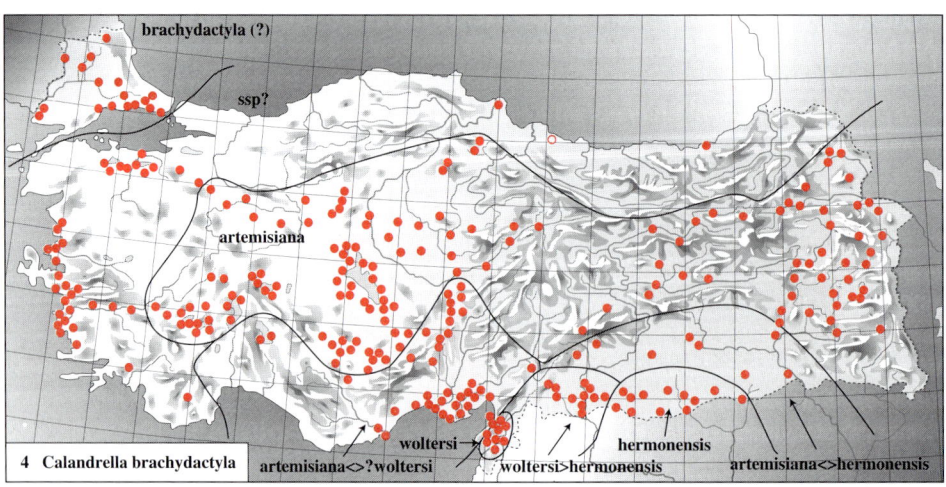

4 Calandrella brachydactyla

Geographical Variation

Subspecies described or recorded in the region:

(W) **C. b. brachydactyla** (Leisler), 1814, Montpellier (France). [Upperparts rufous-brown (when fresh) to cinnamon-buff or pink-buff (when worn), feathers with rather broad dull black centres, crown usually contrastingly intense rufous, but latter more often so in the south-western part of its south European range than in the south-east; ground-colour of breast buff to cream. Wing of ♂ from Italy, former Yugoslavia, and Greece 96.2 (92-102) (n=37), of ♀ 90.1 (86-94) (n=17), bill of both sexes 14.2 (13.4-15.2) (n=13) (Roselaar in Cramp 1988).]

(T) **C. b. woltersi** Kumerloeve, 1969, Amik Gölü (southern Turkey). [Upperparts pale grey with slight pink tinge and broad contrastingly black streaks, paler, greyer, and more heavily streaked than in *hermonensis* (see below); streaks heavier and more contrasting than in *longipennis*; ground-colour not as rufous as in *brachydactyla*: ground-colour of breast pale cinnamon-cream; cap like upperparts, but usually with smaller or larger rufous spots. See also Kumerloeve (1969b). Wing of ♂ from Amik Gölü 96.2 (94-98.5) (n=4), of ♀ 91.2 (89-92) (n=3), bill of both sexes 14.1 (13.3-14.7) (n=7) (CSR).]

(S) **C. b. hermonensis** Tristram, 1864, Mount Hermon (Lebanon). [A rufous subspecies; upperparts pink- or sandy-cinnamon, marked with rather narrow ill-defined dark brown-olive streaks; colour of crown usually not contrasting with the remainder of the upperparts; breast more cinnamon than in other subspecies. Wing of ♂ from Israel, Lebanon, Jordan, and Syria 93.2 (91-97) (n=8), of ♀ 89.8 (86-92) (n=7), bill of both sexes 13.8 (13.1-14.4) (n=9) (CSR).]

(E) **C. b. artemisiana** Banjkovski, 1913, Tbilisi (Georgia, western Transcaucasia). [Upperparts well-marked with black streaks, as in *woltersi*, but ground-colour slightly darker, grey with pink or buff tinge; cap rufous, or grey with rufous dots, but occasionally similar to the remainder of the upperparts; breast cream to off-white; bill long, tip laterally compressed. Wing of ♂ from north-east Turkey (Malatya, Van area, Erçis, Erzurum, Igdir, Agri, Ani, Rize) 96.2 (94.5-99) (n=9), of ♀ 91.7 (88-94) (n=6), bill 14.6 (13.6-15.5) (n=13); a few examined from the Caucasus and Transcaucasia similar to these (CSR).]

(E) **C. b. longipennis** (Eversmann), 1848, Dzhungaria (north-west Sinkiang, China). [Upperparts sandy-grey, dark streaks narrow and ill-defined; cap similar to upperparts; breast cream or off-white; bill small. Wing of ♂ from the plains of the lower Volga and the Crimea east to Lake Zaisan 96.3 (93-101) (n=12), of ♀ 91.6 (87-96) (n=12), bill 13.3 (12.4-13.9) (n=21) (CSR).]

Subspecies recognized in Turkey: variation obscured by strong influence of wear, marked individual variation, widespread occurrence of wintering *longipennis* throughout Turkey, and absence of clear natural boundaries between many subspecies. All races recorded in the region (see above) are breeding, except for *longipennis*, but quite a number of populations combine characters of several races or are variable. Nominate *brachydactyla* occurs in Thrace; wing here 96 and 93 (♂) and 90 (♀) (Rokitansky & Schifter 1971, CSR). This subspecies may occur also in coastal Western Anatolia, where the species is common, but the few spring birds examined from this area were apparently migrant *longipennis*. *C. b. artemisiana* from the Caucasus area and eastern Turkey occurs west across the Central Plateau to at least Burdur and Eskisehir; wing of ♂ from the Central Plateau 95.4 (93-100) (n=9), of ♀ 93.0 (91-95) (n=10) (CSR). *C. b. hermonensis* is restricted to the area round Ceylanpinar and Urfa in the South-East, contiguous with its distribution further south in the Middle East; wing of ♂ from Ceylanpinar and Urfa 94.4 (92-

96) (n=12), of ♀ 90.2 (89-92) (n=4). *C. b. woltersi* has a very restricted range round Amik Gölü, not occurring outside Turkey. Birds from Birecik and Gaziantep are intermediate between *woltersi* and *hermonensis*, though slightly nearer to *woltersi*; wing of ♂ from this area 95.6 (93-100) (n=10), of ♀ 88.4 (87-90.5) (n=5), bill 14.3 (13.4-15.3) (n=10) (CSR). Birds from the Southern Coastlands from Silifke to Iskenderun and Maras, as well as birds breeding in the Taurus near Beysehir and in the northern foothills of the Taurus north to Karaman, Ulukisla, Ürgüp, and Kayseri, are similar to *artemisiana* but more cinnamon on the upperparts and with a shorter bill; wing of ♂ of these birds 96.8 (94-100) (n=9), of ♀ 92.8 (89-95) (n=3); both these birds and those of Siverek, Diyarbakir, and Siirt can be considered as intermediates between *hermonensis* or *woltersi* and *artemisiana* (CSR).

References Kumerloeve, H. (1967a) Zum Brutverbreitung der beiden *Calandrella* -Arten Kleinasiens. *Bonner Zool. Beitr.* 13, 509-519. Kumerloeve, H. (1969b) Zur Rassenbildung der Kurzzehenlerche, *Calandrella brachydactyla*, im vorderasiatischen Raum. *J. Orn.* 110, 324-325.

CALANDRELLA RUFESCENS
Lesser Short-toed Lark   Küçük Bozkir Toygari

Habitat Flat open ground with scanty tufts of vegetation, often clayey or silty, and generally more open than in *C. brachydactyla*.

Distribution See map 5; also Kumerloeve (1967a). Mainly restricted to plains and valleys in the east and south-east, where locally common on cultivated or uncultivated open gravelly and clayey ground up to at least 2000m. Records of breeding in coastal Western Anatolia (Menemen and Bafa areas) and the Southern Coastlands (Antalya, Göksu delta, Çukurova deltas, Amik Gölü) require confirmation, particularly as it is not clear whether *C. rufescens* or *C. cheleensis* is involved (if misidentification as *C. brachydactyla* can be excluded).

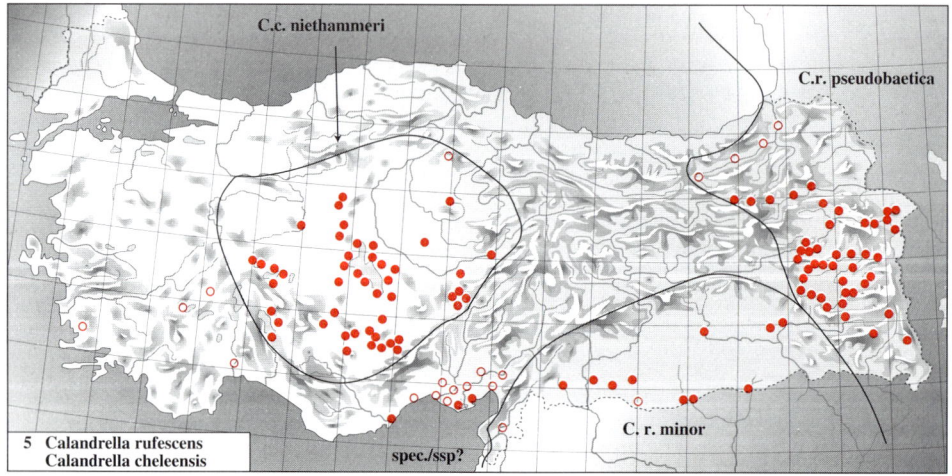

5 Calandrella rufescens / Calandrella cheleensis

Geographical Variation
Subspecies described or recorded in the region:

(S)   ***C. r. minor*** (Cabanis), 1851, El Hammam well (south-west of Alexandria, Egypt). [Upperparts pink-cinnamon with dark olive-brown to black feather centres, but rump and upper tail-coverts of ♂ almost uniform cinnamon; breast pale cream-buff with short black streaks 1-2 (in ♂) or 0.5-1 (in ♀) mm wide. Populations from the Middle East are rather large in size: wing of ♂ from north-east Sinai, Amman (Jordan), and Syria 93.4 (92-95) (n=5), wing of ♂ from Ceylanpinar 93.2 (92-94) (n=5), of ♀ from Ceylanpinar 87, 88 (n=2), tail of ♂ from the Middle East (Ceylanpinar to Sinai) 55.5 (54-58) (n=8), of ♀ 52.5, 53.5 (n=2), bill of both sexes from the Middle East 12.2 (11.4-12.8) (n=12), bill depth 5.2 (5.0-5.6) (n=12) (CSR). Typical *minor* from North Africa is on average smaller: wing of ♂ from Morocco, Algeria, and Tunisia 90.6 (88-95) (n=20), of ♀ 84.5 (82.5-87) (n=13), tail of ♂ 53.0 (50.5-55.5) (n=17), of ♀ 48.6 (46-51.5) (n=13), bill of both sexes 12.1 (10.1-13.3) (n=29), bill depth 5.2 (4.7-5.7) (n=16) (CSR).]

(E) **C. r. pseudobaetica** Stegmann, 1932, Kapa-Siva (at border between Iraq, Iran, and Turkey). [Upperparts rather dark, dark grey-brown or dark olive-grey with slight dull pink wash, feather centres with broad olive-black streaks up to c. 4 mm wide; breast pale cinnamon-buff with narrow and short black streaks 0.5-1 mm wide; rather like *heinei* (below), but ground-colour darker, dark streaks slightly heavier, and bill thicker at base. In the original description, *pseudobaetica* was said to be small, wing 87-95 (n=9) (Stegmann 1932), but birds from eastern Turkey are large, only slightly smaller than *heinei* of central Asia, which is the largest subspecies. Wing of ♂ from the Van area, Erçek Gölü, Diyarbakir, Agri, and the upper Murat valley 97.7 (94.5-102) (n=21), of ♀ 91.0 (88-93) (n=6), tail of ♂ 57.1 (54-59) (n=10), of ♀ 54.0 (52-55) (n=4), bill (both sexes) 12.9 (11.8-14.3) (n=22), bill depth at base (both sexes) 6.0 (5.5-6.3) (n=21) (CSR).]

(E) **C. r. heinei** (Homeyer), 1873, Volga area. [Upperparts grey with slight pink-buff tinge, marked with rather narrow but well-defined black shaft streaks up to c. 2 mm wide; breast pale cream-buff with short black streaks c. 1 mm wide. Wing of ♂ from the lower Volga (south European Russia) east to eastern Kazakhstan 98.5 (95-102) (n=23), of ♀ 93.8 (89-98) (n=13), tail of ♂ 59.7 (57-63) (n=10), of ♀ 56.2 (54-58) (n=7), bill 12.5 (11.3-13.3) (n=33), bill depth at base 5.3 (5.0-5.5) (n=10) (CSR).]

Subspecies recognized in Turkey: unlike *C. brachydactyla*, the races of *C. rufescens* (and its close relative *C. cheleensis*) are well-marked, and each population is quite uniform in appearance (although a reference collection will be needed to identify single specimens). Though isolated from typical *minor* of North Africa (Morocco to north-west Egypt) by the very dark *nicolli* in the Nile delta, birds breeding from Gaziantep to Nusaybin in South-East Turkey, as well as the birds from inland Syria and Israel, Jordan, and the north-east Sinai peninsula, are all indistinguishable from *minor*. On average, birds from the Middle East are slightly larger (see above). The mountain valleys further east (north-east of a line from Askale to Tatvan and Yüksekova) are inhabited by *pseudobaetica*, a subspecies occurring also in southern and eastern Transcaucasia and in north-west Iran. *C. r. heinei* wanders in winter to, e.g., Iraq and Iran, and may winter in Turkey.

References Stegmann, B. (1932) *Calandrella minor pseudobaetica* subsp. nov. Orn. Monatsber. 40, 54.

CALANDRELLA CHELEENSIS

Asian Short-toed Lark

Habitat Level salt and soda plains with a sparse cover of herbs, even more barren than the habitat occupied by *C. rufescens*.

Distribution See map 5. Largely restricted to the barren fringes of salt and soda lakes of the Central Plateau, where locally very common.

Geograpical Variation

Subspecies described or recorded in the region:

(T)   **C. r. niethammeri** Kumerloeve, 1963, Tasköprü-Ortakoy (near Eber and Aksehir Gölü, Central Plateau). [Pale: upperparts light stone-grey (sometimes slightly tinged sandy), feather centres with contrasting black mark 1-2 mm wide; ground-colour of underparts white with slight cream tinge on breast; streaks on breast short, contrastingly deep black, 0.5-1 mm wide. Wing of ♂ from Eber, Tuz, Konya, and Eregli areas 99.3 (97-103.5) (n=11), of ♀ 92, 92.5 (n=2), tail of ♂ 63.8 (60-66) (n=7), of ♀ 58, 60 (n=2), bill of both sexes 13.4 (12.4-14.3) (n=13), bill depth at base 6.3 (6.1-6.7) (n=8) (CSR). See also Kumerloeve (1963b).]

(E)   **C. c. persica** (Sharpe), 1890, Neyriz (southern Iran). [Upperparts very pale (but less so than in *leucophaea* from central Asia, which is greyish-white above), pale sandy-grey with slight pink or cinnnamon tinge, marked by very narrow dark shaft streaks less than 0.5 mm wide; breast cream-white with short black or dark brown streaks c. 0.5 mm wide. Wing of ♂ from the plains and valleys of Iran and Iraq as well as from birds wintering further west in the Middle East 101.8 (97-106.5) (n=22), of ♀ 97.2 (92-103) (n=10), tail of ♂ 62.5 (58-67) (n=14), of ♀ 58.5 (55-62) (n=7), bill 13.6 (12.4-14.7) (n=24), bill depth at base 6.2 (5.7-6.6) (n=10) (CSR).]

Subspecies recognized in Turkey: all breeding birds from the Central Plateau are *niethammeri*. The subspecies *persica* is not yet known to occur in Turkey, but winters widely in Iraq, Kuwait, and Syria and thus may also wander to Turkey. *C. cheleensis* is now usually separated from *C. rufescens* due to the overlap in breeding range of *C. r. heinei* and *C. c. leucophaea* in Transcaspia without apparent interbreeding (Stepanyan 1967, 1990). Which of the subspecies formerly all included in *C. rufescens* belong to *C. cheleensis* and which to *C. rufescens sensu stricto* has not always been clear (see, e.g., Roselaar in Cramp 1988), but examination of most of the races of both groups shows that *niethammeri* and *persica* together with *leucophaea, seebohmi,* and other races further east are to be included in *C. cheleensis*, and *heinei, pseudobaetica, minor,* and other races further west are to be included in *C. rufescens*. No trenchant characters can be given to separate all subspecies of one species from all of the other, but in the area where both species overlap, *C. cheleensis* is the one with a relatively longer tail, with more white in the outer tail feathers, with paler ground-colour of the upperparts, and with narrower dark streaking. *C. cheleensis* is sometimes said to differ from *C. rufescens* by a thicker bill, but at least *leucophaea, seebohmi,* and *cheleensis* of *C. cheleensis* have a slender bill, and the bill of *pseudobaetica* is thicker than that of the other races of *C. rufescens*. Overlap in breeding range of *C. rufescens* and *C. cheleensis* in Syria, as inferred by Hartert (1921-22), has yet to be proven: birds of both species have been collected in El Qaryatein in central Syria, a smaller one inseparable from *C. rufescens minor* (represented, e.g., by a female in the Amsterdam museum, probably locally breeding, collected on 11 May 1914 by J. Aharoni), and a larger paler one, belonging to *C. cheleensis* and separated as *aharonii* Hartert, 1910, which is not certain to breed. Hartert compared the types of *aharonii* (a male with wing 104 and a female with wing 96.5) with the very pale *C. c. leucophaea* from the salt-deserts of central Asia, from which it was stated to differ by slightly darker grey or brown-grey upperparts, slightly broader, blacker, and more conspicuous streaks on the upperparts and breast, less white in the tail, and a thicker bill (Hartert 1921-22, Hartert & Steinbacher 1932-38). As can be deduced from this description, *aharonii* must be near to or similar to *persica*, rather than being an older name for *C. c. niethammeri* of

Central Anatolia. *C. c. persica* breeds from central and eastern Iraq through Iran (except the north) to southern Afghanistan, and eventually it may breed west to central Syria, too.

References Kumerloeve, H. (1963b) *Calandrella rufescens niethammeri*, eine neue Stummellerchenform aus Inneranatolien (Türkei). *Vogelwelt* 84, 146-148. Stepanyan, L. S. (1967) *Calandrella cheleensis* Swinhoe a valid species. *Acta Orn.* 10, 97-107.

GALERIDA CRISTATA
Crested Lark   **Tepeli Toygar**

Habitat On all kinds of open sparsely vegetated soils, usually near human habitation or cultivation, wastelands, roads, etc., generally below *c*. 1500m.

Distribution See map 6. Widespread and very common over much of Turkey, more local in the Black Sea Coastlands, the South-East, and the East, where generally scarce but locally common.

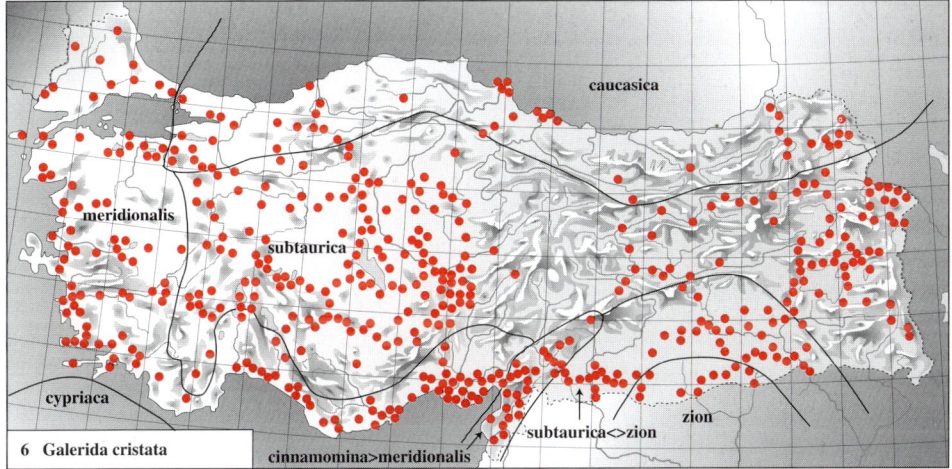

6 *Galerida cristata*

Geographical Variation
Subspecies described or recorded in the region:

(W)   ***G. c. meridionalis*** C. L. Brehm, 1841, Dalmatia (Croatia). [Dark: upperparts dark grey-brown, upperparts and breast marked with rather broad but ill-defined black streaks; a slight cinnamon tinge on the hindneck. Wing of ♂ from Dalmatia, mainland Greece, and Bulgaria 108.7 (106-111.5) (n=9), of ♀ 101.0 (99-102) (n=3); wing of ♂ from Crete 107.3 (104.5-110) (n=19), of ♀ 100.0 (96-103) (n=14) (CSR). Includes *ioniae* (Kollibay), 1912, described from Priene (near the mouth of the Büyük Menderes, Turkey).]

(T)   ***G. c. subtaurica*** (Kollibay), 1912, Eregli. [Pale: upperparts pale olive-grey (when fresh) to sandy isabelline-grey (when worn) with narrow and rather faint brown streaks; underparts cream with rather heavy short black streaks on breast; bill rather long (for bill and wing measurements and plumage characters of this and many other subspecies, see Watson 1962a and Abs 1963). Wing of ♂ from central Turkey (Elmali, Burdur, Beysehir, Ankara, Bolu, Konya, Eregli, Kayseri) 110.2 (108-113) (n=20), of ♀ 103.8 (102-107) (n=10) (Kummerlöwe & Niethammer 1934-35, Rössner 1935, Jordans & Steinbacher 1948, CSR), wing of ♂ from eastern Turkey (Siirt, Van area, Erzurum, Yüksekova) 109.5 (107-113) (n=13), of ♀ 103.3 (101.5-106.5) (n=3) (CSR), wing of ♂ from eastern Iraq and north-west Iran 110.6 (107-117) (n=20), of ♀ 105.4 (100-115) (n=10) (Vaurie 1951). Includes *ankarae* Kummerlöwe & Niethammer, 1934, described from Ankara (Central Plateau, Turkey).]

(S) **G. c. cypriaca** Bianchi, 1907, Cyprus. [A pale subspecies, close to *subtaurica*, but streaks on upperparts and breast slightly larger and blacker. Small: wing of ♂ from Cyprus 104.4 (103-106) (n=5), of ♀ 97.9 (93-100) (n=8) (CSR).]

(S) **G. c. cinnamomina** Hartert, 1904, Mount Carmel (near Haifa, Israel). [A rufous subspecies, rather heavily streaked black-brown on a cinnamon-rufous ground-colour; size as in *zion*.]

(S) **G. c. zion** Meinertzhagen, 1920, Jerusalem (Israel). [Pale: upperparts rufous-cinnamon (when fresh) to cinnamon-grey (when worn) with rather narrow but sharp brown-black streaks; paler and less heavily streaked than in *cinnamomina*; streaks on breast short and narrow but sharply defined. Wing of ♂ from Ceylanpinar (south-east Turkey), Jordan, and north-west Iraq 108.6 (106-113) (n=7), of ♀ 104.5 (103-105.5) (n=3) (CSR).]

(E) **G. c. caucasica** Taczanowski, 1887, Lagodekhi (Georgia, western Transcaucasia). [Dark: upperparts dark brown-grey, slightly tinged buff when fresh; upperparts and chest marked with short and sharp black streaks; close to *meridionalis* in colour, but slightly greyer and with more contrasting dark streaks, less pink-cinnamon on hindneck when worn. Wing of ♂ from the Caucasus 105.0 (102-107) (n=4), of ♀ 102.3 (101-104) (n=3) (CSR), or of ♂ 108.6 (103-113) (n=13) (Abs 1963).]

Subspecies recognized in Turkey: difficult to summarize. Within a given area, most birds are usually quite constant in appearance, but small differences occur between areas. These are slight but just recognizable when series of skins are compared, but do not deserve subspecific names, unless one wants to split up the species in hundreds of subspecies. No marked difference in size is observed in the region, apart from the small *cypriaca*. The large size of *subtaurica*, stressed by, e.g., Kumerloeve (1961a), is an average difference of a few mm at most. Two main groups occur in Turkey: dark brown-grey birds (*meridionalis*, *caucasica*) and pale sandy-grey birds (*subtaurica*, *zion*); although the pale and small *cypriaca* occurs as near Turkey as Karpathos and Rodhos, no such birds were found among the Turkish birds examined. Dark birds are restricted to the coastal zone of the Black Sea Coastlands, Thrace, Western Anatolia, and the Southern Coastlands, pale birds to Central Anatolia, the East, and the South-East. The dark populations from the Black Sea Coastlands (from Kars west to at least Kastamonu) are considered to be *caucasica*, the dark ones of Thrace, Western Anatolia (Manyas Gölü to Aci Gölü and Mugla) as well as those from Kas and Antalya east to the Çukurova deltas and north to the neighbouring parts of the Taurus mountains are referred to as *meridionalis*; in view of the minor difference between both subspecies, this division as well as any other suggested is purely arbitrary. Wing of ♂ from Western Anatolia 107.7 (101-111) (n=16), of ♀ 99.7 (95-104) (n=10) (Weigold 1913-14, CSR), wing of a single ♂ from Thrace 109 (Rokitansky & Schifter 1971). Inland, the pale *subtaurica* inhabits a large range, from Kütahya, Burdur, and Elmali through the northern foothills of the Taurus and Çorum east to Igdir, the Van area, Siirt, and Yüksekova; for measurements, see 'Subspecies described', above. The only other pale Turkish race, *zion*, is restricted to Ceylanpinar in the extreme South-East. Birds from Gaziantep, Birecik, Urfa, and Elazig are intermediate between *zion* and *subtaurica*; for these populations, the name *weigoldi* (Kollibay), 1912, described from Urfa, is available, but in view of the slight differences between *zion* and *subtaurica*, the naming of an intermediate bird is not warranted; wing of ♂ of this population is 110.0 (107.5-113) (n=11), of ♀ 101, 107 (n=2) (CSR). The birds from the Amanus area (Antakya, Amik Gölü, Haruniye, Osmaniye) do not fit in the pattern of paler and darker birds described above: they show heavy black or dark brown streaks on the upperparts, like those of *meridionalis*, combined with rather pale rufous-cinnamon feather fringes, rather like those of *zion*. Thus, the same contrasting plumage pattern is found in this area as that of the subspecies of *Melanocorypha calandra* and *Calandrella brachydactyla* of the same area. As the populations of the localities mentioned differ somewhat mutually, naming does not seem appropriate, and the Amanus population can probably best be considered as intermediate between *meridionalis* and *cinnamomina* (a subspecies restricted to the coastal strip of north-west Israel and south-west Lebanon), or between *meridionalis* and *zion* . The wing of this Amanus population is rather short: wing of ♂ 106.5 (103-110) (n=24), of ♀ 98.9 (96.5-101) (n=8) (CSR).

References Watson, G. E. (1962a) A revision of Balkan, Aegean, and Anatolian Crested Larks. *Bull. Brit. Orn. Club* 82, 9-18. Abs, M. (1963) Zur Evolution der *Galerida*-Arten. *Bonner zool. Beitr.* 14, 1-128.

LULLULA ARBOREA
Wood Lark   Orman Toygari

Habitat Stony hills, fields, and mountain plateaux covered with open forest, heathland, or with short scanty vegetation interspersed with scattered shrubs or isolated trees, often at the border of the tree line. Occurs at 0-1200m in Thrace and Western Anatolia, but at 1200-2600m in the Taurus and the Ilgaz Daglari and at 1800-3000m in the South-East and East.

Distribution See map 7. Occurs on all hills and mountains except for some isolated ones on the Central Plateau and in the South-East, where suitable habitat is lacking; also, rather local in the Black Sea Coastlands where the forest is sometimes too dense.

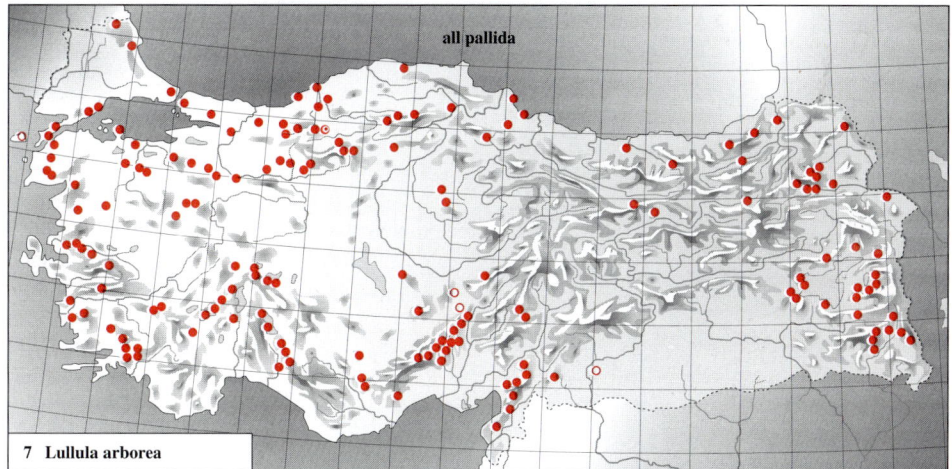

7 Lullula arborea

Geographical Variation
Subspecies described or recorded in the region:

(W) ****L. a. flavescens**** Ehmcke, 1903, Romania. [Upperparts contrastingly streaked yellow-buff and brown-black; breast tinged buff; upperparts less cinnamon-brown as in the subspecies *arborea* (Linnaeus), 1758, described from Sweden; breast and flanks less tawny.

	Wing of ♂ from Romania, Bulgaria, and mainland Greece 98.0 (96-102) (n=8), of ♀ 93, 94 (n=2) (CSR), close to the wing of *arborea* (see Roselaar in Cramp 1988).]
(W)	***L. a. wettsteini*** Niethammer, 1943, Crete. [Like *flavescens*, but ground-colour of upperparts greyer. Small: wing of ♂ from Crete 94.2 (92-96.5) (n=8), of ♀ 90.5, 92.5 (n=2) (CSR).]
(E)	***L. a. pallida*** Zarudny, 1902, 'Transcaspian hills' (cf. in Turkmenistan). [Upperparts pale and grey, as *wettsteini*, less buff as *flavescens*. Wing of ♂ from Iran 98.3 (96-101) (n=3), of ♀ 93.0 (90-95) (n=3) (Paludan 1938, 1940, Schüz 1959).]

Subspecies recognized in Turkey: the birds examined, mainly from the breeding season, are pale, greyish, and rather large, similar to *pallida*. Wing of ♂ from northern Asia Minor (excluding 4 birds in juvenile plumage) 99.0 (94-101.5) (n=6), of ♀ 97 (n=1) (Kummerlöwe & Niethammer 1934-35, Rössner 1935, CSR). Birds examined from Thrace by Rokitansky & Schifter (1971) (wing of ♂ 95 and 97, of ♀ 93) are also referred to as *pallida*. The type of *flavescens* from Romania is a bird in fresh plumage; worn birds from Romania and the Balkan countries are distinctly greyer, inseparable from *pallida* and *wettsteini*. The separation of *wettsteini* on its small size seems hardly justified either, and probably all pale southern and eastern birds, including those of Turkey, should be united under the oldest available name, *pallida*.

ALAUDA ARVENSIS

Eurasian Skylark   **Tarlakusu**

Habitat Grassy slopes of undulating hills or on plateaux, moist short-grass lake shores, meadows or marshes with low grassy vegetation; usually less a species of cultivated fields (unlike the situation in central and western Europe), but in the west of the country also on arable land. In general, mainly at 0-1500m in the west and centre, but at 1000-3100m in the east.

Distribution See map 8; also, Kumerloeve (1971). Reaches the coast in Thrace and Western Anatolia, but largely absent from the dry parts of the Southern Coastlands and the wooded parts of the Black Sea Coastlands.

Geographical Variation

Subspecies described or recorded in the region:

(W)	***A. a. cantarella*** Bonaparte, 1850, Italy. [Upperparts pale and greyish, dark shaft streaks rather broadly bordered by ill-defined olive-brown; upperparts without pink-cinnamon tinge or with traces of cinnamon on hindneck and mantle only; ground-colour of breast pale cinnamon to buff (when fresh) or cream-white (when worn), not extending to the unstreaked part of the breast. Wing of ♂ from the Balkan countries and Greece 115.7 (113-119) (n=5), of ♀ 103.3 (102-106) (n=3), bill of ♂ 15.5 (15.3-15.9) (n=5), of ♀ 13.0 (12.8-13.1) (n=3); wing of ♂ from the Caucasus 116.1 (115-117) (n=4), of ♀ 107.5 (106-109) (n=3) (CSR).]
(E)	***A. a. armenicus*** Bogdanov, 1879, Akhalzyk near Tbilisi (Georgia, western Transcaucasia). [Upperparts pink-cinnamon, grey only when heavily worn; black streaks on upperparts broader than in *cantarella*, but only narrowly bordered by olive, upperparts appearing cinnamon with heavy black streaks on cap and rump and black spots on mantle and scapulars; breast tinged rufous-cinnamon (when fresh) to pale cinnamon (when worn), the cinnamon extending to the unstreaked part of the breast. Large: wing of ♂ from South-East Turkey (Van and Erçek areas) 121.2 (117-127) (n=13), of ♀ 110.1 (108-112) (n=6), bill of ♂ 16.8 (15.5-18.5) (n=13), of ♀ 15.2 (14.8-15.5) (n=6) (CSR).]

Larks

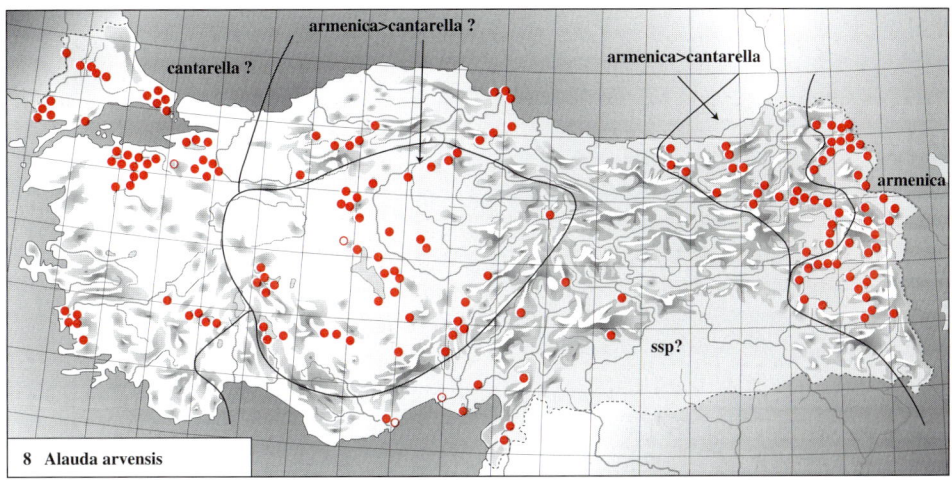

Subspecies recognized in Turkey: all breeding birds from the South-East are *armenica*, but the birds from further north (upper Murat valley, Agri, Erzurum, Kars) are similar to *cantarella* in colour of plumage and in bill size, though the wing averages longer: wing of ♂ from the north-east 118, 121.5, of ♀ 110, 110, bill of ♂ 15.2, 15.6, of ♀ 14.0, 14.0 (n=2 for each sex). Further west, the situation is obscure; though the species is widespread, only a single breeding bird has been examined, a probable ♂ from the Eregli area, which is more or less intermediate between *cantarella* and *armenica* in size, though nearer to the latter in breeding plumage. As *cantarella* is the subspecies occurring in Bulgaria and Greece, one may expect this form to breed in at least Thrace and Western Anatolia. Winter birds from the latter area as well as from the north-east are all *cantarella*: wing of wintering ♂ from Eregli, Haruniye, Erzurum, and Ispir 115.7 (113-118) (n=6), of ♀ 102, 105 (n=2), bill of ♂ 14.9 (13.8-15.5) (n=6), of ♀ 13.0, 15.4 (n=2) (CSR). In winter, *arvensis* from central and northern Europe may also occur (a subspecies similar to *cantarella*, but with browner upperparts and warmer buff-brown breast and flanks), as well as *dulcivox* from the Volga steppes and Kazakhstan (which is similar to *cantarella*, but with the upperparts mainly grey, the olive borders along the black of the streaks reduced, the streaks apppearing narrower, and with the dark streaks on the breast narrower and reduced in extent, with the throat often unspotted).

References Kumerloeve, H (1971) Zur Brutverbreitung der Feldlerche in Kleinasien. *Ardea* 59, 61-63.

EREMOPHILA ALPESTRIS
Horned Lark   Kulakli Tarlakusu

Habitat Generally on alpine meadows or arid hillsides strewn with scattered rocks and boulders at 1900-4000m, but at least locally also recorded breeding in open wastelands and even fallow cultivated land in valley bottoms at 1000-1500m (see, e.g., Vauk 1973 and Schubert 1979a). Perhaps part of the population breeds at low levels early in spring (before many birdwatchers are active), retreating to the mountains when the valleys dry up.

Distribution See map 9; also, Kumerloeve (1961a). Recorded from many mountains (and, locally, their foothills: see above), from the mountains near Abant Gölü and the Taurus eastward; also, the Ulu Dag and (at least formerly) the Boz Dag in Western Anatolia. Only a few data are available from between the Ilgaz Daglari and Sumela in the north and between the Erciyas Dag and Erzincan in central Turkey. Though these areas are under-recorded, the species may really be absent, as, for instance, Vielliard (1968) could not find the species in the Munzur Daglari.

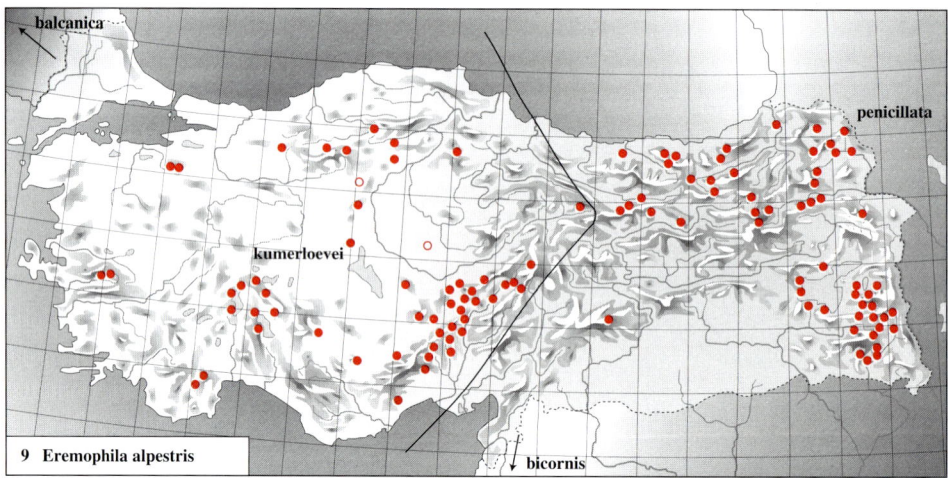

9 Eremophila alpestris

Geographical Variation
Subspecies described or recorded in the region:
(W)   ***E. a. balcanica*** (Reichenow), 1895, Stara Planina (Bulgaria). [Like *penicillata* (see below), but upperparts of ♂ purer grey, of ♀ more dull sandy-grey with heavier black streaks; face and throat slightly deeper yellow. Wing of ♂ from the Balkan countries 115.9 (111-120) (n=25), of ♀ 109.4 (106-112) (n=5) (Stresemann 1920, Niethammer 1943, Bub & Herroelen 1981, CSR).]
(T)   ***E. a. penicillata*** (Gould), 1838, Erzurum. [Black of cheek connected with black of chest, unlike subspecies in northern Eurasia and America; upperparts of ♂ dull pinkish grey-buff with vinous-pink hindneck and with distinct though fine dark streaking on mantle and scapulars; upperparts of ♀ grey-buff with more distinct streaking; face and throat light yellow; lesser coverts grey with slight pink tinge. Wing of ♂ from eastern Turkey 120.4 (117-124) (n=10), of ♀ 109.4 (108-110.5) (n=4); wing of ♂ from the Caucasus and Trans-caucasia 121.3 (118-126) (n=12), of ♀ 109.5 (107-112) (n=5) (CSR).]
(T)   ***E . a. kumerloevei*** Roselaar, Bereketlü (Taurus, Turkey): see chapter V. [As *penicillata*, but mantle and scapulars of ♂ greyish-pink, hardly buff, without dark streaks (near to *bicornis*, but hindneck contrastingly vinous-pink, like *penicillata*); upperparts of ♀ pinkish sandy-buff, streaking as in *penicillata*, face pale yellow (paler than in *penicillata*). Wing of ♂ from the Ulu Dag, Elma Dag (near Ankara), and the Taurus mountains 118.4 (114-121) (n=8), of ♀ 107.6 (106-110.5) (n=4) (CSR).]
(S)   ***E. a. bicornis*** (C. L. Brehm), 1842, Lebanon. [Upperparts of ♂ uniform vinous-buff, without contrasting hindneck and upper mantle; upper tail-coverts buff-cinnamon, not pink-grey as in the subspecies above; upperparts of ♀ sandy-buff with pink tinge, faintly streaked; face and throat of both sexes whitish-yellow. Wing of ♂ from Lebanon 115.8 (112-120) (n=23), of ♀ 107.6 (105-110) (n=10) (CSR).]
Subspecies recognized in Turkey: birds from eastern Turkey (Verçenik, Erçek, Van, Baskale, Hakkari) are *penicillata*, as are those of the Caucasus, Transcaucasia, and western and

northern Iran east to Gorgan. Birds from the Ulu Dag, the Ankara area, the Taurus, and probably other mountain ranges surrounding the Central Plateau are *kumerloevei*. See chapter V.

References Bub, H., & P. Herroelen (1981) Lerchen und Schwalben. Neue Brehm Bücherei 540. Wittenberg Lutherstadt.

RIPARIA RIPARIA
Sand Martin   Kum Kirlangici

Habitat Breeds in soft banks, cliffs, and earth-mounds, natural or artificial, virtually always close to water, foraging aerially in open spaces nearby. The altitude of breeding is largely governed by the occurrence of suitable breeding habitat; the species is mainly confined to streams and lakes at the lower altitudes of the coastlands and plateaux, but it breeds up to 2000m in the east, e.g., it is very common in the Van area.

Distribution See map 10; only localities where the species is known to breed are given or where birds were seen in numbers in June or early July over water near banks and where the species thus probably nests. Likely to occur everywhere in suitable habitat, but probably severely underrecorded; the distribution is obscured by the occurrence of migration up to early June and from July onwards.

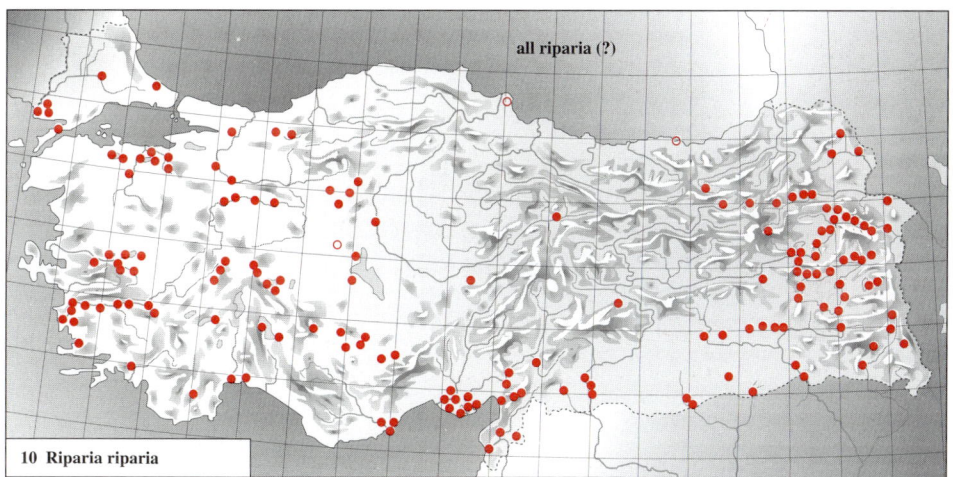

10 Riparia riparia

Geographical Variation
Subspecies described or recorded in the region:

(W) **R. r. riparia** (Linnaeus), 1758, Sweden. [Upperparts and breast band drab-brown, throat white. Large: wing in north-west Europe 106.9 (103-111) (n=41) (CSR).]

(S) **R. r. eilata** Shirihai & Colston, 1992, Elat (southern Israel, on migration). [Upperparts and wing darker, dark earth-brown or (on cap, mantle, and wing) sooty brown; throat grey-brown with white mottling, hardly contrasting with the dark grey-brown breast band. Small: wing 94.0 (87.5-98.6) (n=105) (Shirihai & Colston 1992). Breeding grounds not known.]

Subspecies recognized in Turkey: birds from the Ankara and Van areas are attributed to *R. r. riparia*, and this is supported by plumage characters and measurements: e.g., wing of ♀ 105.6 (103-107.5) (n=4) (Kummerlöwe & Niethammer 1934-35, CSR). *R. r. eilata* is a migrant in Elat, Israel, breeding somewhere to the north of Elat, but probably south of the range of the northern subspecies *R. r. riparia* (elsewhere in the range of the Sand Martin, smaller subspecies also breed south of larger ones). Thus, it may probably breed either in the Balkan area, Turkey, the Caucasus area, or Iran, and breeding birds of these areas should be carefully investigated. In this respect, it is interesting to note that a single bird collected by G. Nikolaus on 27 September 1974 in Kayseri had the plumage characters of *eilata*, being markedly darker than other Turkish birds examined, though the wing (103 mm) was too long for *eilata*, and more in agreement with the equally dark *ijimae* of eastern Asia. Also, note that migrant birds from Herceg Novi, on the Dalmatian coast of southern Croatia, described by Tschusi in 1912 as *fuscocollaris*, are also said to be darker than *R. r. riparia*, especially on cap, ear, breast band, and wing, though showing a white throat (Hartert 1921-22). Obviously, more data on breeding birds of the southern Palearctic are needed.

References Shirihai, H., & P. R. Colston (1992) A new race of the Sand Martin *Riparia riparia* from Israel. *Bull. Brit. Orn. Club* 112, 129-132.

PTYONOPROGNE RUPESTRIS
Crag Martin   Kaya Kirlangici

Habitat As a breeder, confined to steep crags, gorges, and rock faces near open ground or water and thus mainly confined to mountains, but breeds locally in similar habitat near sea-level, and occasionally nests on houses, as e.g. in Adana. Breeds up to 3000m, occurring even higher during foraging trips.

Distribution See map 11. Apparently absent (or underrecorded?) Thrace and north-west Anatolia, and scarce or almost absent in the Black Sea Coastlands, the shores of the Mediterranean Sea, and the plains of the South-East.

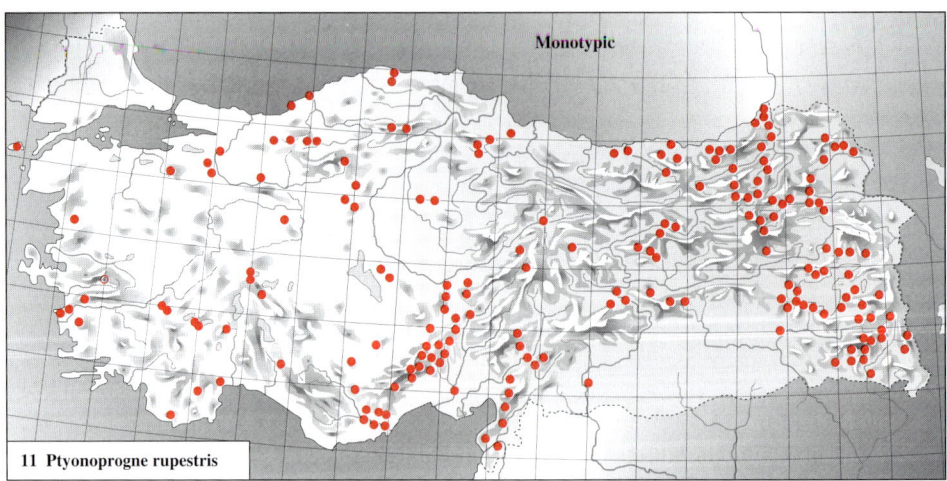

Geographical Variation None. *P. rupestris* (Scopoli), 1769, originally described from Tirol (Austria or Italy) and occurring throughout the mountains of the southern Palearctic region from Morocco to China, is considered to be monotypic (Roselaar in Cramp 1988). Birds from the Alps and the Balkan countries have wing 132.2 (128-135) (n=18: 12 ♂, 6 ♀) (CSR), those of Central Asia average larger, e.g., wing in Tibet 136.0 (128-145)

(n=58) (Vaurie 1972). Wing length of 8 ♂ and 4 ♀ from Turkey is 130.5 (125.5-136) (CSR), close to wing of European birds, but Rössner (1935) recorded a ♂ with wing of 122 in the Bolu area.

References Vaurie, C. (1972) Tibet and its birds. London.

HIRUNDO RUSTICA

Barn Swallow   Is Kirlangici

Habitat Open fields or parkland, usually near water and preferably near grazing cattle, with sheds, houses, or barns available which provide shelter for nesting. This habitat occurs mainly at lower levels, but locally it breeds up to at least 2000m.

Distribution See map 12. Everywhere common, the lack of dots on the map being mainly due to observers simply stating 'breeds everywhere' without mentioning localities. Absent from mountain tops and apparently rare or absent in the valleys of the Kurdish Alps.

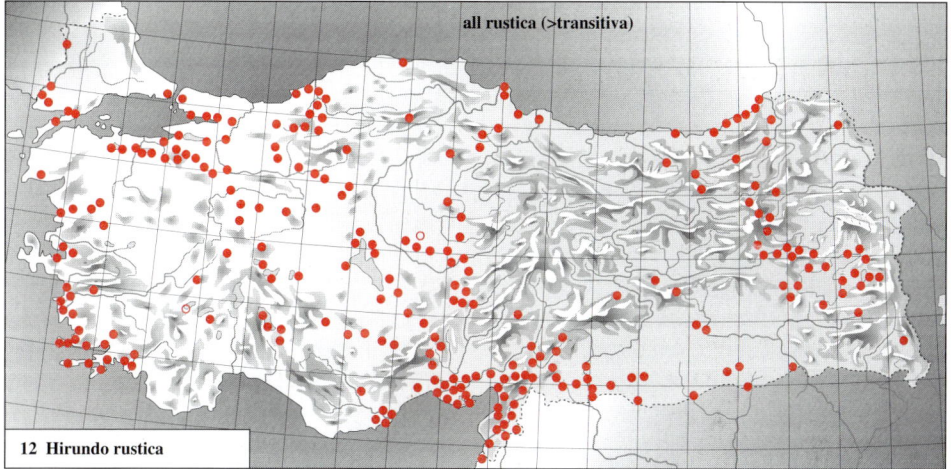

Geographical Variation

Subspecies described or recorded in the region:

(W/E) **H. r. rustica** Linnaeus, 1758, Sweden. [Belly cream-white, pink-cream, or (in some birds) rufous, under tail-coverts slightly paler. Large: adult wing 123.6 (118-129) (n=90) in western and central Europe, 123.9 (120-130) (n=14) in the Balkan countries (CSR).]

(S) **H. r. transitiva** (Hartert), 1910, Plain of Esdraelon ('Emeq Yizre'el, northern Israel). [Belly to under tail-coverts uniform deep rufous in all birds; size as in *rustica* (larger than rufous-chestnut-bellied subspecies *savignii* from Egypt). Wing of birds from Lebanon, Israel, and Jordan 123.6 (119-128) (n=18) (CSR).

Subspecies recognized in Turkey: a number of birds in the Balkan countries and Turkey show a rufous belly, and the proportion of birds with this character is apparently higher than in the remainder of Europe, especially in Turkey: of 19 Turkish birds, 12 have a more or less rufous-pink belly (Jordans & Steinbacher 1948, CSR). However, in none of the Turkish populations does every bird appear to be uniform deep rufous below, in contrast to *transitiva*, and the under tail-coverts of Turkish rufous-bellied birds are generally not as dark rufous as in that subspecies. Therefore, Turkish birds are included in *H. r. rustica*, though tending somewhat to *transitiva*. Birds from the Balkan countries and northern Syria are also intermediates between *rustica* and *transitiva*. Wing in Turkey 123.8 (118-134.5) (n=11), the largest wing being that of an unusually large ♂ from Zonguldak (CSR). Wing of migrants Beysehir 123.0 (116-131) (n=66) (Vauk 1973).

HIRUNDO DAURICA
Red-rumped Swallow   Kizil Kirlangiç
Habitat Breeds in cracks and niches of rocks and caves, below bridges, in ruins, or in deserted or occupied buildings, with open fields, parkland, or water for aerial foraging nearby. Mainly in lowlands and valleys, not in mountains.

Distribution See map 13; also, Kumerloeve (1961a). Virtually absent from northern Thrace and the Central Plateau, as well as from entire northern and eastern Turkey, but apparently recently spreading into the western part of the Black Sea Coastlands. Widespread and locally common only in areas within influence of the Mediterranean Sea and the Sea of Marmara.

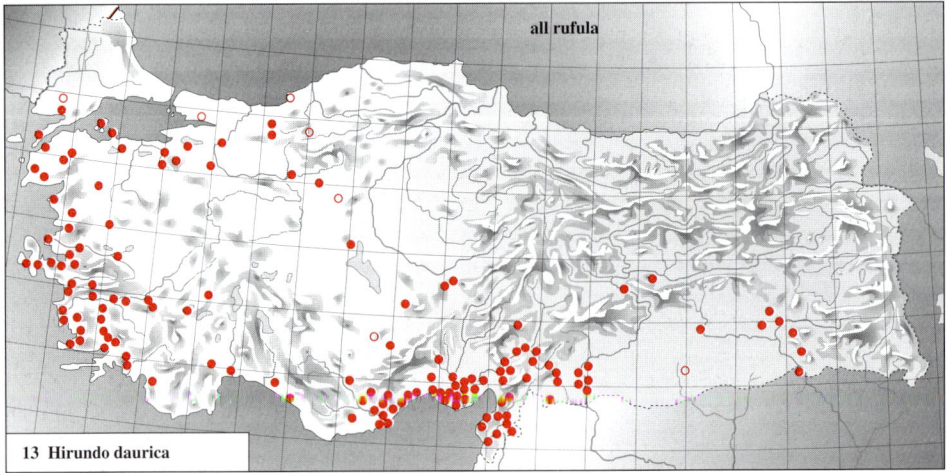

13 Hirundo daurica

Geographical Variation
Subspecies described or recorded in the region:

(W/S) ***H. d. rufula*** Temminck, 1835, 'Africa, Egypt, Sicily, etc. '. [A large pale subspecies; a rufous collar round the hindneck; rump unstreaked rufous-cinnamon and cream; streaks on underparts narrow, less than c. 0.3 mm wide, hardly extending to lower flanks and vent. Adult wing in southern Europe and north-west Africa 123.0 (117-128) (n=23), in Cyprus, Egypt, and the Levant 123.2 (118-127) (n=32) (CSR).]

Subspecies recognized in Turkey: all birds are attributable to *rufula*, as was to be expected because other subspecies occur far away in the Afrotropics and in southern and eastern Asia. Wing of adult Turkish birds 123.8 (120-128.5) (n=7) (CSR).

DELICHON URBICA
House Martin   Pencere Kirlangici
Habitat Requirements as in *Ptyonoprogne rupestris* (see above), but makes more intense use of man-made structures at lower altitudes, such as buildings in towns; however, breeding on natural rock faces in mountains is far more common than nesting on man-made structures, especially in central and eastern Turkey. In the west of the country, breeds up to at least 2000m, in the east to at least 3000m.

Distribution See map 14. Largely absent from the Central Plateau as well as from much of the East and South-East, where breeding is noted for a few mountains or towns only, though the species is probably underrecorded; it is more widespread east from a line running from Erzurum through Siirt to Baskale. Common in Thrace, Western Anatolia, and the the Black Sea Coastlands, where underrecorded.

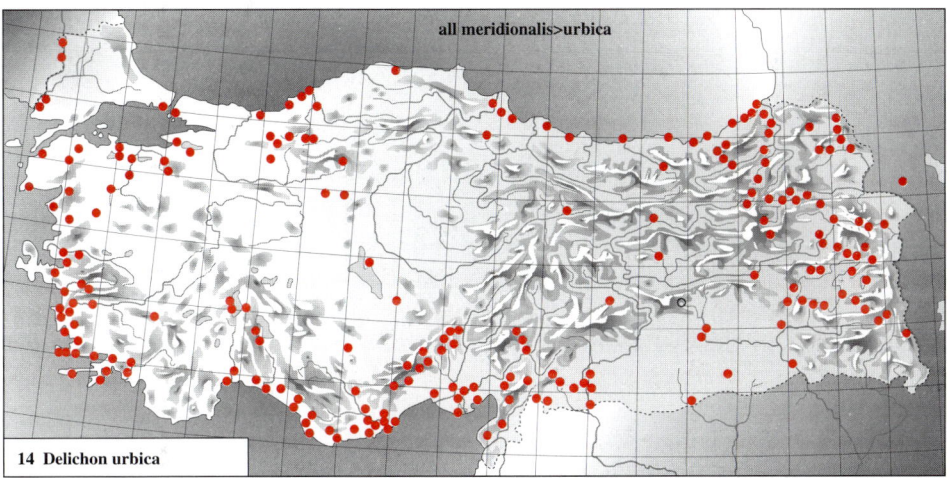

14 *Delichon urbica*

Geographical Variation
Subspecies described or recorded in the region:
- (W) ***D. u. urbica*** (Linnaeus), 1758, Sweden. [Larger, wing in Sweden 102-118 (n=67) (Svensson 1992), in the Netherlands 110.6 (105-116) (n=63) (Roselaar in Cramp 1988).]
- (S) ***D. u. meridionalis*** Hartert, 1910, Hammam R'Hira (north-east Algeria). [Smaller, wing in north-west Africa 103.9 (99-108) (n=32) (Roselaar in Cramp 1988), in Spain (including Balearics) 104.5 (101-108) (n=20) (CSR), in Iran 102.0 (98-106) (n=18) (Witherby 1903, Hartert 1903-10, Vaurie 1951, Diesselhorst 1962). No overlap in wing-length between *urbica* from northern and north-central Europe and *meridionalis* from the Mediterranean basin east to Iran when sex and age are taken into account, but the populations breeding in south-central and south-east Europe are intermediate. For these latter, the name *fenestrarum* (Brehm), 1831, described from central Germany, is sometimes in use, but the populations here are better considered as intermediate between *urbica* and *meridionalis*.]

Subspecies recognized in Turkey: birds examined are intermediate in size between both subspecies, as are those of south-central and south-east Europe. Wing of breeding birds from Turkey is 105.8 (101-109) (n=3) (CSR), slightly nearer to *meridionalis* than to *urbica*, but the sample is very small.

ANTHUS CAMPESTRIS
Tawny Pipit   Kir Incirkusu
Habitat Arid cultivated or uncultivated steppe, fields, slopes, or plateaux, barren or covered with short vegetation alternated with patches of bare sandy or clayey soil, sometimes with some scattered trees or shrubs. From sea-level up to at least 2500m (in the Taurus) or 3000m (on the Nemrut Dag in the east).

Songbirds of Turkey

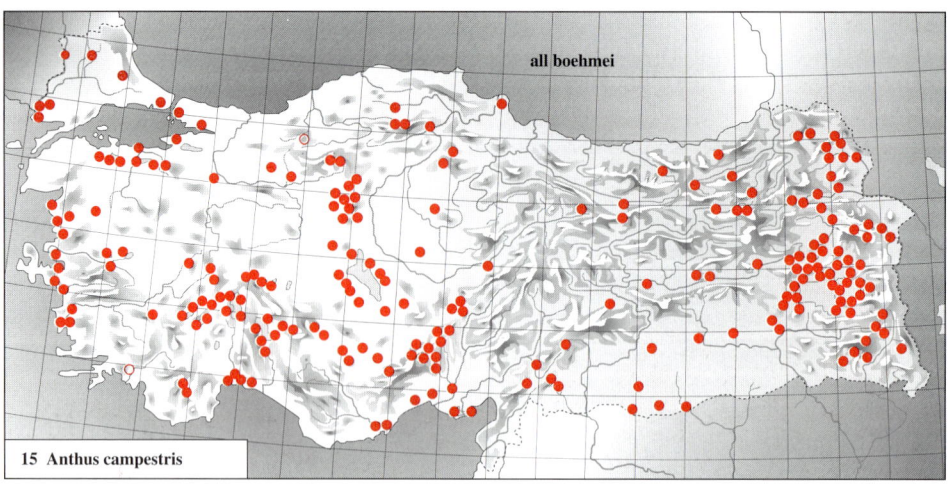

15 *Anthus campestris*

Distribution See map 15. Widespread, but sometimes locally absent in apparently suitable habitat; absent from the coastal zone of the Black Sea Coastlands.

Geographical variation

Subspecies described or recorded in the region:

(W) **A. c. campestris** (Linnaeus), 1758, Sweden. [Rather large; upperparts tinged light sandy-brown. Wing of ♂ from central Europe 91.1 (88-94) (n=17), of ♀ 87.3 (84-90) (n=11), adult bill (both sexes) 18.3 (16.8-19.3) (n=20) (CSR).]

(W/E) **A. c. boehmei** Portenko, 1962, Plyura valley near Kiara (south slope of Damavand mountains, northern Iran). [Large, wing longer and bill heavier than in *campestris* (Portenko 1962); colour of upperparts intermediate between *campestris* and *griseus* (feathers light sandy-brown with rather small pale grey-brown centres). Wing of ♂ from Iran 95.6 (93-98) (n=10), of ♀ 88.5 (87-90) (n=4) (Stresemann 1928, Paludan 1938, 1940, Schüz 1959, Diesselhorst 1962, Érard & Etchécopar 1970); wing of ♂ from the Caucasus and northern Iran east to Khorasan 94.0 (88-100) (n=31), of ♀ 87.4 (80-92) (n=15) (Portenko 1962); wing of ♂ from Greece and Bulgaria 95.9 (93-98) (n=11), of ♀ 87.8 (86-90) (n=3), bill in Greece and Bulgaria 18.8 (18.0-19.5) (n=10) (CSR).]

(E) **A. c. griseus** Nicoll, 1920, 'Tyshkan', former Soviet Turkestan. [Rather large; upperparts grey, feathers with narrow pale sandy-brown fringes. Wing of ♂ from Tien Shan (Central Asia) 92.5 (89-96) (n=16), of ♀ 87.0 (84-90) (n=6), bill 18.2 (17.7-19.0) (n=13) (CSR).]

Subspecies recognized in Turkey: birds examined are inseparable in both colour and size from birds from Iran, the Caucasus area, and Greece, and thus can be attributed to *boehmei*; wing of ♂ from Turkey 95.8 (92-101, once 89.5) (n=28), of ♀ 89.7 (87-94, once 85) (n=17), bill 19.0 (17.8-20.7) (n=31), no variation within Turkey (CSR). *A. c. griseus* occurs only east from north-east Iran and southern Transcaspia, and does not reach Turkey as far as known. The boundary between *boehmei* and *campestris* in the Balkan area needs to be settled. One may doubt whether a bird intermediate in plumage characters between *campestris* and *griseus* deserves recognition, but *boehmei* shows a feature not shared with either of these, viz., large size.

References Portenko, L. A. (1962) [New subspecies of the passerine birds] (Aves, Passeriformes). II. *Trudy zool. Inst. Akad. Nauk* 30, 385-394.

ANTHUS TRIVIALIS

Tree Pipit Ağaç İncirkuşu

Habitat Open forest of evergreen *Quercus* or of coniferous trees (e.g., *Juniperus*), clearings and glades with scattered shrubs, forest edges, and burnt sites, mainly on slopes or tops of hills and mountains, at c. 1000-2000m in the west but to the upper tree limit at c. 2500m in the east.

Distribution See map 16. Restricted to the hills and mountains of Thrace, the Black Sea Coastlands, and the Upper Aras valley; may breed elsewhere in the East (recorded near Tatvan on 12 July and 3 August).

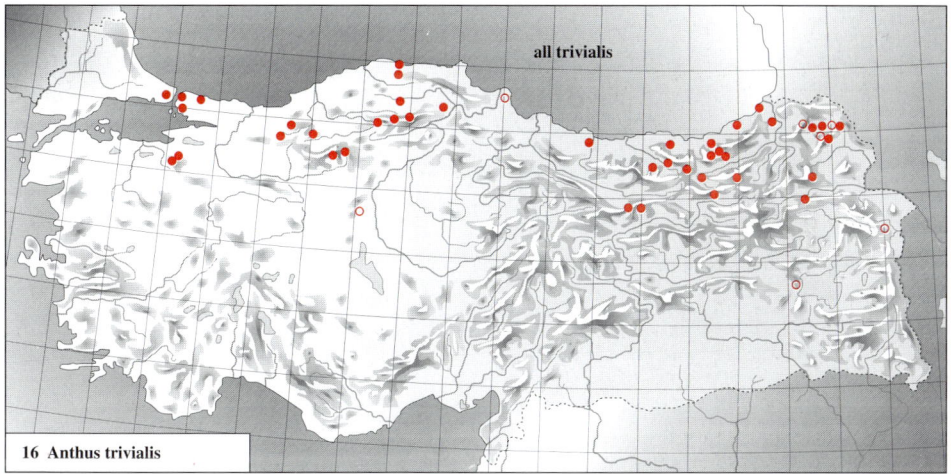

16 Anthus trivialis

Geographical Variation

Subspecies described or recorded in the region:

(W) **A. t. trivialis** (Linnaeus), 1758, Sweden. [Upperparts dark olive-buff with narrow dusky streaking on cap, mantle, and scapulars; throat and breast pale cream-buff with dark brown streaking, remainder of underparts white with fine dark streaks on flanks. Wing of ♂ from central and northern Europe 88.0 (85-94) (n=95), of ♀ 84.9 (82-88) (n=38), bill (both sexes) 15.2 (14.3-16.4) (n=53) (CSR).]

(T) **A. t. differens** Clancey, 1987, Sarikamis (north-east Turkey). [Stated to have the upperparts more brown-olive, with the dark streaks broader and blacker, throat and breast warmer reddish-buff, more heavily streaked with deeper black, and with the ground-colour of the remainder of the underparts tinged yellow, less white (Clancey 1987). Wing of breeding ♂ from Turkey (Bolu Dag, Rize, above Eleskirt, and at Sarikamis) 89.1 (86-91.5) (n=4), of ♀ 83.8 (82-85) (n=5), bill 15.7 (15.2-16.2) (n=7) (CSR).]

Subspecies recognized in Turkey: breeding birds in slightly worn plumage from northern Turkey (including the type specimen of *differens*) do not show any obvious difference in colour from those of central and northern Europe, and *differens* is here considered to be a synonym of *trivialis*. Moderately worn breeding birds from Turkey are greyish-olive above and very whitish below, the dark streaks appearing more extensive than in freshly moulted plumage and contrasting more with the ground-colour, but *trivialis* from northern Europe is similar in this plumage, and a number of heavily worn breeding specimens from northern Europe were inseparable from similarly plumaged Turkish birds. Birds with freshly moulted body from Rize (still in wing moult and undoubtedly breeding locally) are also inseparable in colour from birds from northern Europe in equally fresh plumage. Wing of migrating ♂ of *trivialis* from Turkey 88.9 (85-94) (n=11), of ♀ 84.2 (82-91) (n=5) (Weigold 1912-13, Kummerlöwe & Niethammer 1934-35, Rössner 1935, Rokitansky & Schifter 1971, CSR).

References Clancey, P. A. (1987) The Tree Pipit *Anthus trivialis* (Linnaeus) in southern Africa. *Durban Mus. Novit.* 14, 29-42.

ANTHUS SPINOLETTA
Water Pipit   Dere Incirkusu

Habitat Moist alpine meadows, partly covered with rocks, herbs, low scrub, or stunted trees, from the upper tree line to the snow line. At 1500-2100m on the Ilgaz Daglari, over 2000m on the Ulu Dag, at 1800-3500m in the Taurus, at 2000-3000m in the east, and perhaps at 1200m or over in the mountains above Burdur Gölü.

Distribution See map 17. Probably breeds on many mountain tops, but the species is either underrecorded, or some tops are too bare to afford suitable habitat; occurrence in the mountains above Izmir requires confirmation.

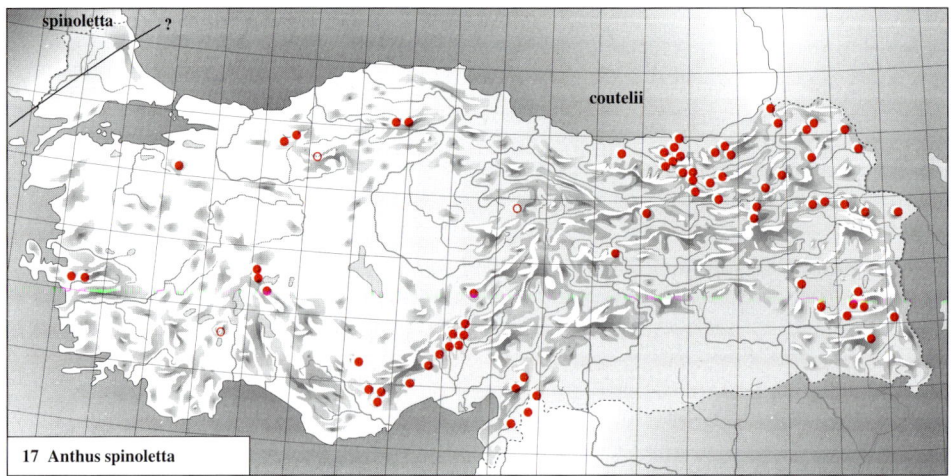

17 *Anthus spinoletta*

Geographical Variation
Subspecies described or recorded in the region:

(W)   **A. s. spinoletta** (Linnaeus), 1758, Italy. [Rather large; in breeding plumage, upperparts rather pure medium or dark grey, underparts vinous-pink; in winter, upperparts olive-brown, underparts extensively streaked and spotted dark grey or sooty on pale cream ground-colour. Wing of ♂ from Iberia, the Alps, and the western Balkan countries 91.5 (88-96) (n=61), of ♀ 85.4 (82-90) (n=36), bill of both sexes 16.7 (15.3-17.8) (n=55) (Roselaar in Cramp 1988).

(E)   **A. s. coutellii** Audouin, 1828, Egypt (in winter). [Smaller; in breeding plumage, upperparts darker olive-grey with more distinct streaks than in *spinoletta*; underparts warmer buff, often partly streaked black; in winter, upperparts dark brown-olive, less brown than *spinoletta*; underparts rather extensively streaked dark on pale buff ground-colour. Wing of ♂ from the Caucasus area and Iran 88.1 (85.5-91) (n=12), of ♀ 80.5 (79-82) (n=6), bill (both sexes) 16.6 (15.6-17.2) (n=8) (Stresemann 1928, Paludan 1938, Schüz 1959, Diesselhorst 1962, CSR). *A. s. coutellii* is often included in the subspecies *blakistoni* Swinhoe, 1863, the latter described from the banks of the Yangtze River (China) and occurring in central Asia (e.g., by Vaurie 1959); *blakistoni* is paler in plumage than *coutellii*, with greyer upperparts and with more creamy and less streaked underparts, and the wing of the ♂ is generally over 90 and the ♀ is generally over 84: see Roselaar in Cramp (1988). In case *coutellii* is not a valid name (it may have been based on a wintering bird of *spinoletta* or *blakistoni*), *caucasicus* Laubmann, 1915, described from the Kuban valley in the north-west Caucasus, is an alternative name sometimes in use.]

Subspecies recognized in Turkey: birds from north-east Turkey and the Ulu Dag, collected from June to early September, are similar to birds from the Caucasus and northern Iran; measurements are also rather similar, wing of ♂ from Turkey 89.7 (88-91.5) (n=12), of ♀ 81.0 (79-84) (n=3), bill 16.6 (15.5-17.6) (n=13) (CSR). Thus, birds breeding in Turkey are *coutellii*. Autumn and winter birds collected near Ankara, Emir Gölü, and Izmir do not differ from *spinoletta*; wing of these 89 and 92 (♂) and 87 (♀), bill 16.9 and 17.6 (♂) (Kummerlöwe & Niethammer 1934-35, Kumerloeve 1961a, CSR).

MOTACILLA FLAVA
Yellow Wagtail   Sari Kuyruksallayan

Habitat Damp meadows and marshes grazed by cattle at borders of lakes, streams, and coastal lagoons; sometimes in salt marsh. Occurs mainly below 100m in the west, but on the Central Plateau, in the Taurus, and in the east inhabits wetlands at 1000-2100m altitude.

Distribution See map 18. Widespread in all suitable habitats throughout Turkey, though apparantly uncommon in the Black Sea Coastlands (except for some deltas) and in the plains of the South-East.

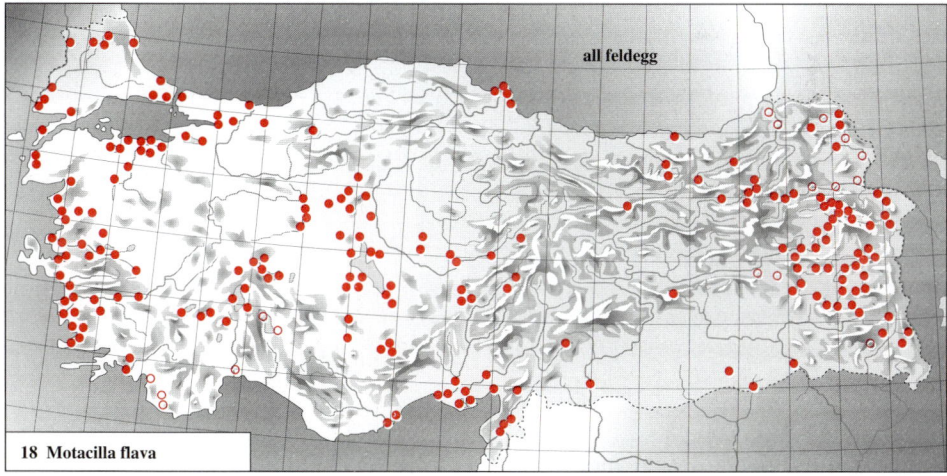

18 Motacilla flava

Geographical Variation
Subspecies described or recorded in the region:
(W/E) **M. f. feldegg** Michahelles, 1830, southern Dalmatia (Croatia). [Cap and sides of head of adult ♂ in summer deep black, chin and throat bright deep yellow, similar to the remainder of the underparts; no white supercilium and no white at the border of black and

yellow on the lower cheek; head of ♀ in summer dark olive-green, without a pale supercilium or with a vestigial one only. Size similar to various other subspecies from Europe, but bill rather long: wing of ♂ from the Balkan and Caucasus areas 83.0 (80-87) (n=17), of ♀ 79.6 (77-82) (n=5), bill of both sexes 16.7 (16.0-18.0) (n=30), against bill 15.9 (14.5-17.2) (n=250) in other subspecies (CSR).]

Subspecies recognized in Turkey: all birds examined and those recorded in the literature as breeding in Turkey are *feldegg*, like those of the Balkan and Caucasus areas and the Levant, west, south, and east of Turkey. During migration, a number of European and west Siberian subspecies occur, of which males in spring are usually distinct but females and autumn birds less so: see Roselaar (in Cramp 1988). Wing of Turkish ♂ 83.8 (82-86.5) (n=5), of ♀ 82 (n=1) (CSR). Occasionally, birds with aberrant head patterns occur: see Kumerloeve (1964a).

MOTACILLA CITREOLA

Citrine Wagtail   **Karaense Kuyruksallayan**

Habitat Damp meadows and marshes with short vegetation.

Distribution See map 19; also Kasparek (1992). Restricted to some wetlands in the east, which have either been recently colonized or the species was previously overlooked. Pairs observed in the Sultan Marshes and at Tuzla Gölü (in the latter area, seen carrying nest material) may indicate a spread further west, and there are an increasing number of spring records in the west of Central Anatolia, e.g., near Gölbasi and in the Hotamis marshes (Kirwan 1993, Schekkerman & Van Roomen 1993).

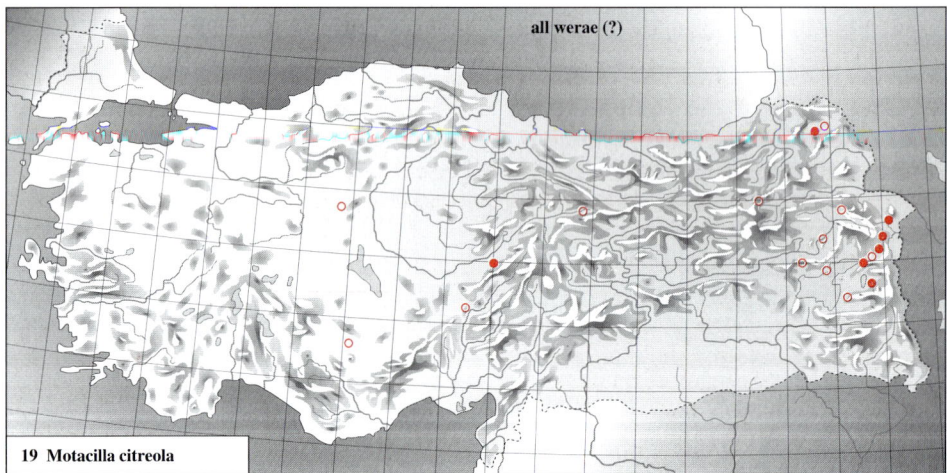

19 *Motacilla citreola*

Geographical Variation

Subspecies described or occurring in the region:

(E?)   ***M. c. citreola*** Pallas, 1776, east Siberia. [A large brightly yellow and grey-coloured subspecies, with a broad black shawl across the hindneck in adult ♂. Wing of ♂ from northern European Russia 87.9 (85-90) (n=17), of ♀ 82.3 (80-85) (n=7) (CSR).]

(E?)   ***M. c. werae*** (Buturlin), 1907, Promzino, Sura valley (south-east European Russia). [Smaller; upperparts on average paler grey, black shawl of adult ♂ narrower; yellow of head and underparts slightly less deep; yellow of sides of breast and flanks washed pale grey. Wing of ♂ from southern European Russia 82.4 (80-84) (n=17), of ♀ 78.5 (76-82) (n=5) (CSR); also, tail, bill, and tarsus lengths distinctly shorter: see Roselaar in Cramp (1988).]

Subspecies recognized in Turkey: no certain breeding specimens from Turkey have apparently been collected, and neither *citreola* nor *werae* is known to breed in an area close to Turkey, and thus it is not known which subspecies may breed in Turkey. However, a single ♀, collected at Beysehir Gölü on 12 May by Vauk (1973) and presumed to be on spring migration, is *werae* according to its measurements, and *werae* is also known to migrate through northern Iran (Diesselhorst 1962). The species is a common migrant through northern Turkey (see Hustings 1994), and as these birds are probably *werae*, this subspecies is more likely to settle as a breeding bird in Turkey than *citreola*, of which no proof of occurrence in the area yet exists.

MOTACILLA CINEREA
Grey Wagtail   Dag Kuyruksallayani

Habitat Brooks and streams with boulders, rock slabs, or gravelly shores, bordered by crags and slopes covered with trees or dense vegetation, and thus mainly confined to forested slopes of hills and mountains. Mainly at 200-1800m in the west and at 1000-2000m in the centre, but up to at least 2500m in the east.

Distribution See map 20. Confined to the hills of Thrace and north-west Anatolia, the Black Sea Coastlands, the foothills and slopes of the Taurus and Anti-Taurus, and the mountain slopes of the East and South-East.

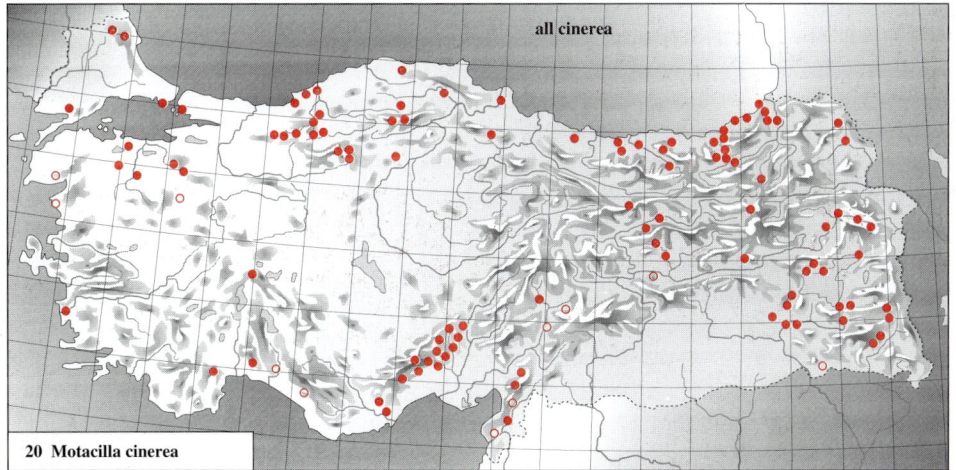

20 Motacilla cinerea

Geographical Variation
Subspecies described or occurring in the region:

(W) ***M. c. cinerea*** (Tunstall), 1771, Wycliffe (Yorkshire, England). [Large; wing of both sexes in western and central Europe 83.7 (80-89) (n=64), tail 95.7 (90-104) (n=62) (CSR); wing of both sexes from the Balkan countries and Greece 82.4 (78-86) (n=53), tail 96.7 (90-105) (n=38) (Stresemann 1920, CSR).]

(E) ***M. c. caspica*** (S. G. Gmelin), 1774, Enzeli (Bandar-e Pahlevi, north-west Iran). [Said to be smaller; wing in eastern Transcaucasia and northern Iran 82.9 (80-86) (n=11), tail 93.2

(92-103, once 82) (n=7) (Stresemann 1928, Paludan 1940, Schüz 1959, Diesselhorst 1962, CSR); no difference from *cinerea* in colour.]

(E) **M. c. melanope** Pallas, 1776, Dauria (Transbaikalia, Russia). [Like *cinerea* and *caspica*, but tail shorter; wing in central Asia (Pamir to Transbaikalia) 81.7 (78-87) (n=48), tail 88.2 (83-95) (n=45); tail in most birds below 92, against over 92 in above races (CSR).]

Subspecies recognized in Turkey: difference in measurements between *caspica* and *cinerea* are too small to consider *caspica* a valid race (see above), and birds from the Caucasus area as well as from the Balkan countries are all united in *cinerea*. Turkish birds between these areas are thus also *cinerea*, as supported by measurements: wing 82.9 (80-87) (n=19), tail 94.5 (88.5-102) (n=18) (Weigold 1912-13, Kummerlöwe & Niethammer 1934-35, CSR). There is no evidence of the occurrence of *melanope* in Turkey.

MOTACILLA ALBA
Pied Wagtail   Ak Kuyryksallayan

Habitat Open fields or parkland, usually near cultivation, virtually always near water, up to 2000m in the west and centre, to 2300m on the Ulu Dag and in the east.

Distribution See map 21. Widespread in small numbers in the western and northern half of the country, rather common in the valleys of the east, virtually absent from the steppe in the South-East.

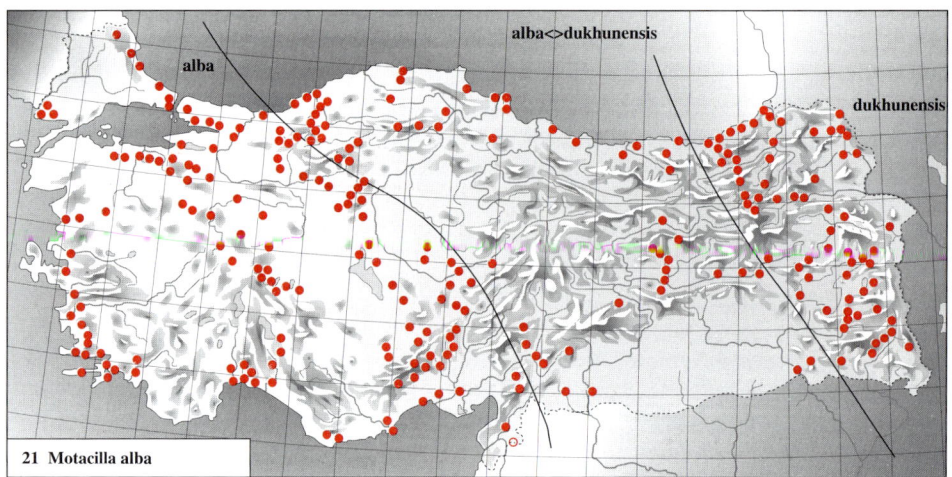

21 Motacilla alba

Geographical Variation
Subspecies described or occurring in the region:

(W) **M. a. alba** Linnaeus, 1758, Sweden. [Upperparts medium grey; white tips of median and greater upper wing-coverts rather narrow, forming two white wing-bars. Wing of ♂ in northern and central Europe 90.3 (84-96) (n=111), of ♀ 86.5 (80-92) (n=55) (CSR).]

(E) **M. a. dukhunensis** Sykes, 1832, Deccan (central India, in winter). [Upperparts paler grey; white on upper wing-coverts more extensive, in ♂ often forming a single large mirror on each wing. Breeds in Siberia between the Urals and the Yenisey, winters in India. Wing of migrant ♂ from Afghanistan 90.0 (86-96) (n=22), of ♀ 86.2 (84-92) (n=11) (Paludan 1959).]

Subspecies recognized in Turkey: clearly, two groups breed, (1) birds with narrow white wing-bars and darker grey upperparts, inseparable from *alba*, occurring in western Turkey east to Bolu (in the north) and Haruniye (in the south), and (2) birds with a large white wing-mirror (♂) or broad white wing-bars (♀), as well as with a variable tinge of grey on the upperparts, occurring in eastern Turkey west to Sumela and Sivrikaya (above Ikizdere, in the north) and Hakkari (in the south). Birds from the area in between, e.g., from Daday (near Kastamonu) and Zonguldak in the western Black Sea Coastlands and from Tirebolu to Cizre in the south appear to be variable in both the amount of white in the wing and in the grey of the upperparts. Large white wing patches as found in the birds from central and eastern Turkey occur also in *dukhunensis* from northern Asia (which has a paler grey back), in *persica* from the plateau of Iran, and in *personata* from the mountains of west-central Asia (both of the latter have much black on lower face and side of neck). Usually, Turkish white-winged birds are called *dukhunensis*, but more likely they are intergrades between *alba* and *personata*, without contact with *dukhunensis*, though the upperparts of some (but not all) birds are rather pale. These intergrades are closer to *alba* than to *personata*, unlike *persica* which is an intergrade nearer to *personata* than to *alba*. Birds examined from the Caucasus mostly resemble birds from eastern Turkey in showing much white on the wing, combined with medium grey upperparts. They may deserve recognition as a separate subspecies, for which the name *intermedia* Domaniewski, the type of which could be rather similar to Caucasus birds and which is an intergrade between *alba* and *dukhunensis* rather than between *alba* and *personata*, preferably should not be used. The situation in northern Iran (north of the Elburz) and on the western slopes of the Zagros mountains of western Iran needs further research; birds from here are called *dukhunensis* (Paludan 1938, 1940), though they are not connected with true *dukhunensis* in northern Asia. Wing of ♂ from western and central Turkey 89.5 (86-93) (n=11), of ♀ 85.5 (84-88) (n=6) (Weigold 1912-13, Kummerlöwe & Niethammer 1934-35, Jordans & Steinbacher 1948, CSR), of ♂ from eastern Turkey and the Caucasus 93.5 (91.5-97) (n=7), of ♀ 86.2 (84-88) (n=3) (CSR). Wing of ♂ from Transcaucasia and western Iran 93.1 (90-96.5) (n=10), of ♀ 86.8 (84-90) (n=3) (Paludan 1938, 1940, CSR).]

PYCNONOTUS XANTHOPYGOS
White-spectacled Bulbul Gri Bülbül
Habitat Slopes and valleys with dense bushes and trees at 0-1000 (-1500) m.

Distribution See map 22; also Kumerloeve (1957b, 1961a). Restricted to the Southern Coastlands, from Kemer and Antalya eastward to the western foot of the Anti-Taurus, the Amanus area, and to Amik Gölü, with occasional records of pairs north of Pozanti and near Kilis. Recent records further west may point to an extension of the range, but confirmation is required: seen in the Dalyan area near Köycegiz Gölü, September 1988 (Snow 1988), and at Kisla mountain above Bodrum, May 1989 (Brock 1989).

Geographical Variation. None. *P. xanthopygos* (Hemprich & Ehrenberg), 1833, originally described from the Lebanon (Bates 1935a, Stresemann 1962) and occurring in the Middle East from southern Turkey, the Levant, and the Sinai south throughout the Arabian peninsula is considered to be monotypic (Roselaar in Cramp 1988); birds described from Haifa (Israel) as *vallombrosae* by Bonaparte, 1856, and said to be greyer above and below than birds from Asia Minor and to show a deeper black head and throat, appear to be similar to birds from the Lebanon and Turkey at a similar stage of plumage wear. Wing of ♂ from the Sinai and Gaza 100.4 (97.5-101.5) (n=5), of ♀ 93.4 (91-97) (n=9), of ♂ from Israel, Jordan, and Lebanon 99.4 (96-104) (n=18), of ♀ 93.0 (89.5-95) (n=19) (CSR), of ♂ from

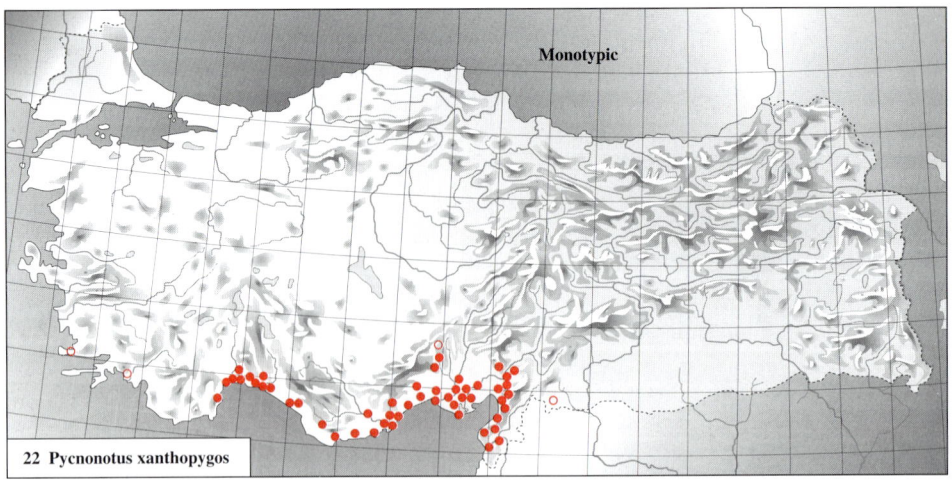

22 Pycnonotus xanthopygos

Monotypic

Turkey 93.8 (91-98) (n=6), of ♀ 91.5 (n=1) (Kumerloeve 1961a, CSR). From these data, a cline of decreasing size towards the north seems apparent, but perhaps the Turkish sample contains a number of wrongly sexed birds.

References Bates, G. L. (1935a) On the type-locality of *Pycnonotus x. xanthopygos* (Hemprich & Ehrenberg). *Bull. Brit. Orn. Club* 55, 118-119. Kumerloeve, H. (1957b) Zur Verbreitungsgrenze des Gelbsteissbülbüls in Kleinasien. *Anz. orn. Ges. Bayern* 4, 574-576. Stresemann, E. (1962) Hemprich und Ehrenberg zum Gedenken. *J. Orn.* 103, 380-388.

CINCLUS CINCLUS
White-throated Dipper   Su Karatavugu

Habitat Clear shallow streams and torrents on lower slopes of mountains, strewn with boulders and rock slabs, up to 2200m in the west and centre and to 2500m in the east.

Distribution See map 23. Restricted to mountain regions with suitable running water, largely avoiding Western Anatolia, the Central Plateau, and the steppe country of the South-East.

Geographical Variation

Subspecies described or recorded in the region:

(W)   *C. c. aquaticus* Bechstein, 1803, Germany. [Cap umber-brown, remainder of upperparts grey with narrow black-brown scalloping; breast rufous-chestnut or deep chestnut, rather contrasting with the blackish flanks, belly, and vent. Breeds from central Europe to north-east Spain and to the north of former Yugoslavia. *C. c. orientalis* Stresemann, 1919, described from Han-Abdipasa (Makedonija, former Yugoslavia) as a slightly darker subspecies and occurring in the eastern Balkan countries and Greece is probably best included in *aquaticus*, as the difference in colour between both is an average one only: the cap of *'orientalis'* is on average greyer-brown, the scalloping of the remainder of the upperparts is sometimes slightly broader and blacker, and the breast is slightly duller rufous-brown. Wing of ♂ *'orientalis'* from the Balkan countries and Greece 93.9 (91-97) (n=19), of ♀ 86.7 (85-89) (n=7) (Stresemann 1920, CSR).]

(S)   *C. c. olympicus* Madarász, 1903, Troödos Mountains (Cyprus). [Dark, like *caucasicus* (see below), but cap slightly paler grey-brown, upperparts (especially mantle) slightly more extensively grey, and breast dull tawny-brown; bill not laterally compressed (but the difference in bill shape between all races is subtle). Intermediate between *'orientalis'* and *caucasicus*.]

Dippers

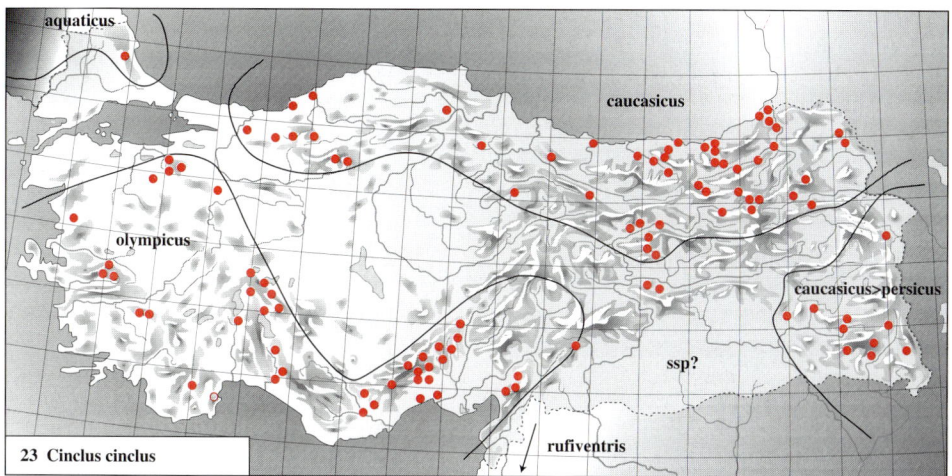

23 Cinclus cinclus

(S) **C. c. rufiventris** Tristram, 1884, Litani River (Lebanon). [Pale, like *persicus* (see below); cap and hindneck dull tawny-brown, black scalloping of the remainder of the upperparts narrow, breast to mid-belly deep rufous-chestnut (brighter and more rufous than in *aquaticus*); size perhaps smaller, wing of ♂ from 'Syria' and Lebanon c. 92.5 (n=1), of ♀ 84, 87 (n=2) (Hartert 1903-10, Greenway & Vaurie 1958, Vaurie 1959, CSR).]

(E) **C. c. caucasicus** Madarász, 1903, Vladikavkaz (Ordzhonikidze) and Pyatigorsk (northern Caucasus, Russia). [Cap dusky grey-brown or chocolate-brown, brown extending further backwards over mantle than in *aquaticus* and *rufiventris*, dark fringes of scapulars and back broad, grey feather centres rather small; breast, belly, and vent uniform dusky brown with a slight tawny tinge, usually without a rusty tinge on the upper breast; underparts not unlike those of the subspecies *cinclus* from Scandinavia, but colour less deep black and without a rusty bar across the upper breast; bill slender, tip laterally compressed. Wing of ♂ from the Caucasus 94.5 (91-98) (n=14), of ♀ 86.7 (83-89) (n=9) (Vaurie 1951, CSR). In Iran, birds from Azarbaijan are intermediate in size between *caucasicus* and *persicus*, but the colour is as *caucasicus*; in the Hamadan area, both colour and size are intermediate. Wing of ♂ from Iranian Azarbaijan 97.5 (95-101) (n=8), of ♀ 91.0 (90-92) (n=3) (Vaurie 1951). *C. c. amphitryon* Neumann & Paludan, 1937, described from Varsambeg (Verçenik, south of Rize, Turkey), is a synonym of *caucasicus* (see below).]

(E) **C. c. persicus** Witherby, 1906, Malamir (Izeh, Bakhtiari, south-west Iran). [Like *caucasicus*, but much paler and larger; cap to mantle tawny grey-brown, remainder of upperparts with more grey visible; breast and belly bright tawny-cinnamon. Wing of ♂ from the Zagros mountains (south-west Iran) 100.5 (98-104) (n=11), of ♀ 92.1 (90-94) (n=8) (Vaurie 1951).]

Subspecies recognized in Turkey: birds from north-east Turkey are dark, like *caucasicus*, and birds further west in the Bolu area are also indistinguishable from *caucasicus* (and thus browner than the black-bellied subspecies *cinclus* from Scandinavia); in all northern birds, the brown tinge of the cap reaches almost as far back on the mantle as in *caucasicus*, and the bill is rather compressed at the tip, similar to *caucasicus*. These northern birds are best included in *caucasicus* rather than using the name *amphitryon* (Neumann & Paludan 1937), as the characters of the type series of *amphitryon* are close to those of *caucasicus*. Wing of ♂ from Verçenik in the north-east 95.4 (94-98) (n=6), of ♀ 88, 92 (n=2) (CSR), of ♂ from the Bolu area 92.7 (91-94) (n=5), of ♀ 86, 89.5 (n=2) (CSR;

Rössner's data of these birds from 1935 with wing of ♂ 89.6 (89-90) (n=5) and of ♀ 87, 87 (n=2) were probably obtained with a different measuring technique). Birds examined from the Ulu Dag are closely similar to those of Verçenik and Bolu (thus, the brown of the mantle extends further down than in *aquaticus*), but the breast is paler, more tawny (less rusty than in *aquaticus*), tending towards *rufiventris*, though the plumage is less light than that subspecies; wing of ♂ 91, 94 (n=2) (CSR). The birds from the Central Taurus and Osmaniye are closely similar to those of the Ulu Dag, though the breast and belly tend to be more light tawny-brown, less pure tawny; wing of these birds 94.7 (93-97.5) (n=5) in ♂, 85.5 (84-87) (n=3) in ♀ (Kumerloeve 1961a, CSR). The birds from the Ulu Dag, the Central Taurus, and Osmaniye are easily separable from *caucasicus* from the Black Sea Coastlands; they are rather intermediate in characters between *caucasicus* and *'orientalis'*. Provisionally, they are best included in *olympicus*, being less pale than what one may expect of intermediates between *caucasicus* and *rufiventris*, which is another option. Birds occurring in the Hakkari area are probably similar to the birds from Iranian Azarbaijan (for characters of which see above); this was collected here by Woosnam, but no specimen has been examined.

References Neumann, O., & K. Paludan (1937) Zwei neue geografische Rassen aus Klein-Asien. *Orn. Monatsber.* 45, 15-16. Greenway, J. C., & C. Vaurie (1958) Remarks on some forms of *Cinclus* (Aves). *Breviora* 89, 1-10.

TROGLODYTES TROGLODYTES
Northern Wren   Çitkusu
Habitat Low scrub or tangled low growth in open mature deciduous, coniferous, or mixed forest, from sea shores to above tree line, at 0-2500m.

Distribution See map 24. Proved breeding only in Thrace, the Black Sea Coastlands, and Western Anatolia. Birds singing in spring in the Taurus mountains are supposed to be lingering migrants (e.g., Kumerloeve 1970b), but the number of records of birds still present in late April and May is large, and the species is locally recorded throughout summer, thus breeding is highly likely, especially as it is also known to breed further south on Rodhos, on Cyprus, and in the Lebanon. Records after 29 April in the southern mountains are shown in black.

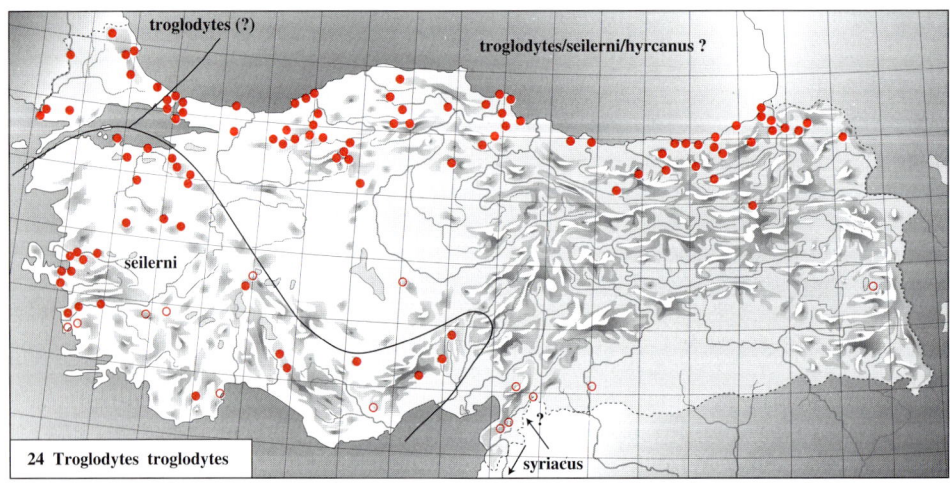

24 Troglodytes troglodytes

Geographical Variation

Subspecies described or recorded in the region:

(W) **_T. t. troglodytes_** (Linnaeus), 1758, Sweden. [Upperparts rufous-brown, closely barred with poorly contrasting dull black bars from the scapulars backwards; lower flanks and vent closely barred black, rufous-brown, pale buff, and cream-white; bill rather short. Wing of ♂ in north-west and central Europe 49.8 (47-52) (n=79), of ♀ 46.9 (45-50) (n=30), bill of ♂ 13.9 (13.1-14.6) (n=31), of ♀ 13.3 (12.2-14.2) (n=21); wing of ♂ from Romania, Bulgaria, and Greece 48.9 (47-51) (n=5), of ♀ 45.5 (n=1), bill of ♂ 13.9 (13.4-14.4) (n=4), of ♀ 13.6 (n=1) (CSR).]

(W) **_T. t. stresemanni_** Schiebel, 1926, Crete (Greece). [Upperparts olive-brown or olive-grey, less rufous than in _troglodytes_; underparts pale buffish-grey; black barring more extensive, extending to upper mantle on upperparts and to upper flanks and chest on underparts, throat finely speckled black. Wing of ♂ from Crete 49.9 (48-53) (n=7, including the type specimen), of ♀ 48.5 (n=1), bill of ♂ 14.2 (13.8-14.6) (n=6), of ♀ 14.4 (n=1)]

(W) **_T. t. seilerni_** Sassi, 1937, Rodhos (Greece). [Barring as in _cypriotes_ (below), but ground-colour of body greyer on upperparts and whiter on underparts.]

(S) **_T. t. cypriotes_** (Bate), 1903, Cyprus. [Ground-colour of body as in _troglodytes_ or very slightly paler and greyer; dark bars more contrasting, extending up to the mantle on the upperparts and up to the breast and upper flanks on the underparts; bill on average c. 1.0mm longer than in _troglodytes_.]

(S) **_T. t. syriacus_** Meinertzhagen, 1933, Levant. [Body well barred, bars extending up to the throat on the underparts, as in _stresemanni_, but ground-colour of upperparts distinctly paler and greyer.]

(E) **_T. t. hyrcanus_** Zarudny & Loudon, 1905, Caspian provinces of northern Iran. [Similar to _troglodytes_, but cap and upper mantle darker, more fuscous, and the rufous of the remainder of the upperparts duller brown; dark barring slightly more extensive. Wing of ♂ from the Caucasus, Transcaucasia, and northern, western, and south-western Iran 50.0 (47-53) (n=54), of ♀ 47.3 (46-49) (n=22), bill of ♂ 14.9 (13.2-16.0) (n= 39), of ♀ 14.3 (13.2-15.0) (n=21) (Stresemann 1928, Vaurie 1951, Schüz 1959, CSR).]

Subspecies recognized in Turkey: difficult to establish. In all western Palearctic populations, a strong influence of bleaching and wear on the ground-colour of the body is apparent, the upperparts grading from rufous-brown (when fresh, in September) through olive-brown and olive-grey (when moderately worn) to grey (when heavily worn, in July), the underparts from buff-brown to dirty white; this variation is found in central Europe and the Balkan countries, as well as in Turkey and on the Mediterranean islands, and differences in ground-colour between subspecies cited above appear not to be valid, except perhaps for a slightly more rapid bleaching to grey in spring in southern populations than in those of north. The contrast of the dark barring depends on the tinge of the ground-colour (more contrast showing when the plumage is worn to grey) and is not a valid character either. Only the dull dark colour of fresh _hyrcanus_ seems to be valid for a diagnosis. Consequently, the only differences between the subspecies are in extent of barring (limited in _troglodytes_ of central Europe and the Balkan countries, intermediate in _seilerni_, _cypriotes_, and _hyrcanus_, extensive in _stresemanni_ and apparently in _syriacus_) and in bill length (long in _hyrcanus_ and _cypriotes_, short in the others, but no data for _syriacus_). _T. t. syriacus_ may be either a synonym of _cypriotes_ or of _stresemanni_. A fair series of birds from the Ulu Dag area of Turkey was examined, as well as a single bird from Pozanti in the Taurus, but no birds from the Black Sea Coastlands. The Ulu Dag series was highly variable in ground-colour, similar in this respect to *troglodytes* from the Balkan countries, but the extent of the barring was intermediate between *troglodytes* and *stresemanni*. This series, as well as the Taurus bird (which was similar), is best included in *seilerni* (also intermediate in extent of barring and with a short bill). Wing of ♂ from the Ulu Dag area and the Taurus 48.6 (47.5-50) (n=8), of ♀ 46, 47 (n=2), bill of ♂ 13.9 (13.2-14.6) (n=6), of ♀ 13.0 (n=1) (CSR). In the Black Sea Coastlands, birds from the Ilgaz Dagi and near Inebolu are called *erwini* by Kummerlöwe & Niethammer (1934-35) (a synonym of *hyrcanus*: Stepanyan 1990), but a bird from Sumela (near Maçka) is considered to be *troglodytes* by Witherby (1910), as are birds from the Bolu area (Rössner 1935) and a juvenile from Abant Gölü (Rokitansky & Schifter 1971); in reality, they may belong to either *troglodytes*, *seilerni*, or *hyrcanus*, or to intermediates between these. Birds from Thrace are likely to belong to *troglodytes*, as this subspecies breeds in nearby Bulgaria.

PRUNELLA MODULARIS
Dunnock   Bozbogaz
Habitat Undergrowth of open mixed or coniferous forest, forest edges, or scrub at the border of the tree line; recorded at 1500-2500m on the Ulu Dag and at 1000-2500m in the Black Sea Coastlands, but may occasionally breed lower down.

Distribution See map 25. Restricted to scattered localities in the Black Sea Coastlands; also, the Ulu Dag and apparently Thrace. Records up to early May further south possibly refer to late migrants.

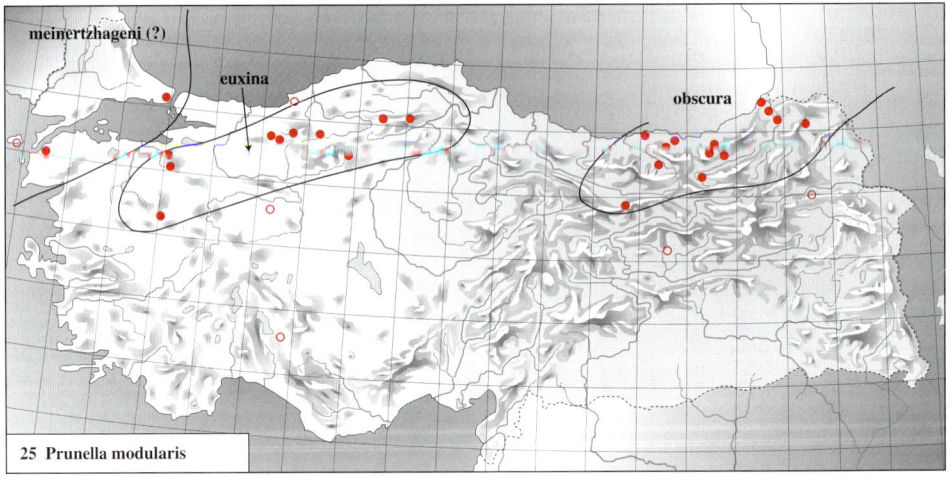

25 *Prunella modularis*

Geographical Variation
Subspecies described or recorded in the region:

(W)   *P. m. modularis* (Linnaeus), 1758, Sweden. [Mantle and scapulars rufous-brown (when fresh) to dark olive-brown (when worn), marked with sharply defined short black streaks 3-4 mm wide; supercilium, cheeks, throat, and chest medium to dark grey, faintly mottled olive-brown and white when plumage fresh; flanks streaked rufous- or fuscous-brown. Wing of ♂ in central Europe 71.1 (69-74) (n=78), of ♀ 68.6 (65-72) (n=49), bill of both sexes 14.5 (13.6-15.6) (n=50) (Roselaar in Cramp 1988).]

(W)   *P. m. meinertzhageni* Harrison & Pateff, 1937, Beglik (Bulgaria). [Rufous-brown of upperparts slightly darker than in *modularis*, black streaks slightly wider, *c.* 4 mm; grey of face and underparts slightly darker; streaks on flanks black, sharply defined. Wing of ♂ from Bulgaria 69.4 (64-74) (n=7), of ♀ 67.0 (65-68) (n=3) (Mauersberger 1971); wing of ♂ from Bulgaria and from Mount Olimbos (Greece) 71.7 (70-73) (n=3), of ♀ 68 (n=1), bill 13.4 (13.1-13.8) (n=4) (CSR).]

(T) **P. m. euxina** Watson,1961, Ulu Dag (Western Anatolia, Turkey). [Streaks on upperparts narrow but sharply defined, 1.5-2 mm wide, dark brown on mantle and inner scapulars, black on outer scapulars; grey of face and underparts dark, as in *meinertzhageni*, but slightly tinged olive and with traces of white feather fringes when fresh; streaks on flanks reduced in extent, dark rufous-brown. See also Watson (1961) and Mauersberger (1971). Wing of ♂ from the Ulu Dag 68, 69 (n=2) (Watson 1961); wing of ♂ from the Ulu Dag and from winter specimens of this subspecies from Osmaniye and Antakya 69, 71 (n=2), of ♀ 67, bill of both sexes 14.1 (13.5-15.2) (n=3) (CSR).]

(E) **P. m. obscura** (Hablizl), 1783, Samamisian Alps (Gilan, north-west Iran). [Streaks on mantle and scapulars rather broad, *c.* 2-3 mm wide, but ill-defined, dark brown or fuscous rather than black; supercilium brown rather than grey; grey of face and underparts rather pale, tinged olive, broadly fringed or spotted off-white on breast when plumage fresh; streaks on flanks rufous-brown, ill-defined. Wing of ♂ from the Caucasus area and northern Iran 69.2 (67-73) (n=43), of ♀ 67.6 (65-70) (n=42) (Marien 1951, Mauersberger 1971).]

Subspecies recognized in Turkey: two distinct subspecies occur, *euxina* on the Ulu Dag and *obscura* in the eastern Black Sea Coastlands, west to at least Tirebolu. Wing of ♂ of *obscura* from Erzurum (October!) 70, bill 14.2; wing of ♀ from Verçenik (8 September) 69, bill 14.0 (CSR). It is not known what subspecies inhabits the western Black Sea Coastlands (Abant Gölü to the Ilgaz Daglari), but as *euxina* occurs in winter as far east as Osmaniye and Antakya, probably *euxina* is involved. Birds breeding in Thrace and Çanakkale in the north-west (if any) are likely to be *meinertzhageni*. *P. m modularis* may occur in winter.

References Marien, D. (1951) Notes on the bird family Prunellidae in southern Eurasia. *Am. Mus. Novit.* 1482, 1-28. Watson, G. E. (1961) Aegean bird notes. I. Descriptions of new subspecies from Turkey. *Postilla* 52, 1-15. Mauersberger, G. (1971) Über die östlichen Formen von *Prunella modularis* (L.). *J. Orn.* 112, 438-450.

PRUNELLA OCULARIS

Radde's Accentor   Sürmeli Çitcerçesi

Habitat Gravelly slopes and gullies strewn with scattered boulders and partly covered by a dense layer of low scrub or clumps of herbs (often spiny); breeds only above the tree line, occurring at 1900-2700m in the Taurus and in the eastern Black Sea Coastlands and at (2100-) 2500-3000 (-4000)m in the East and South-East. See also Loskot (1988).

Distribution See map 26. Restricted to mountain tops, from the Central Taurus east to the mountains of the eastern Black Sea Coastlands, the East, and the South-East.

Geographical Variation None. *P. ocularis* (Radde), 1884, was originally described from the Kiz Yurdi mountain in the Talysh area of Azerbaijan (south-east Transcaucasia). The species is restricted to the Caucasus and to the mountains of Turkey, Transcaucasia, Iran, and Turkmenistan. It is considered to be monotypic (Roselaar in Cramp 1988, which see for discussion of relationships). Wing of ♂ from Iran (mainly from the south-west) 76.7 (74-80) (n=24), of ♀ 74.1 (71-78) (n=10) (Marien 1951); wing of ♂ from northern Iran 78.2

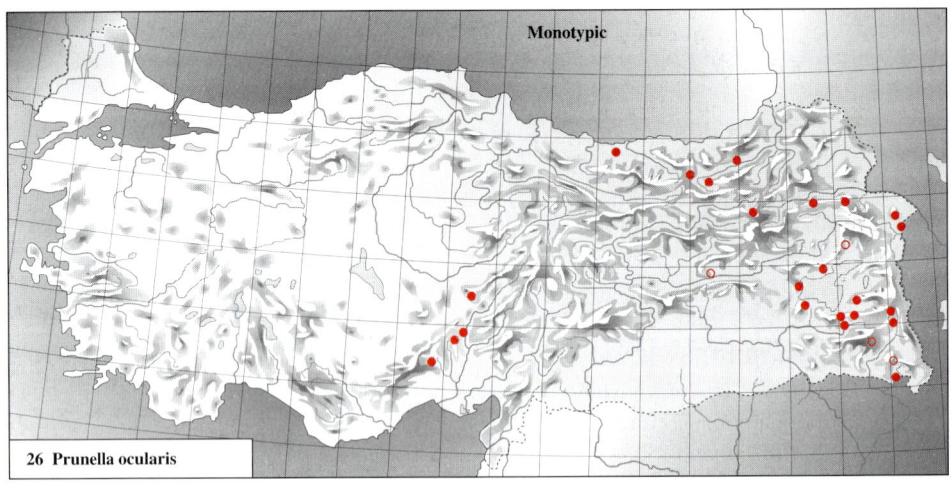

26 Prunella ocularis

(77-79) (n=5), of ♀ 73.5 (73-74) (n=3) (CSR). Wing of ♂ from Erzurum (Turkey) 77.5 (n=1) (CSR).

References Loskot, V. M. (1988) [New data on the distribution and life history of the Brown Accentor (*Prunella ocularis* Radde).] *Trud. zool. Inst. Akad. Nauk SSSR* 182, 89-115.

PRUNELLA COLLARIS
Alpine Accentor   Alp Serçesi

Habitat Alpine grasslands alternated with areas with scree, crags, and boulders, generally devoid of bushes, between the tree line and the snow line; over 2000m in the west, at 2400-3500m in the Taurus, and at 2400m to over 4200m in the east.

Distribution See map 27. Restricted to mountains which reach to over 2000m, and thus mainly confined to the Taurus and the mountains of the eastern third of the country, with isolated occurrences on the Ulu Dag, the Ilgaz Daglari (Schweiger 1965), and the Boz Dag above Sardis (for which latter confirmation is required).

Geographical Variation

Subspecies described or recorded in the region:

(W)   *P. c. subalpina* (C. L. Brehm), 1831, Dalmatia (Croatia). [Upperparts and chest grey, narrowly but distinctly streaked dusky on mantle and scapulars, outer scapulars pale rufous or olive-brown; white throat patch narrowly barred black; flanks narrowly streaked pale rufous; bill longer. Bill of both sexes from the Balkan countries 17.6 (16.7-18.3) (n=19) (all save two 17.3 and over), from Crete 17.9 (17.7-18.0) (n=4) (CSR); for wing, see discussion below.]

(E)   *P. c. montana* (Hablizl), 1783, Gilan (northern Iran). [Upperparts olive-grey with distinct buff tinge on mantle and scapulars, very faintly streaked dusky; throat heavily barred black; flanks extensively deep rufous, rufous extending to sides of breast and to under tail-coverts; bill shorter. Bill of both sexes from the Caucasus and northern Iran 16.9 (16.4-17.5) (n=9), from Dihok (north-east Iraq, at the border of Turkey) 16.8 (16.3-17.6) (n=5) (CSR).]

Subspecies recognized in Turkey: birds from the eastern third of Turkey are *montana* according to plumage and size, e.g., bill of both sexes from the north-east 16.4 (16.0-17.2) (n=7) (CSR). Measurements of birds from the Ulu Dag are similar to *montana*, e.g., bill of ♀ is 16.9 (16.5-17.4) (n=4) (CSR), but the plumage of the birds examined is too worn to esta-

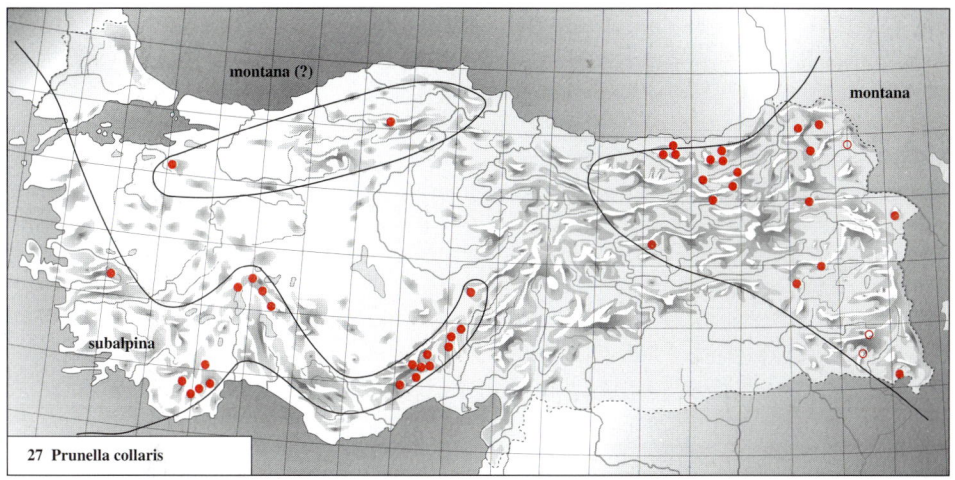

27 Prunella collaris

blish to what subspecies they belong; according to Kumerloeve (1961a), they are *montana*, but they appear to be nearer to *subalpina*. Taurus birds are similar to those of Crete and thus belong to *subalpina*, agreeing with this subspecies both in colour and size, e.g., bill of ♀ 17.5 (17.4-17.6) (n=3) (CSR), though the upperparts have a slight buff tinge, more so than typical *subalpina* from the Balkans. The variation in wing length is slight: wing of ♂ of *subalpina* from former Yugoslavia and Greece (including Crete) 103.1 (100-107) (n=19), of ♀ 97.6 (94-100) (n=18), of ♀ of *subalpina* from the Taurus 95.7 (93-98) (n=3), of ♀ of *montana* from the Ulu Dag 97.4 (94-101) (n=4), of ♂ of *montana* from north-east Turkey and further east 103.4 (100-107) (n=18), of ♀ 97.1 (93-102) (n=16) (CSR).

CERCOTRICHAS GALACTOTES
Rufous-tailed Scrub-Robin   **Kizil Çalibülbülü**

Habitat Scrub, bushes, hedges, gardens, palm groves, and other not too tangled growth (e.g., of tamarisk or oleander), bordered by open ground, along fields, roads, ditches, and canals, at 0-700m in the west, to 900m on the Ala Dag, to 1100m in the Amanus mountains, and mainly to *c.* 1100m in the east, though occasionally up to 1700 (-1800) m.

Distribution See map 28. Locally common in southern Thrace, Western Anatolia, the Southern Coastlands, and the South-East, penetrating to the Central Plateau (mainly at its fringe) and to some valleys of the East and South-East.

Geographical Variation

Subspecies described or recorded in the region:

(W/S) ***C. g. galactotes*** (Temminck), 1820, Algeciras (southern Spain). [Upperparts rufous, similar in colour to the rump; tail longer, 68.2 (64-73) (n=34); white on tail-tips 15-20mm long; wing-tip rounder, outermost primary 3-5.5 mm shorter than the longest, innermost 14-19 mm (Roselaar in Cramp 1988). Breeds from south-west Europe through northern Africa to Levant, north to southern Syria.]

(W/S) ***C. g. syriacus*** (Hemprich & Ehrenberg), 1833, Lebanon. [Upperparts drab-brown, contrasting with the rufous rump and outer tail feathers; supercilium whiter, contrasting with a blackish stripe through the eye; underparts paler; tail shorter, 63.2 (59-66) (n=20); white on tail-tips 8-15 mm long; wing-tip more pointed, outermost primary 1.5-3 mm shorter than the longest, innermost 17-21 shorter (Roselaar in Cramp 1988). Breeds in the Balkan

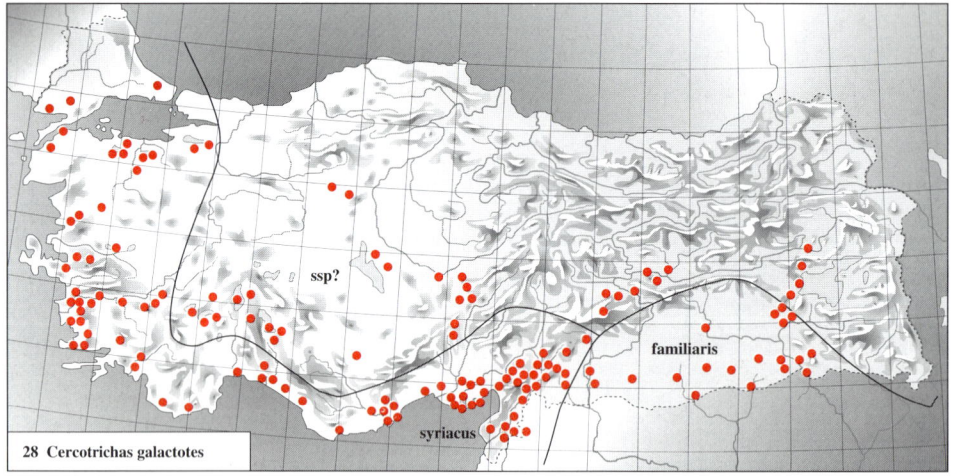

28 Cercotrichas galactotes

countries, Greece, and in the northern part of the Levant, south to Lebanon.]

(E) **C. g. familiaris** (Ménétries), 1832, Sal'yany (Azerbaijan, eastern Transcaucasia). [Similar to *syriacus*, but drab-brown of upperparts paler, light drab-grey; rufous of rump paler; underparts almost white. Breeds from Transcaucasia and Iraq to west-central Asia.]

Subspecies recognized in Turkey: no *galactotes* is found in Turkey, as this rufous subspecies does not extend further north in the Middle East than to southern Syria. All Turkish birds belong to one of the grey-backed subspecies *syriacus* or *familiaris*. The difference between these is often difficult to establish, because many birds appear to be intermediate in colour of upperparts; also, individual variation is large and the colour is in part influenced by bleaching and wear. All birds from Thrace, Western Anatolia, and the coastal zone of the Southern Coastlands east to Gaziantep are clearly drab-brown above, like *syriacus* from the northern part of the Levant and from the Balkan area. East from Urfa, the paler *familiaris* is found. However, a few (but not all) birds from Gaziantep, Amik Gölü, and even a bird from as far west as Burdur Gölü are as pale as *familiaris*. Thus, the border between both subspecies is apparently not as sharp as, e.g., Weigold (1912-13) suggested. Wing of ♂ of *syriacus* from western Turkey (Izmir, Aydin, Solak) is 88.6 (87-91) (n=8), of ♀ 85.4 (84-86) (n=4); tail of ♂ 63.4 (61-65.5) (n=8), of ♀ 62.1 (61-64) (n=4) (CSR); wing of *familiaris*-like males from Burdur, Amik Gölü, and Urfa 88.1 (86-91) (n=4), tail 66.4 (64-68) (n=4) (Weigold 1912-13, CSR). The subspecies occurring between Malatya and Siirt to Tatvan and that breeding along the fringe of the Central Plateau has yet to be established.

ERITHACUS RUBECULA
European Robin   **Kizilgerdan**

Habitat Deciduous, mixed, and coniferous forests and plantations, usually near damp spots at borders of glades or at the forest edge; at 350-600m in Western Anatolia, at 1000-1500m on the Ulu Dag, and up to 2000m in the north and east.

Distribution See map 29. Confined to the area north of 40ºN, as well as north-west Anatolia south to the Büyük Menderes river, with isolated occurrences further south (Beynam Forest, Elazig area); birds singing in the Taurus in late April are probably late migrants.

Geographical Variation
Subspecies described or recorded in the region:

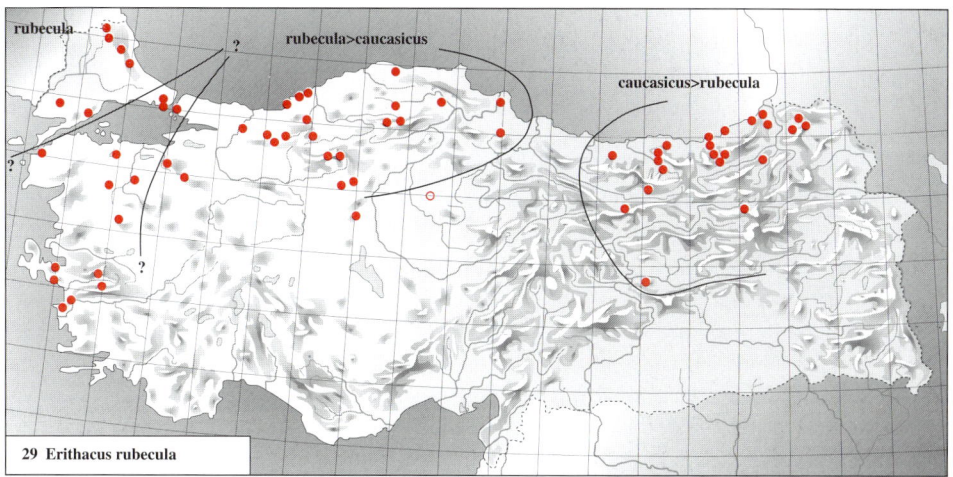

29 *Erithacus rubecula*

(W) ***E. r. rubecula*** (Linnaeus), 1758, Sweden. [Upperparts including upper tail-coverts olive-green, breast tawny-orange. Wing of ♂ from northern and central Europe 72.8 (70-77) (n=91), of ♀ 71.0 (68-74) (n=40), bill of both sexes 14.3 (13.5-15.6) (n=127) (Roselaar in Cramp 1988).]

(T) ***E. r. balcanicus*** Watson, 1961, Boz Dag, above Ödemis (Western Anatolia, Turkey). [Upperparts olive-grey, upper tail-coverts grey, throat pale orange, flanks brownish. Wing of one ♂ and 2 ♀ from Boz Dag 70, 70, 68, and bill 15, 15.2, 14.2, respectively (Watson 1961). *E. r. xanthothorax* Salvadori & Festa, 1913, described from wintering birds collected Rodhos (Greece), is characterized by greyish upperparts, a pale orange-yellow throat, and some rufous on the tail; wing 70.5 (69-72), bill 14.4 (14.5-15) (n=4) (Watson 1961). Birds which agree in characters with this description occur in northern Turkey, and perhaps *xanthothorax* is an older name for *balcanicus*.]

(E) ***E. r. caucasicus*** Buturlin, 1907, Caucasus and Transcaucasia. [Similar to *rubecula*, but longer upper tail-coverts and base of tail usually tinged rufous, face and breast deeper rufous-orange, and bill longer on average. Wing of ♂ from the Caucasus and Transcaucasia 72.8 (71-75) (n=13), of ♀ 73.3 (70.5-76) (n=5), bill 16.3 (14.8-17.5) (n=18) (Vaurie 1955, CSR).]

(E) ***E. r. hyrcanus*** Blanford, 1874, Gilan (northern Iran). [Upperparts and fringes of flight feathers rufous-brown instead of olive-brown; longer upper tail-coverts and basal half of tail rufous-chestnut; face and breast deep orange-chestnut; flanks buff-brown instead of grey-olive; under tail-coverts buff instead of whitish. Wing of ♂ from northern Iran and of wintering birds from Iraq 74.2 (72-78) (n=15), of ♀ 71.8 (70-76) (n=11), bill 16.2 (15.3-17.2) (n=26) (CSR).]

Subspecies recognized in Turkey: spring birds with greyish upperparts and pale orangey throat similar to the characters described for *balcanicus* (and *xanthothorax*) were examined from western Turkey, Greece, Bulgaria, eastern and southern Romania, and the south and centre of former Yugoslavia. However, similar birds also occur commonly among breeding *E. r. rubecula* from further north in Europe, and *balcanicus* is therefore not considered to be a valid race. Breeding birds from the Ulu Dag often show some rufous on the longer upper tail-coverts and at the tail-base, but some birds from central Europe have a similar amount of rufous. Thus, all birds from the Balkan countries, Greece, and western Turkey are similar in plumage to *E. r. rubecula*, as suggested by, among others, Stresemann (1920), Kummerlöwe & Niethammer (1934-35), Vaurie (1959), and Rokitansky & Schifter (1971). Wing of ♂ of *rubecula* from western Turkey (Thrace, Ulu Dag area, Bolu, Boz Dag, Ankara, Abant Gölü, and Mersin, partly from winter) 71.9 (69-74.5) (n=5), of ♀ 69.8 (68-72) (n=6), bill of both sexes 15.0 (14.1-15.5) (n=4) (Kummerlöwe & Niethammer 1934-35, CSR): though the plumage (including the colour of face and breast) is similar to *rubecula*, the bill is on average longer, tending slightly to *caucasicus*. Further east, birds from Giresun, Erzurum, Rize, and Borçka are similar to those of the western Black Sea Coastlands, but the face and breast are deep rufous-orange, like *caucasicus*, and the bill averages still longer, the latter perfectly intermediate between *rubecula* and *caucasicus*: wing of ♂ 72.5 (71.5-73) (n=5), of ♀ 70.5 (n=1), bill (both sexes) 15.3 (15.0-15.5) (n=5) (CSR; though partly collected in winter, the birds are considered to be local breeding birds). Watson (1961) suggests *caucasicus* breeds as far west as Trabzon or Samsun, but whether this statement is based on examination of specimens is not clear. In my opinion, birds from the eastern Black Sea Coastlands are best considered intermediate: though similar to *caucasicus* in plumage, the bill averages shorter. In winter, *hyrcanus* from south-east Transcaucasia and northern Iran occurs commonly in north-west Iran as well as in Iraq (many examined; it occurs as close to Turkey as Dihok), and this subspecies is also likely to reach Turkey.

LUSCINIA MEGARHYNCHOS
Rufous Nightingale   Bülbül

Habitat Dense deciduous growth in valleys, gullies, and on slopes, or in thick undergrowth of open woodland, usually near water or in damp places; mainly at 0-1000 (-1700) m.

Distribution See map 30. Very common in Thrace, the Black Sea Coastlands, and Western Anatolia, more local elsewhere in the country where suitable habitat exists, absent from the steppe on the Central Plateau and in the South-East and from the mountains of the east. The distribution pattern in the south is influenced by many birds singing when temporarily stopping-over during migration: these are frequently considered to be breeding locally by many observers, but probably are not.

Geographical Variation
Subspecies described or occurring in the region:

(W)   **L. m. megarhynchos** C. L. Brehm, 1831, Germany. [Upperparts umber-brown, breast and flanks extensively olive-buff. Rather small: wing of ♂ from western Europe 83.9 (81-87) (n=28), of ♀ 81.8 (78-85) (n=23), tail 63.5 (58-68) (n=49); wing/tail ratio 1.31 (n=48) (Roselaar in Cramp 1988).]

(W)   **L. m. baehrmanni** Eck, 1975, Veles (Titov Veles, Makedonija, former Yugoslavia). [Upperparts olive-brown, slightly paler and greyer than in *megarhynchos*, upper tail-coverts more yellow, tail and wing-tip longer (Eck 1975a,b). Wing of ♂ from Makedonija 86.2 (83-90) (n=53), of ♀ 83.1 (81-85) (n=7) (Stresemann 1920); tail 68.4 (62-76) (n=98) (Loskot 1981); wing/tail ratio 1.26 (n=47) (Eck 1975a,b, 1990).]

(E)   **L. m. africana** (Fischer & Reichenow), 1884, Little Arusha near the Kilimanjaro (north-east Tanzania, in winter; breeds in the Caucasus area). [Upperparts grey-brown or dull brown-grey, less rufous than in *megarhynchos*, less olive than in *baehrmanni*, and less sandy-grey than in *hafizi*; underparts rather pale, breast grey-brown; size as in *baehrmanni* but tail longer. Wing of ♂ from the Caucasus area and northern Iran 86.2 (82-92) (n=100), of ♀ 84.4 (80-87) (n=18), tail 74.2 (67-85) (n=118) (Stresemann 1928, Loskot 1981); wing/tail ratio 1.16 (1.13-1.21) (n=7) (Stresemann 1928).]

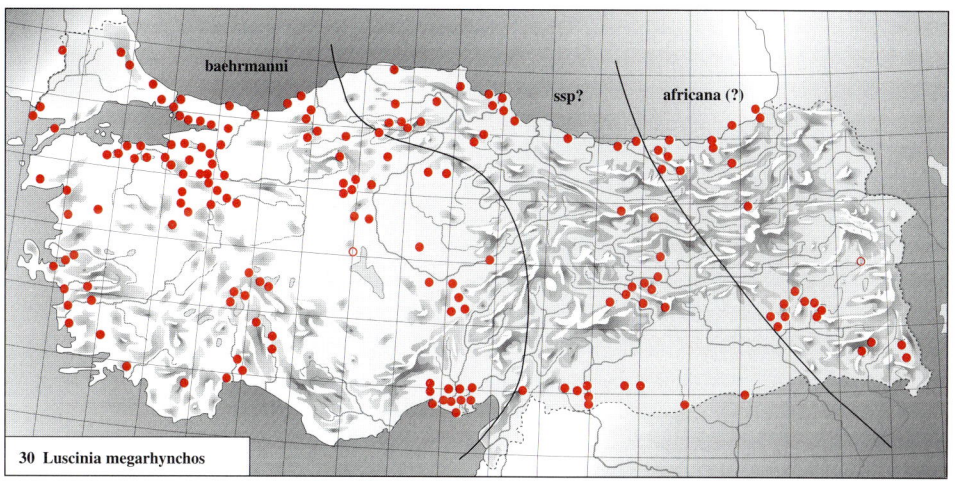

30 Luscinia megarhynchos

(E) ***L. m. hafizi*** Severtzov, 1873, Turkestan. [Upperparts distinctly paler than other subspecies, sandy-grey; supercilium whitish; underparts white with restricted sandy-buff tinge on breast; fringes of flight feathers sandy-grey; underwing cream; large, especially tail. Wing of ♂ from central Asia and Mongolia 91.5 (85-99) (n=130), of ♀ 90.1 (84-97) (n=51), tail 80.9 (74-88) (n=150); wing/tail ratio 1.12 (n=170) (Loskot 1981, Roselaar in Cramp 1988).]

Subspecies recognized in Turkey: *L. m. baehrmanni* occurs east to at least a line from Zonguldak over Ankara to the Çukurova deltas; wing of ♂ of *baehrmanni* from western Turkey 86.4 (83.5-89) (n=7), of ♀ 86 (n=1), tail 68.3 (65.5-72) (n=6) (Kummerlöwe & Niethammer 1934-35, CSR). As *africana* is the subspecies inhabiting the Caucasus area and Iranian Azarbaijan, one may expect this race to occur in eastern Turkey; a single bird from Varsambeg (Verçenik, eastern Black Sea Coastlands), collected in September by Jordans & Steinbacher (1948), has a relatively long tail, like *africana*, but it may have been a migrant from elsewhere. Spring migrants from the Urfa area have rather short tails and are probably *baehrmanni*, intermediates between *baehrmanni* and *africana*, or a mixture of both races (CSR): wing of both sexes 82.0 (78-87) (n=12), tail 68.0 (62-75.5) (n=12), wing/tail ratio 1.21 (1.13-1.32) (n=12) (Weigold 1912-13). The validity of *baehrmanni* of the Balkan area is sometimes doubted and birds from western Turkey are sometimes referred to as *megarhynchos*, but most measurements of *baehrmanni* average clearly larger than in *megarhynchos*, and *baehrmanni* is easily separable in colour when series of both subspecies can be compared. Yet, the boundaries between the subspecies are not sharp, *baehrmanni* probably grading into *megarhynchos* and *africana*, and *africana* into *hafizi*. The song of the Thrush Nightingale *Luscinia luscinia* (Benekli Bülbül) is frequently heard in spring, sometimes up to early June, but breeding in Turkey has not been proved.

References Eck, S. (1975a) Eine neue subspezies der Nachtigall, *Luscinia megarhynchos* (Aves, Turdidae). *Zool. Abh. Staatl. Mus. Tierkde. Dresden* 33, 223-224. Eck, S. (1975b) Über die Nachtigallen (*Luscinia megarhynchos*) Mittel- und Südosteuropas. *Beitr. Vogelk.* 21, 21-30. Loskot, V. M. (1981) [On the subspecies of the Nightingale (*Luscinia megarhynchos* Brehm)]. *Trudy Zool. Inst. Akad. Nauk SSSR* 102, Filogeniya i sistematika ptits, 62-71. Eck, S. (1990) Über Maße mitteleuropäischer Sperlingsvögel. *Zool. Abh. Staatl. Mus. Tierkde. Dresden* 46, 1-55.

LUSCINIA SVECICA

Bluethroat   Bugdaycil

Habitat Low tangled scrub and herbs on stony slopes near water, as well as bushes in damp subalpine meadows; above the tree line, at (1700-) 2000-2600m.

Distribution See map 31. Confined to the mountains east of a line from Trabzon and Refahiye to Baskale. May occur in the Hakkari area, where suitable habitat exists (Helbig 1984).

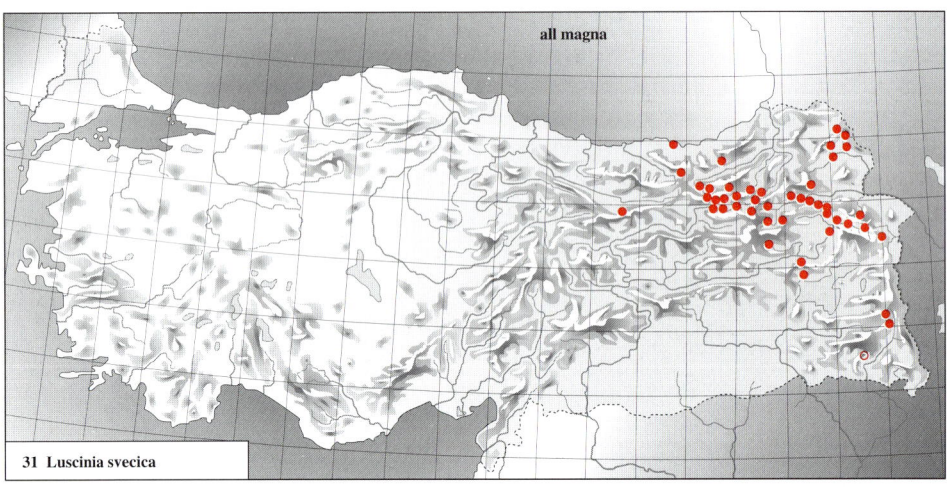

31 Luscinia svecica

Geographical Variation

Subspecies described or occurring in the region:

(E)   ***L. s. magna*** (Zarudny & Loudon), 1904, 'Bidesar, Arabistan' (=Khuzestan, south-west Iran, in winter). [A large subspecies; ♂ usually with uniform blue throat, bordered below by a distinct red band, without the rufous or white spot on the mid-throat shown by most other subspecies, but occasionally with a trace of a white spot; matched by the unnamed Spanish population only, which is somewhat smaller, however. Wing of ♂ breeding in Iran and of migrants in the Middle East 82.0 (79-84) (n=16), of ♀ 78.9 (76-81) (n=5), bill 17.1 (16.4-18.0) (n=19) (CSR); in other races, of which *svecica*, *volgae*, and *pallidogularis* occur in Turkey on migration or in winter, wing is below 79 in ♂, below 75 in ♀, and bill is below 16.8 in both sexes (Roselaar in Cramp 1988).]

Subspecies recognized in Turkey: only *magna* is breeding, as far as can be judged from 2 birds collected on the Tahir Geçidi and at Çildir Gölü in eastern Turkey. Wing of these 80 (♂) and 76.5 (♀) (Kumerloeve 1968, CSR). This race breeds also in the Caucasus area southeast to northern Iran.

IRANIA GUTTURALIS

White-throated Robin   Akgerdan

Habitat Stony or boulder-strewn slopes or gullies, fairly densely covered with scrub or with stunted deciduous or coniferous forest (e.g., *Quercus, Berberis, Pinus, Juniperus*), mainly at 500-1200m in the west and centre, up to 2300m in the Taurus, but mainly at 2000-3000m in the mountains of the east.

Distribution See map 32; also, Kumerloeve (1964c, 1966b). Occurs on dry hills and slopes east from Pamukkale and Pazarcik (and perhaps Bodrum, where recorded singing in May: Brock 1989), avoiding the moister Black Sea Coastlands and the flat areas of the Central

Plateau and the South-East. Very local and scarce in the northern part of the Turkish range. Apparently extinct on the slopes above Izmir, where recorded in the 19th century (Kumerloeve 1964c).

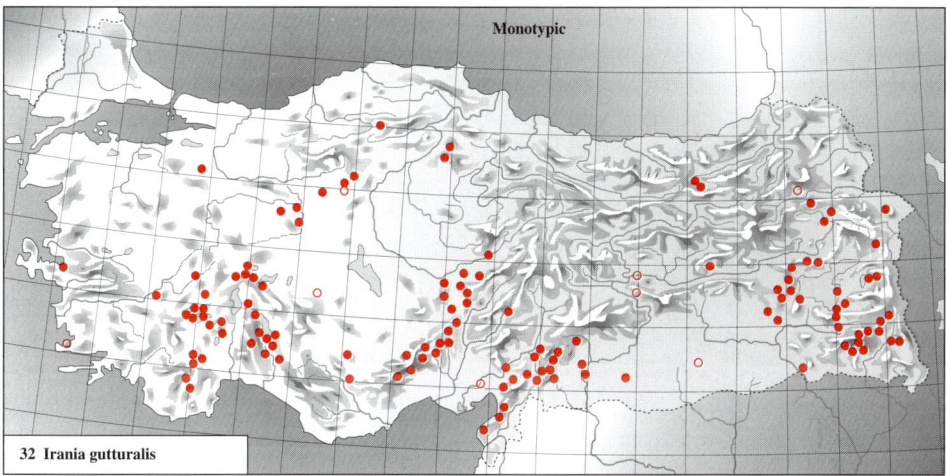

32 *Irania gutturalis*

Geographical Variation None. *I. gutturalis* (Guérin-Méneville), 1843, originally described from a bird wintering in Ethiopia, breeds from Turkey and Transcaucasia east through Iran to Kyrgyzstan, Tajikistan, and Afghanistan. It is considered to be monotypic. *I. albigularis* (Pelzeln), 1863, described from near Izmir, is a synonym. For individual variation in throat colour and pattern, see, e.g., Grote (1943). Wing of ♂ from Iran and Iraq 95.8 (92-101) (n=11), of ♀ 94.3 (90-99) (n=15) (Stresemann 1928, Paludan 1938, 1940, Érard & Etchécopar 1970, CSR); wing of ♂ from Turkey 95.1 (93-99.5) (n=10), of ♀ 90 (n=1) (CSR).

References Grote, H. (1943) Zur Kenntnis von *Irania gutturalis* (Guérin). *Orn. Monatsber.* 40, 148-149. Kumerloeve, H. (1964c) Sur la distribution d'*Irania gutturalis* (Guérin) en Asie Mineure et dans les régions voisines. *Alauda* 32, 97-104. Kumerloeve, H. (1966b) Zusätzliche Bemerkungen zur Verbreitung des Weisskehlsängers, *Irania gutturalis* (Guérin) in Kleinasien. *Alauda* 34, 153-156.

PHOENICURUS OCHRUROS
Black Redstart   Ev Kizilkuyrugu

Habitat Rocky boulder-strewn slopes and peaks, down to the open forest at the border of the tree line in the east, but sometimes lower down on houses or ruins in the west; mainly at 1200-1500m in the western part of the Black Sea Coastlands, at (1400-) 1700-3650m in Western Anatolia and the Taurus, and at 1200m to over 3500m further east, but lower where adapted to man-made structures.

Distribution See map 33. Widespread and often common on higher slopes and peaks of mountains in the entire country, absent lower down except where living close to humans as in Western Anatolia and Thrace.

Geographical Variation

Subspecies described or recorded in the region:

(W)   ***P. o. gibraltariensis*** (J. F. Gmelin), 1789, Gibraltar. [Upperparts of ♂ largely grey, apart

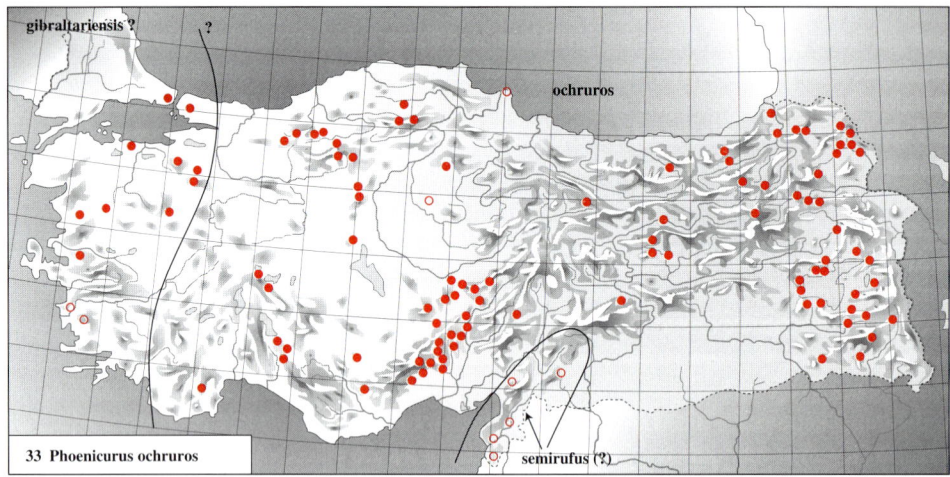

33 Phoenicurus ochruros

from rufous rump to tail; face to breast black, flanks grey, lower belly and vent grey-white, under tail-coverts rufous; outer fringes of outer tertials and inner secondaries white, forming contrasting patch; underwing mixed grey and black; ♀ largely dark brownish-grey. Rather large: wing of ♂ from west and central Europe 86.3 (82-91) (n=67), of ♀ 82.9 (79-90) (n=35) (Roselaar in Cramp 1988).]

(S) **P. o. semirufus** (Hemprich & Ehrenberg), 1833, Lebanon. [Upperparts of ♂ grey with varying amount of black (especially in adult, often hardly so in first year), black sometimes extending up to cap; belly to under tail-coverts rufous; flanks and underwing rufous, sometimes mixed with some grey; wing with white patch, as in *gibraltariensis*; ♀ warm buff-brown, tinged pale tawny on belly, less dark and grey as ♀ of *gibraltariensis*. Size small, wing of ♂ 81-84, of ♀ 75-77 (n=2 in each) (CSR).]

(E) **P. o. ochruros** (S. G. Gmelin), 1774, mountains of Gilan (north-west Iran). *P. o. erythroprocta* Gould, 1855, Erzurum (eastern Turkey), is a synonym. [Similar to *semirufus*, but upper flanks and under wing-coverts dark grey (thus, upper border of rufous on belly n-shaped instead of u-shaped), sometimes slightly mixed rufous; black on mantle and scapulars of adult ♂ (if any) sometimes more restricted; tertials and secondaries fringed blue-grey, generally without white; ♀ as in *semirufus*. Size intermediate, wing of ♂ from the Caucasus and northern Iran 85.3 (83-87) (n=5), of ♀ 81.1 (79-83.5) (n=4) (CSR). Males rather variable, some like *gibraltariensis* (especially in the northern Caucasus), others almost as black above and as rufous below as *semirufus*; see, e.g., Stegmann (1928).]

Subspecies recognized in Turkey: only a series from Varsambek (Verçenik, above Rize in the north-east) examined, as well as some breeding birds from the central Taurus and Ürgüp, and a few wintering birds from Izmir, Bolu, and Haruniye. Those from Varsambek, the Taurus, and Ürgüp are quite uniform, as described for *ochruros* above (apart for the age-related differences in the amount of black on the upperparts). Birds from Erzurum are *ochruros* also, but those from Artvin in the extreme north-east are more variable (Stegmann 1928). Males from the Taurus and Ürgüp are similar to those of Varsambek, but show some white on the inner flight feathers; they can be considered to be *ochruros* with some influence of *semirufus* from the Levant. *Ochruros*-like birds occur west to at least Abant Gölü (Kumerloeve 1964a) and Sultan Dagi. The wintering birds examined from Izmir, Bolu, and Haruniye are all *gibraltariensis*. It cannot be excluded that these birds are local breeding birds, and thus *gibraltariensis* may be the subspecies breeding at low level near human habitation in Thrace and Western Anatolia. Wing of ♂ of *ochruros* from central and eastern Turkey 84.0 (81-86.5) (n=9), of ♀ 79.5 (n=1), bill (both sexes) 14.9 (14.1-16.1) (n=8); wing of wintering *gibraltariensis* 84.8 (81-88) (n=3), bill 15.3 (15.1-15.4) (n=3) (CSR).

References Stegmann, B (1928) Zur Systematik der Rotschwänze. *J. Orn.* 76, 496-503.

PHOENICURUS PHOENICURUS

Common Redstart   Bahçe Kizilkuyrugu

Habitat Open deciduous, mixed, or coniferous woodland, plantations, or parks, usually with a limited amount of undergrowth, or at the edges of grassy glades, mainly at 900-2100m, sometimes lower in the west or higher in the east.

Distribution See map 34. Locally common in the Istranca Daglari (Thrace), the Marmara region of Western Anatolia, and in the inland parts of the Black Sea Coastlands; scarce in the Taurus, the Amanus mountains, and in the valleys of the East and South-East. Apparently absent from large parts of Thrace, Western Anatolia, the western Taurus, the Anti-Taurus, and from the entire Central Plateau and the steppe of the South-East.

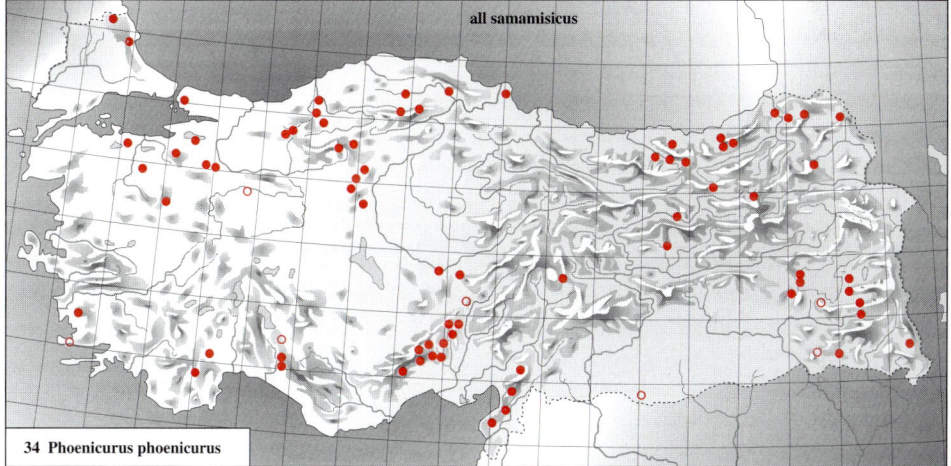

34 Phoenicurus phoenicurus

Geographical Variation

Subspecies described or recorded in the region:

(W)   ***P. p. phoenicurus*** (Linnaeus), 1758, Sweden. [In ♂, upperparts from mantle grey, underparts below chest rufous-cinnamon, fringes of flight feathers grey; in both sexes, outermost functional primary usually longer than 5th outermost, and wing-tip formed by 2nd outermost. Wing of ♂ from Fenno-Scandinavia 81.7 (76-86) (n=109) (Blondel 1967); wing of ♂ from the Netherlands and Germany 80.0 (76-84) (n=67), of ♀ 78.1 (74-82) (n=50) (CSR).]

(E)   ***P. p. samamisicus*** (Hablizl), 1783, Samamisian Alps (Gilan, north-west Iran). [Upperparts of ♂ darker grey than in *phoenicurus*, underparts deeper rufous, but main difference is presence of white fringes along flight feathers, which form conspicuous white patch on inner wing unless the plumage is heavily worn; in some birds ('*incognita*' morph), the upperparts show a variable amount of black (see Stegmann 1928); in both sexes, the outermost functional primary is usually shorter than the 5th outermost, and the wing-tip is formed by the 2nd-4th outermost. Only the adult ♂ has distinct white fringes along the outer webs of the outer tertials and the secondaries, forming a large white patch; juvenile ♂, first year ♂ up to 1-year-old, as well as adult ♀, either have traces of white only, forming a small patch which is generally worn away when 1-year-old, or show no white at all; juvenile and 1st adult ♀ do not show white. Thus, in the breeding season, approximately half of the males (the 1-year-olds) cannot be recognized in the field. Average wing of ♂ 79.9 (n=60) (Blondel 1967); wing of ♂ from Bulgaria, the Levant, and northern Iran 80.4 (78-82.5) (n=20), of ♀ 78, 79 (n=2) (Stresemann 1928, Paludan 1938, 1940, Schüz 1959, CSR).]

Subspecies recognized in Turkey: all Turkish breeding birds are referable to *samamisicus*, all adult ♂ examined showing a white patch on the wing. Frequently, males without a patch are recorded, but these are either migrating *phoenicurus* from further north or heavily worn first year *samamisicus* (see above). *P. p. samamisicus* also occurs in the Levant, Iran, the Caucasus area, and in the Balkan countries west to the south of former Yugoslavia. Wing of ♂ of *samamisicus* from Turkey 78.7 (77-80) (n=5), of ♀ 78 (n=1) (Kummerlöwe & Niethammer 1934-35, CSR); wing of ♂ of migrant *phoenicurus* from Turkey 79.7 (77-82) (n=14), of ♀ 76, 81 (n=2) (Weigold 1912-13, Kummerlöwe & Niethammer 1934-35, Rössner 1935, CSR).

References Blondel, J. (1967) Étude d'un cline chez le rouge-queue à front blanc, *Phoenicurus phoenicurus phoenicurus* (L.): la variation de la longueur d'aile, son utilisation dans l'étude des migrations. *Alauda* 35, 83-105, 163-193.

SAXICOLA RUBETRA
Whinchat   **Çayir Taskusu**

Habitat Extensive damp meadowland covered with scattered shrubs and clumps of herbs (e.g., Umbelliferae) at or above the tree line, at *c.* 1700-2300m.

Distribution See map 35; also, Kumerloeve (1969c). Proved breeding only in the Ardahan area, the upper Çoruh, Aras, Kara, and Kelkit valleys, and in the valleys north-east, east, and south-east of Van and Erçek Gölü. Records on other sites in eastern Turkey in May and August may in part indicate local breeding. No certain breeding records for Thrace and the western part of the Black Sea Coastlands, but may occur (e.g., at Abant Gölü?).

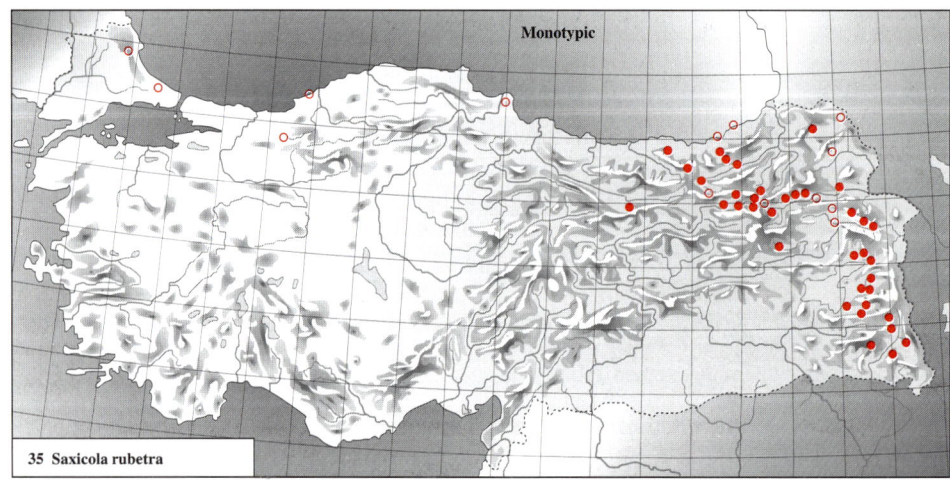

35 *Saxicola rubetra*

Geographical Variation

Subspecies described or recorded in the region:

(W) **S. r. rubetra** (Linnaeus), 1758, Sweden. [Upperparts sandy-buff with rather broad black streaks. Wing of ♂ from western and central Europe 77.2 (74-81) (n=42), of ♀ 75.9 (73-78) (n=20), bill of both sexes 14.8 (13.9-15.5) (n=48) (CSR).]

(W) **S. r. spatzi** (Erlanger), 1900, Gafsa (Tunisia, on migration; said to breed in the south-west and south of former Yugoslavia). [Upperparts slightly paler and more greyish, dark streaks slightly narrower. Wing of ♂ from the south-west and south of former Yugoslavia, as well as from Bulgaria, Greece, and Romania 77.8 (75-81) (n=28), of ♀ 75.6 (74-78) (n=16), bill 14.9 (14.1-15.3) (n=15) (Stresemann 1920, CSR).]

(T) **S. r. senguni** Kumerloeve, 1969, Yüksekova, Van, and Ermaniz Gölü near Van (South-East Turkey). [Darker than other subspecies; cap virtually uniform black, mantle and scapulars black-brown with narrow sandy-brown streaking; smaller than *noskae*. Wing of adult ♂ 77.5, of adult ♀ 74, of juvenile '♂' 74, bill of adult 14.9, 15.3 (n=2) (CSR).]

(E) **S. r. noskae** (Tschusi), 1902, Laba valley (north-west Caucasus, Russia). [Upperparts pale, as *spatzi*, but streaks broader, as *rubetra* and *senguni*; 'larger than *rubetra*, wing up to 82 rather than up to 81' (Hartert 1903-10). *S. r. margaretae* (Johansen), 1903, described from Tomsk (south-west Siberia), is probably a synonym. Birds from south-west Iran show colour characters comparable to *noskae*, and size of these slightly larger than in *rubetra*, apparently as in *noskae* : wing of ♂ 79 (78-82) (n=10), of ♀ 75.8 (74-78) (n=4) (Vaurie 1955); not certain to breed in south-west Iran, however, but does so in northern Iran.]

Subspecies recognized in Turkey: examination of many specimens throughout the breeding range of the species shows that geographical variation is very slight or absent, and that none of the subspecies cited above can be recognized. Thus, *S. rubetra* is considered to be monotypic. The differences seen in plumage are all caused by influence of bleaching and wear: in fresh plumage, the feather fringes of the upperparts are buff-brown; due to bleaching, these fringes gradually become creamy or off-white, and due to abrasion they gradually become narrower, the black streaks on the upperparts then becoming more prominent. Black-capped birds with heavily black-streaked upperparts like the adults of 'senguni' occur also in central and northern Europe, the only difference being that this stage of plumage is reached a few weeks earlier in Turkey than in northern Europe, perhaps as a consequence of an earlier breeding season in Turkey. The slightly worn juvenile of 'senguni' is not different from birds from central Europe in the same plumage. Wing of migrant ♂ from Turkey 76.6 (75-79.5) (n=14), ♀ 74.8 (73-77) (n=6), bill 14.5 (14.1-15.5) (n=11) (Weigold 1912-13, Kummerlöwe & Niethammer 1934-35, Rössner 1935, Rokitansky & Schifter 1971, CSR).

References Kumerloeve, H. (1969c) Zur Brutverbreitung des Braunkehlchens (*Saxicola rubetra*) in der Türkei. *J. Orn.* 110, 220-221.

SAXICOLA TORQUATA

Common Stonechat   **Taskusu**

Habitat Dry plains and hillsides covered with rough grasses and heather mixed with open ground, scattered scrub, stunted trees (e.g., pine or juniper), boulders, or stone walls, at edge of thickets, forest, or cultivated fields. At 100-900m in Western Anatolia, at 1000-1500m in the western part of the Black Sea Coastlands, at 1000-2000m in the central Taurus and the north-east, and at 1600-2400 (-3000)m in the Van area.

Distribution See map 36. Widespread but local in Thrace, the Black Sea Coastlands, Western

Songbirds of Turkey

Anatolia, and on the mountain slopes in the east, scarce in the Taurus, largely absent from the Central Plateau, and apparently entirely absent from the southern coastlands, the Anti-Taurus, and the steppe of the South-East.

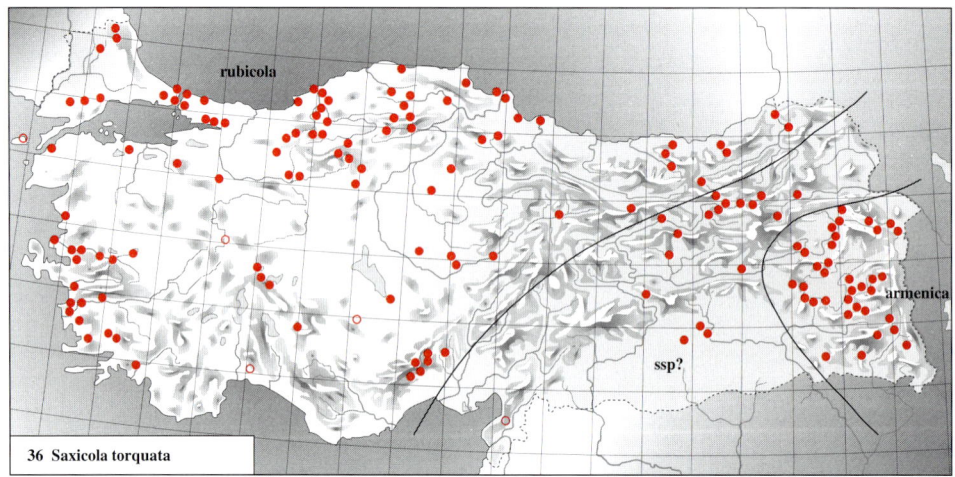

36 Saxicola torquata

Geographical variation

Subspecies described or recorded in the region:

(W) **S. t. rubicola** (Linnaeus), 1766, Seine Inférieure (France). [Feathers of upperparts fringed buff-brown in fresh plumage; upper tail-coverts of ♂ mixed black-and-white; tail-base black; if worn, chest of ♂ deep rufous, sides of belly and flanks gradually paler rufous. Wing of both sexes in western Europe 66.0 (63-68) (n=44), bill 15.0 (13.7-15.8) (n=39) (Roselaar in Cramp 1988). *S. t. graecorum* Laubmann, 1927, described from Corfu (Kerkira, Ionian Islands, Greece), is a synonym; it was described as being smaller than *rubicola*, but the difference is not convincing, as the wing of both sexes from Makedonija, Romania, and Greece is 64.7 (62-69) (n=91), bill 14.9 (14.3-15.5) (n=20) (Stresemann 1920, CSR), close to typical *rubicola*. Other synonyms are *gabrielae* Neumann & Paludan, 1937, described from the Ulu Dag (Western Anatolia, Turkey), and *amaliae* Buturlin, 1929, described from Vladikavkaz (Ordzhonikidze, northern Caucasus, Russia); both the latter are said to be paler or darker rufous than *rubicola* on the breast of the male, but this character is somewhat variable in all populations. Wing of ♂ from the northern Caucasus 63.1 (61.5-67) (n=9), bill of ♂ 15.0 (13.7-15.7) (n=9) (Buturlin in Molineux 1930-31).]

(E) **S. t. armenica** Stegmann, 1935, 'Adshafana, Kurdistan' (near the border of Iran, Iraq, and Turkey). [A subspecies of the Asiatic *maura* group, in which worn males differ in general from the ♂ of *rubicola* and other races of the European *rubicola*-group by showing uniform white upper tail-coverts, a smaller rufous patch on the breast, and a longer wing, but a shorter bill, legs, and feet. In *armenica*, the feather fringes of the upperparts are paler in fresh plumage than in *rubicola*, the upper tail-coverts of the ♂ are white, forming a large white patch; about one-quarter of the base of the outer 5 tail feathers is white; the breast has a rather restricted amount of deep rufous-chestnut, the remainder of the underparts is contrastingly white. Rather large: wing of ♂ in the Zagros mountains (Iran) 76 (74-81) (n=15) (Vaurie 1959), but wing of ♂ in western Transcaucasia and northern Iran 70.2

(68-72) (n=6), bill 14.2 (13.5-14.5) (n=6) (CSR).]
(E) **S. t. variegata** (S. G. Gmelin), 1774, Shemakha (Azerbaijan, eastern Caucasus). [Male as ♂ of *armenica*, but half to two-thirds of the tail-base is white, except for the central pair, resulting in an *Oenanthe*-like tail; white patch on side of neck and on tertials and inner greater upper wing-coverts larger; rufous of breast slightly paler; ♀ with c. 1 cm of pale buff at tail-base. Wing of ♂ from north-western shore of Caspian Sea (Guriev to the eastern Caucasus) and of wintering birds in Sudan 71.9 (70-74) (n=7), of ♀ 70.2 (68-72) (n=3), bill (both sexes) 14.2 (13.7-15.0) (n=8) (CSR). Further north, this subspecies is replaced by *maura* in western and central Siberia and north-east Europe, which differs from *variegata* mainly in the virtual absence of white on the tail-base, slightly smaller size, and smaller white patches on the neck and wing; see Roselaar in Cramp (1988).]

Subspecies recognized in Turkey: all birds examined from the western half of Turkey and from the Black Sea Coastlands east to Borçka are inseparable from *rubicola*; wing of ♂ of these 65.1 (64-68.5, once recorded as 60) (n=15), of ♀ 64.1 (63-65.5) (n=5), bill 14.8 (14.3-16.1) (n=8) (Weigold 1913-14, Kummerlöwe & Niethammer 1934-35, CSR); this subspecies breeds also in the western Caucasus and in western Transcaucasia. Birds from the Van area (Van, Erçek, Yüksekova, Agri) are *armenica*; wing of ♂ 73.5, 75, of ♀ 72, 73.5, bill 13.5, 15.1 (n=2 in each) (CSR). The boundary between both subspecies has yet to be established, e.g., it is unclear what race breeds in the Aras valley and it is unknown how far *rubicola* extends to the east in the southern half of the country. In the eastern Caucasus area, subspecies of both groups are said to show some overlap without apparent interbreeding, separated by habitat: *rubicola* breeds on higher, more stony and scrubby ground, *armenica* and *variegata* on flatter, more humid, and more grassy ground (Buturlin 1929 in Molineux 1930-31). There is no evidence of occurrence of *variegata* or *maura* in Turkey, but both may visit the country during migration.

OENANTHE ISABELLINA
Isabelline Wheatear **Toprak Renkli Kuyrukkakan**

Habitat Arid open ground, especially loamy and sandy steppe and forest-steppe, either flat, hilly, or gently sloping, covered with scattered vegetation of, e.g., *Artemisia*. The altitudinal range extends from plains near the sea to high mountain plateaux, at 0-1300m in the west and centre of the country, but at 700-2300m in the east.

Distribution See map 37; also, Grote (1937) and Kumerloeve (1975b). Occurs everywhere except for the coastal zone along the Black Sea and the Mediterranean (but not avoiding the coastlands of the Aegean Sea); absent from mountain tops, where replaced by *O. oenanthe*. Numerous on the steppe of the Central Plateau and of the South-East, as well as in open valleys of the East in the Upper Aras area and near Van Gölü.

Geographical Variation Very slight and no subspecies recognized. *O. isabellina* (Temminck), 1829, was originally described from a bird wintering in Nubia (northern Sudan). In birds examined (sexes combined), western birds are smaller than those from Central Asia, but size apparently decreases again east from the Altai mountains and Mongolia: wing in Greece, North Africa, the Levant, and in the area of the lower Volga 97.1 (94-100) (n=13), bill 19.2 (16.4-20.5, in only 3 birds over 20) (n=13) (CSR); wing in Turkey 94.5 (91.5-98.5) (n=14), bill 19.0 (18.5-19.8) (n=7) (Kummerlöwe & Niethammer 1934-35, Rokitansky & Schifter 1971, CSR); wing in the Tien Shan mountains (Central Asia) 102.1 (96-106.5) (n=18), bill 20.6 (19.0-21.7, only in 3 birds below 20) (n=17) (CSR). In view of the uncertain origin of the type specimen (Nubia, wing 98, bill 20.5: CSR), and the undoub-

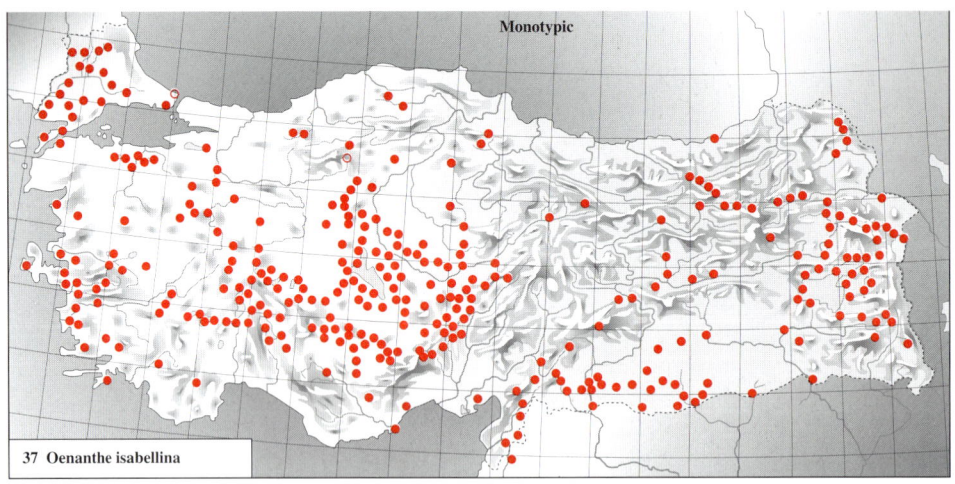

37 Oenanthe isabellina

tedly clinal variation shown by the species, no subspecies are recognized.

References Grote, H. (1937) Die Sommer- und die Winter-Verbreitung von *Oenanthe pleschanka* (Lepechin) und *Oenanthe isabellina* (Cretzschmar). *Orn. Monatsber.* 45, 114-134. Kumerloeve, H (1975b) Zur Verbreitung der Steinschmätzer (*Oenanthe*)-Arten in der Türkei. *Bonner zool. Beitr.* 26, 183-198.

OENANTHE OENANTHE
Northern Wheatear   Kuyrukkakkan

Habitat Stony pastures and mountain slopes with crags and boulders, partly covered with scattered herbs, tolerating presence of isolated bushes and trees; mainly in mountains, locally on isolated crags or boulder-strewn pastures at lower levels; absent from flat or gently undulating featureless landscapes like steppe. At 100-1500m in the west (but at 2000-2300m on the Ulu Dag), (400-) 1100-2700m in the Taurus and the Black Sea Coastlands, and at 1000-4000m further east.

Distribution See map 38; also, Kumerloeve (1975b). Breeds everywhere, except for the coastal zone of the Black Sea and the Mediterranean. Absent from the open plains of the Central Plateau and from the steppe of the South-East where its habitat is scarce or lacking.

Geographical Variation

Subspecies described or recorded in the region:

(W/E) ***O. o. oenanthe*** (Linnaeus), 1758, Sweden. [Plumage darker, feather fringes of upperparts olive-brown, warm brown, or slightly rufous, black of tail-tips rather extensive (but less so than in subspecies *leucorhoa* from Canada, Greenland, and Iceland); ♂ with a limited amount of white on forehead, upperparts light blue-grey when worn, breast cream-buff to warm buff; breast of ♀ cinnamon-buff; bill shorter. Wing of ♂ from Scandinavia 96.4 (94-99) (n=19), of ♀ 91.8 (89-95) (n=6) (Salomonsen 1934), bill of migrants in western Europe (both sexes) 17.3 (16.3-18.8) (n=88); wing of ♂ from Bulgaria, Romania, and mainland Greece 95.9 (92-97.5, once 101) (n=9), of ♀ 93.0 (88.5-94.5, once 98) (n=5), bill 18.1 (17.2-18.9) (n=15) (CSR).]

(W) ***O. o. virago*** Meinertzhagen, 1920, Mount Ida (Idhi Oros), Crete (Greece). [Similar to *libanotica* (below), but wing on average shorter. Wing of ♂ from Crete 94.3 (91-97) (n=14), of ♀ 88.5, 89 (n=2), bill 18.5 (18.2-18.9) (n=16) (CSR). Occurs also on Kikhlades (Greece): Steinbacher (1937).]

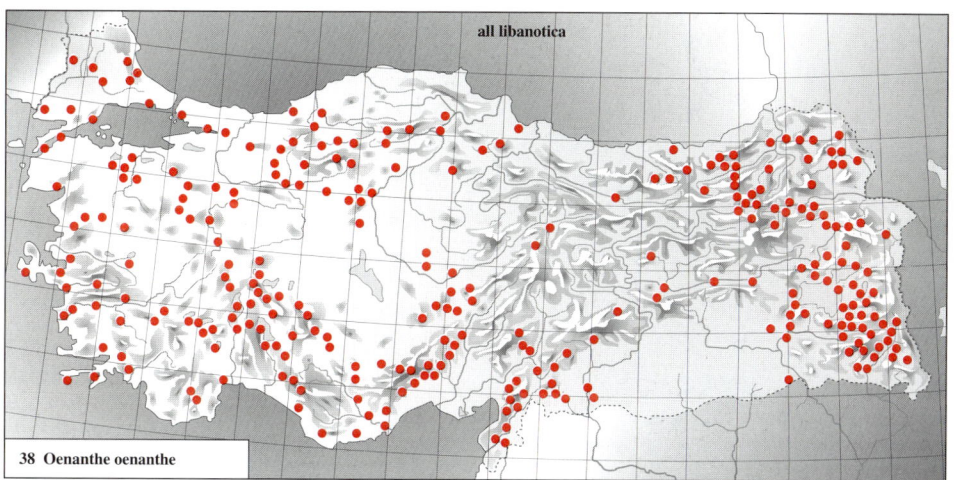

38 Oenanthe oenanthe

(S) ***O. o. libanotica*** (Hemprich & Ehrenberg), 1833, Lebanon. [Plumage paler than in *oenanthe*, independent of abrasion (Ticehurst 1931, CSR); feather fringes in fresh plumage paler grey-brown or sandy-grey, tail-tips with less black, fringes of flight feathers paler; upperparts of ♂ paler ash-grey, forehead more extensively white, chest paler cream; cheeks, throat, and breast of ♀ paler cream-buff; adult ♀ often resembles ♂ in showing grey upperparts, whitish forehead, and blackish face-mask and wing. Wing length as in Scandinavia; bill long, as on Crete. See also Mackworth-Praed & Grant (1951).]

(E) ***O. o. rostrata*** (Hemprich & Ehrenberg), 1833, Beni Suef (northern Nile Valley, Egypt, on migration). [Colour as in *oenanthe*, not as pale as *libanotica*; bill longer on average than in other subspecies. Wing of migrant ♂ from Egypt 101.4 (100-103) (n=5), of ♀ 96.1 (94-98) (n=4), bill 18.5 (17.6-20.0) (n=9); breeding grounds of these migrants unknown (CSR).]

Subspecies recognized in Turkey: all breeding birds examined from Turkey are pale, similar to *libanotica* and *virago*. As *virago* and *libanotica* do not differ in colour and the difference in size between both is slight, *virago* is better not recognized. Wing of breeding ♂ from Turkey 97.0 (93-99) (n=16), of ♀ 94.1 (92.5-96.5) (n=4), bill 18.2 (17.3-19.0) (n=18) (CSR). Wing of migrant ♂ from Turkey 96.0 (93-101.5) (n=19), of ♀ 93.8 (88-99) (n=5), bill 17.2 (15.8-20.6) (n=7) (Weigold 1913-14, Kummerlöwe & Niethammer 1934-35, CSR); according to colour and size, this latter sample is mainly *oenanthe*, but it apparently contains a few *rostrata*.

References Ticehurst, C. B. (1931) Notes on Egyptian birds. *Ibis* (13)1, 575-578. Salomonsen, F. (1934) La variation géographique et la migration du Traquet motteux (*Oenanthe oenanthe* (L.)). *Oiseau* 4, 223-237. Steinbacher, F. (1937) *Oenanthe oenanthe virago* Meinertzh. auf den Kykladen. *Orn. Monatsber.* 45, 204-205. Mackworth-Praed, C. W., & C. H. B. Grant (1951) On the races of the Wheatear *Oenanthe oenanthe* (Linnaeus) occurring in eastern Africa. *Ibis* 93, 234-236.

OENANTHE PLESCHANKA
Pied Wheatear   Alaca Kuyrukkakan

Habitat Stony slopes and hilly pastures in steppe country, covered with scattered boulders or rocky outcrops; also, river banks, gullies, crags, and cliffs. Breeding grounds usually partly covered with fairly lush patches of scrub and low vegetation. The terrain inhabited is generally more rugged and broken than that of *O. hispanica*. Breeds at 800-2000m.

Distribution See map 39; also, Grote (1937, 1939) and Kumerloeve (1975b). Restricted to the area east and north of Van Gölü, as well as near Bingöl (Dott 1967). Reports of *O. pleschanka* further west (and likely some in the east, too) either do not refer to breeding birds or are wrongly identified: *O. pleschanka* is easily confused with heavily worn birds of *O. finschii* (Kumerloeve 1969d, 1970b, 1975b). Recorded 'SSW of Van Gölü' on 25 June 1905 (Witherby 1907). Occurrence in eastern Turkey in the same area as its close relative *O. hispanica* is remarkable, as both species largely exclude each other elsewhere in the breeding range, frequently hybridizing where they come in contact. They meet each other in Iran (Haffer 1977, Vuilleumier 1977), Mangyshlak Peninsula in Kazakhstan (Panov & Ivanitskii 1975, Panov 1986), eastern Bulgaria (Baumgart 1971, Haffer 1977), and perhaps former Yugoslavia (Siegner 1983). Both species probably favour different habitats in Turkey, thus avoiding contact, but confirmation of difference in habitat requirements and research on extent of overlap in breeding range in Turkey is urgently needed. The breeding ranges in Turkey are apparently not fully separated, as a hybrid has been reported (Ullman 1992). Possible occurrence in southern Turkey (as, e.g., in the Gaziantep area, where reported by Warncke 1964) cannot entirely be dismissed, as the species (or, more probably, its close relative *O. cypriaca*) may breed in the Qaryatein-Palmyra area of central Syria (Aharoni 1932; Kumerloeve 1969d; *Zool. Middle East* 6, 13-19, 1992). Aharoni (1931) knew the nest-sites of this species well, and described the eggs as 'almost rounded, bright green-blue with red-brown dots, the smallest of all *Oenanthe* in Syria; closely similar to those of *hispanica*, but smaller'.

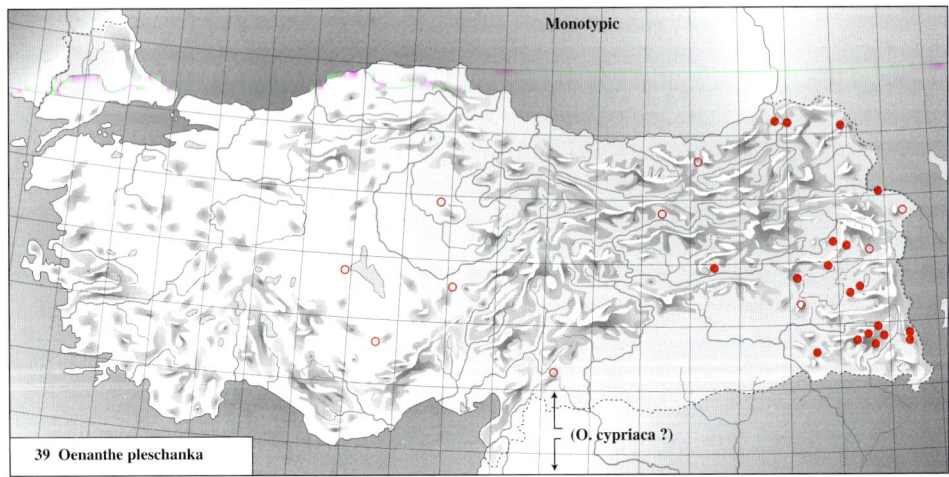

39 Oenanthe pleschanka

Geographical Variation Slight and no subspecies recognized. *O. pleschanka* (Lepechin), 1770, was originally described from Saratov (south-east European Russia). *O. cypriaca* (Homeyer), 1884, an endemic taxon occurring on Cyprus (and perhaps Syria: see above), is now considered to be a separate species rather than a subspecies of *O. pleschanka*, both taxa differing in plumages, size, voice, and behaviour (Christensen 1974, Sluys & Berg 1982, Oliver 1990); for characters and measurements of *cypriaca*, see Roselaar in Cramp (1988). Thus, *O. pleschanka* is a monotypic species, occurring through the steppes from eastern Bulgaria, eastern Romania, and Turkey east to China. Within the species, wing length slightly and gradually increases towards the east, probably as a consequence of the longer migration route of the eastern populations (all winter in Africa). Wing length of males is 92.6 (89-95) (n=12) in eastern Romania and in the steppe along the lower Volga (CSR), 92.7 (88-100) (n=13) in Iran (Sluys & Berg 1982), 95.8 (90-101) (n=13) in Tibet (Vaurie 1972), 95.8 (91-101) (n=27) in Mongolia (Piechocki & Bolod 1972, Piechocki et al. 1982), and 96.8 (94-100) (n=8) in Kansu (China) (Stresemann et al. 1937). The overlap in size is too large and too gradual to separate subspecies. No birds appear to have been collected in eastern Turkey, but they are likely not to differ in size from birds collected in Romania, the lower Volga, and Iran.

References Stresemann, E., W. Meise, & M. Schönwetter (1937) Aves Beickianae. Beiträge zur Ornithologie von Nordwest-Kansu nach den Forschungen von Walter Beick in den Jahren 1926-1933. *J. Orn.* 85, 375-576. Grote, H. (1939) Ist *Oenanthe pleschanka* (Lepech.) eine Rasse von *Oenanthe hispanica* (L.)? *Orn. Monatsber.* 47, 54-57. Dott, H. E. M. (1967) Le Traquet pie, *Oenanthe pleschanka* (Lepechin) en Turquie. *Alauda* 35, 15. Kumerloeve, H. (1969d) On the occurrence of the Pied Wheatear *Oenanthe leucomela* in Asia Minor and adjacent countries. *Ibis* 111, 238-239. Baumgart, W. (1971) Zum Vorkommen von Mittelmeer- und Nonnen-steinschmätzern in der VR Bulgarien. *Beitr. Vogelk.* 17, 449-456. Piechocki, R., & A. Bolod (1972) Beiträge zur Avifauna der Mongolei 2, Passeriformes. *Mitteil. Zool. Mus. Berlin* 48, 41-175. Christensen, S. (1974) Notes on the plumage of the female Cyprus Pied Wheatear *Oenanthe pleschanka cypriaca. Ornis Scandinavica* 5, 47-52. Panov, E. N., & V. V. Ivanitskii (1975) [Evolutionary and taxonomical relations between *Oenanthe hispanica* and *Oenanthe pleschanka.*] *Zool. Zh.* 54, 1860-1873. Haffer, J. (1977) Secondary contact zones of birds in northern Iran. *Bonner zool. Monogr.* 10, 1-64. Piechocki, R., M. Stubbe, K. Uhlenhaut, & D. Sumjaa (1982) Beiträge zur Avifauna der Mongolei 4, Passeriformes. Ergebnisse der Mongolisch-Deutschen Biologischen Expeditionen seit 1962, Nr 124. *Mitteil. Zool. Mus. Berlin* 58, suppl. *Ann. Orn.* 6, 3-53. Sluys, R., & M. van den Berg (1982) On the specific status of the Cyprus Pied Wheatear *Oenanthe cypriaca. Ornis Scandinavica* 13, 123-128. Siegner, J. (1983) Nonnensteinschmatzer *Oenanthe pleschanka* Brutvogel in Jugoslawien. *Verh. orn. Ges. Bayern* 23, 529. Panov, E. N. (1986) [New data on hybridization of *Oenanthe pleschanka* and *O. hispanica.*] *Zool. Zh.* 65, 1675-1683. Oliver, P. J. (1990) Observations on the Cyprus Pied Wheatear *Oenanthe pleschanka cypriaca. Sandgrouse* 12, 25-30. Ullman, M. (1992) Possible hybrid Pied x Black-eared Wheatear *Oenanthe pleschanka* x *O. hispanica. Sandgrouse* 14, 58-59.

OENANTHE HISPANICA
Black-eared Wheatear   Karakulak Kuyrukkakan

Habitat Less confined to open rocky ground and boulders than other *Oenanthe*, requiring scattered shrubs and stunted trees for perching, and consequently legs and feet shorter and more slender than in others, except for its close relative *O. pleschanka.* However, does not avoid boulders and outcrops wherever they occur on the slopes with lush scattered vegetation favoured by the species. Usually at a lower altitude than *O. oenanthe,* but both overlap widely. Breeds mainly at 0-1200m in the west and centre, but up to 1600 (-2150)m in the Taurus and frequently to 2300m in the valleys of the east, some breeding up to 3000m on the Nemrut Dag.

Distribution See map 40; also, Kumerloeve (1975b). Everywhere, except in the coastal zone of

Thrace and the Black Sea Coastlands. Most common on the lower slopes and hills of the west and south, scarce in higher valleys over 2300m in the east, but to over 3000m on the Nemrut Dag.

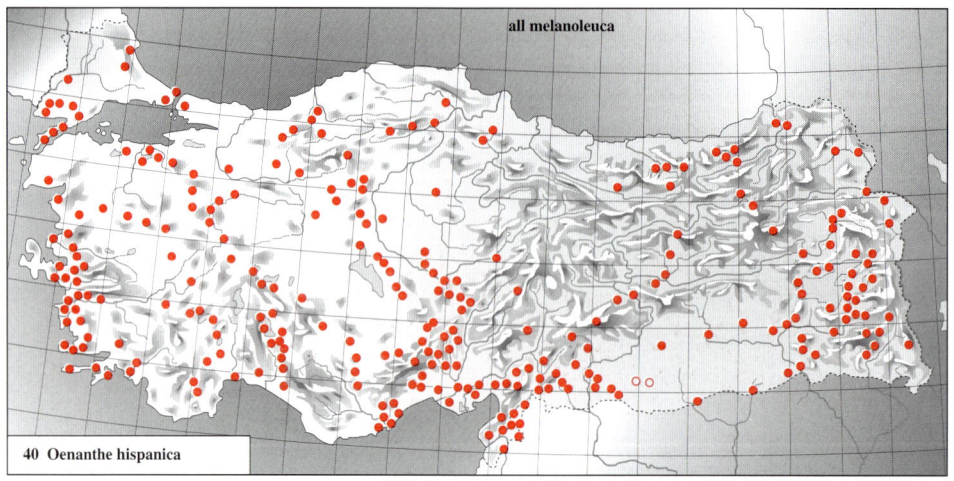

40 Oenanthe hispanica

Geographical Variation

Subspecies described or occurring in the region:

(W/E) ***O. h. melanoleuca*** (Güldenstädt), 1775, Tbilisi (Georgia, western Transcaucasia). [♂ differs from ♂ of *hispanica* of the western Mediterranean basin by the presence of a black stripe of 1-3 mm wide on the forehead, by more brown-grey (less cinnamon) plumage in autumn and whiter (less cream-yellow) plumage in spring, and, in the black-throated morph, by having the black of the throat 21-28 mm long on the middle of the throat instead of 14-21 mm; ♀ buff-brown with sandy tinge (less sandy-cinnamon than ♀ of *hispanica*). Wing of ♂ from southern Italy, former Yugoslavia, and Greece 91.1 (86-95.5) (n=18), of ♀ 88.1 (86-91) (n=4), bill of both sexes 17.0 (15.5-18.3) (n=22) (CSR); wing of ♂ in Iran 92.4 (88-95) (n=10), of ♀ 86.5, 87 (n=2) (Paludan 1938, Schüz 1959, Diesselhorst 1962, Érard & Etchécopar 1970; see also Haffer 1977). As in *O. pleschanka*, two morphs occur in male, pale-throated and black-throated, the proportion of each varying between localities. For this variation in one or both species, see, e.g., Weigold (1913-14), Ticehurst (1927), Vaurie (1949), Mayr & Stresemann (1950), Kumerloeve (1975b), Érard & Etchécopar 1970, Haffer (1977), and Roselaar (in Cramp 1988).]

Subspecies recognized in Turkey: all birds examined are *melanoleuca*, a subspecies occurring from southern Italy and the Dalmatian coast of former Yugolavia east to the Causasus area, northern Iran, and the Mangyshlak Peninsula of western Kazakhstan. Wing of ♂ from Turkey 89.8 (85-93, once 98) (n=25), of ♀ 87.1 (84.5-91) (n=7), bill (both sexes) 16.1 (15.2-16.8, once 17.6) (n=13) (CSR). Thus, the bill is apparently shorter than in the Balkan countries and Greece (see above). The bill from Levant birds is similarly short: 16.1 (15.0-16.9) (n=10) (CSR). Within Turkey, the wing of males from Western Anatolia (Izmir, Aydin, Tire) appears short, 87.7 (85-90) (n=7), against 90.6 (87-93, once 85 and 98) (n=18) in birds from Burdur eastwards; however, this small size is perhaps by chance, as a sample of males from Priene (nearby Izmir, Aydin, and Tire) had wing 90.8 (85-95) (n=36) (Weigold 1913-14). Wing of ♂ from the Urfa area 89.8 (84-94) (n=10), of ♀ 87.5

(85-91) (n=4) (Weigold 1912-13).

References Ticehurst, C. B. (1927) On *Oenanthe hispanica, Oenanthe finschii*, and *Oenanthe picata. Ibis* (12)3, 65-74. Mayr, E., & E. Stresemann (1950) Polymorphism in the chat genus *Oenanthe* (Aves). *Evolution* 4, 291-300.

OENANTHE DESERTI
Desert Wheatear   Çöl Kuyrukkakani

Habitat In Turkey, recorded on a gravelly slope with scattered shrubs at 2200m on the Kuh Dag, south of Erçek Gölü (Gallner 1976).

Distribution No map; no proof of breeding. Found near Amik Gölü on 3 September 1956 (Kumerloeve 1963a) and on the Kuh Dag near Erçek Gölü on 17 August 1973 (Gallner 1976). May breed in Turkey, as recorded breeding on the east side of the Aras valley in Armenia and in Tell Abiad (in Syria, just south of Akçakale), both very close to the Turkish border.

Geographical Variation

Subspecies described or occurring in the area:

(S)   ***O. d. deserti*** (Temminck), 1825, Egypt. [Upperparts greyish-ochre. Wing of ♂ from eastern Egypt (in winter, including the type specimen), Sudan (in winter), the Arabian peninsula, Syria, and north-west Iran 92.3 (89-96) (n=21), of ♀ 89.3 (87-92) (n=8), bill of both sexes 17.4 (16.4-18.2) (n=29) (CSR).]

(E)   ***O. d. atrogularis*** (Blyth), 1847, Agra (India; in winter). [Upperparts buff-brown, less greyish than in *deserti*; white on inner webs of flight feathers does not reach shafts, in contrast to the rather similar subspecies *oreophila* (from the mountains of Central Asia). Larger than *deserti*, wing of ♂ from central and south-east Iran, Turkmenistan, Uzbekistan, Pakistan, and north-west India 98.1 (95-100) (n=12), of ♀ 89, 96 (n=2), bill 17.8 (17.0-18.6) (n=14) (CSR).]

Subspecies recognized in Turkey: none examined, but breeding birds (if any) likely belong to *deserti*, the subspecies occurring both south and east of Turkey. *O. d. atrogularis* occurs from central Iran (south of the Elburz and east of the Zagros mountains) and north-east Iran eastward and may straggle to Turkey.

OENANTHE FINSCHII
Finsch's Wheatear   Kaya Kuyrukkakani

Habitat Steep arid rocky slopes, boulder screes, rocky gullies and ravines, and steep-sided valleys with crags and rocky outcrops at 100-2300m, mainly 500-2000m. Avoids slopes and ravines with dense vegetation, but tolerates scattered scrub and trees.

Distribution See map 41; also, Kumerloeve (1975b). Widespread in suitable habitat east of a line from Kalkan through the Denizli area (Aphrodisias) and Afyon to Kizilcahamam (where recorded by Vermeulen 1987), with occasional records further west. Absent in the steppe country of the Central Plateau and the South-East (except at rocky outcrops and where rivers have made canyons).

Geographical Variation

Subspecies described or recorded in the region:

(S)   ***O. f. finschii*** (Heuglin), 1869, 'Siberia' = Syria? [Cap to mantle of ♂ pale buff-grey (cap white in worn plumage), under tail-coverts pale cream. Cap to mantle of ♀ sandy-grey, ear-coverts brown or black, throat usually with grey or sooty half-collar; underparts pale cream-buff. Smaller, wing of ♂ from Israel, Jordan, Lebanon, and Syria 87.4 (85-91) (n=7), of ♀ 84.3 (82.5-85.5) (n=6) (CSR), tail of ♂ 55.0 (52-58) (n=14) (Vaurie 1949).]

(E) ***O. f. barnesi*** (Oates), 1890, 'Baluchistan and Afghanistan to Persia'. [Cap to mantle of ♂ pinkish cream-buff (cap white in worn plumage), under tail-coverts deep cream-buff. Cap to mantle of ♀ sandy-buff or isabelline-brown, ear-coverts often rufous, throat buff, usually without dark half-collar, underparts buff. Tail-tips of both sexes more extensively black, minimum amount of black on the 2nd to 4th outermost tail feather 9-15 mm (6-12 in *finschii*), maximum on outer web of outermost 12-20mm (5-15 in *finschii*). Larger, wing of ♂ from eastern Iran east to Uzbekistan 90.6 (89-92.5) (n=7), of ♀ 88.5 (86.5-90.5) (n=3) (CSR); wing of ♂ from eastern Iran and Afghanistan 93.1 (89-96.5) (n=23), tail of ♂ 61.2 (58-65) (n=23) (Vaurie 1949); see also Ticehurst (1927).]

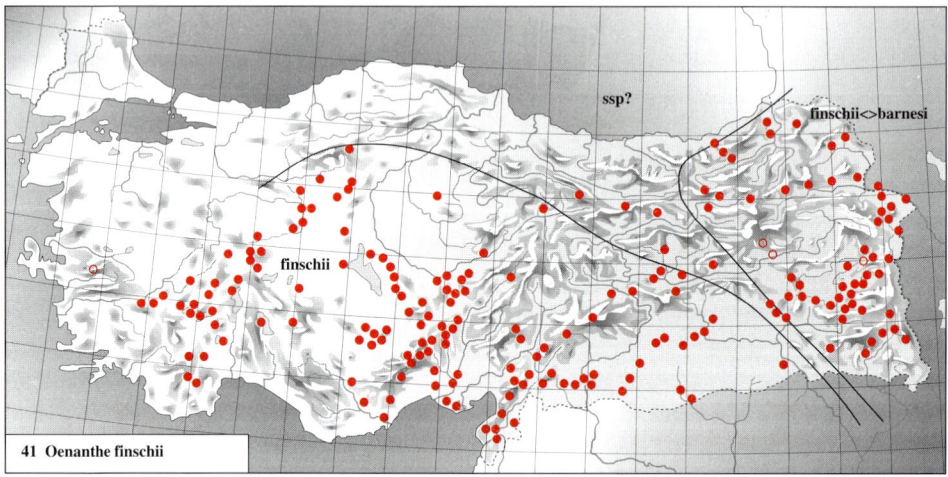

41 *Oenanthe finschii*

Subspecies recognized in Turkey: birds examined from the Central Plateau, the Southern Coastlands, and the South-East east to Ceylanpinar are all similar to *finschii* in plumage and size; wing of ♂ from here 88.9 (87.5-91) (n=10), of ♀ 83.8 (82-86) (n=3) (CSR). Birds from further east in Turkey (examined from Van and Erçek area), as well as those from the eastern Caucasus area and south-west and north-west Iran (east to Gorgan) form a fairly uniform population which is neither *finschii* nor *barnesi*, but combines characters of both: the body colour is similar to *finschii*, but tail pattern and size are similar to *barnesi*. They may be considered to form a separate subspecies rather than an intermediate population, but in view of the slight colour differences between both races (especially in worn plumage), they are perhaps best included in *barnesi*. Wing of ♂ from eastern Turkey 91.4 (89-93) (n=4); wing of ♂ from the eastern Caucasus and north-west and south-west Iran 91.0 (89-95.5) (n=8), of ♀ 86, 86 (n=2) (CSR), tail of ♂ in south-west Iran 60.7 (58-67) (n=9) (Vaurie 1949).

OENANTHE LUGENS

Mourning Wheatear   Karasirt Kuyrukkakan

Habitat Boulder-strewn hills devoid of vegetation, rocky ravines in the desert, and cliffs at the border of dry stony plains.

Distribution No map; no proof of breeding in Turkey. This close relative of *O. finschii* may occur in Turkey, as it nests in the Amadiyah area in northern Iraq, very close to the Turkish border, and in north-west Iran. Recorded as a straggler only: once on 28 February 1914 in Iskenderun, skin formerly in the museum of München (Germany) (Kumerloeve 1975b).

Geographical Variation

Subspecies described or recorded in the region:

(S) ***O. l. lugens*** (Lichtenstein), 1823, 'Deram', Nubia (Egypt or Sudan). [Sexes similar (unlike the subspecies *halophila* from north-west Africa); cap greyish-white; under tail-coverts deep cream-pink; black on tip of outer web of 2nd-4th outermost tail feather 8-15 mm long; white on inner web of flight feathers reaches shaft on middle of secondaries and on bases of primaries. An all-black morph occurs in the basalt deserts of Jordan and southern Syria. Wing of ♂ from Egypt and the Levant 93.8 (89-98) (n=32), of ♀ 90.1 (86-94) (n=23) (Roselaar in Cramp 1988).]

(E) ***O. l. persica*** (Seebohm), 1881, Shiraz (south-west Iran). [As in *lugens*, but cap browner in fresh plumage; under tail-coverts darker rufous-cinnamon; black on 2nd-4th outermost tail feather 13-18 mm long; white on inner web of flight feathers generally does not reach shafts. No black morph recorded. Slightly larger, wing of ♂ from Iran 96.6 (94-99) (n=9), of ♀ 90.6 (88-93) (n=5) (Roselaar in Cramp 1988).]

Subspecies recognized in Turkey: the specimen collected in Iskenderun (see Distribution) has never been identified to subspecies and is now apparently lost (Kumerloeve 1975b). It is most likely to have been *lugens*, which is known to breed from southern Syria southward. The subspecies breeding very close to the Turkish border in northern Iraq and north-west and south-west Iran is *persica*, and this form may occur in the South-East.

OENANTHE XANTHOPRYMNA

Red-tailed Wheatear　　　　　　　　　　　　　　　　**Alkuyruk Kuyrukkakan**

Habitat Dry rocky mountain slopes or valley bottoms, covered with patches of dwarf shrubs and herbs (e.g., *Ferula*), interspersed with bare rocks, scattered boulders, or stunted trees. At c. 1500m in the Yesilce area, at 2000-2100m in the Hakkari area, and at 1000-4100m extralimitally in Iran.

Distribution See map 42; also, Helbig (1984), Kumerloeve *et al.* (1984), and Kasparek (1986, 1992). Proved breeding only in the foothills and mountains bordering the steppe plateau of the South-East, from Yesilce (north-west of Gaziantep) east to the Hakkari area and

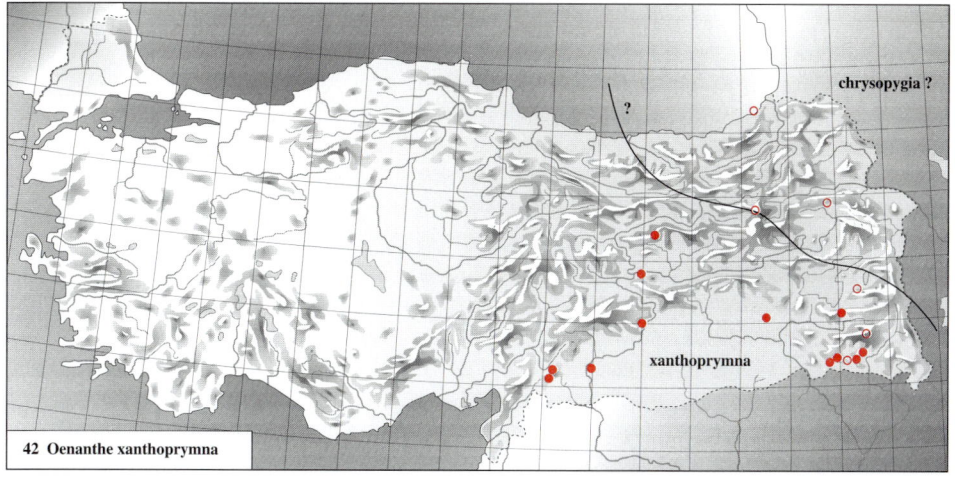

42 Oenanthe xanthoprymna

north to the Munzur Daglari. May breed near Erzurum and Eleskirt as well as in the mountains north-east and east of Van Gölü, thus connecting the known breeding sites in Armenia and Iran with those in Turkey.

Geographical Variation

Subspecies described or recorded in the region:

(E) ***O. x. xanthoprymna*** (Hemprich & Ehrenberg), 1833, Nubia (northern Sudan, in winter). [Both sexes show a male-like plumage. Cap to back dark, dull drab-brown; lower face to throat as well as under wing-coverts in male and most females black or blackish; tail-base white, contrasting with the deep rufous rump and upper tail-coverts (but tail-base rufous in some 1st-year birds). Wing in western Iran and in migrants or wintering birds from the Arabian peninsula and north-east Africa 95.1 (92-97.5) (n=8), of ♀ 89.5 (86-93.5) (n=5) (CSR).]

(E) ***O. x. chrysopygia*** (De Filippi), 1863, Demavend (Damavand, northern Iran). [Both sexes show a female-like plumage. Cap to back sandy-brown; lores and stripe through eye dull black, lower face and throat white, buff towards rear; tail-base always rufous, similar in colour to the rump and upper tail-coverts; under wing-coverts white with some black mottling. Wing of ♂ in Iran 93.2 (90-98) (n=22), of ♀ 90.3 (89-93) (n=13) (Vaurie 1949).]

Subspecies recognized in Turkey: as far as known, all birds observed in Turkey are *xanthoprymna*, although apparently only one specimen has ever been collected (Kumerloeve *et al.* 1984). *O. x. xanthoprymna* has a markedly small breeding range, restricted to Turkey and a narrow zone immediately east of the Turkish border along the western side of the Zagros Mountains in western Iran (between Rezaiyeh and Kermanshah and perhaps further south to above Dezful and Shustar), and probably in extreme north-east Iraq. Further east (in the Zagros proper), as well as further north (in Armenia, e.g., in the Yerevan area) it is replaced by *chrysopygia*. This latter subspecies may occur in eastern Turkey (e.g., on the slopes along the Aras valley, round Dogubayazit and Eleskirt, or even west to Erzurum), but proof is lacking. The relation between both subspecies is unclear; they are said to interbreed (e.g., Vaurie 1949), but the characters found in the supposed hybrid (which has been named *cummingi*) are present in some first year birds of *xanthoprymna* (CSR; see also Bates 1935b and Helbig 1984). Moreover, 'cummingi' and *xanthoprymna* are reported to breed side-by-side (Härms 1925), probably a further proof that this 'hybrid' belongs to *xanthoprymna*. The forms *xanthoprymna* and *chrysopygia* may be separate species, as already advocated by Ivanov (1941). See also Stepanyan (1971).

References Härms, M. (1925) Über *Oenanthe xanthoprymna* (Hemprich & Ehrenberg). *J. Orn.* 73, 390-394. Bates, G. L. (1935b) On *Oenanthe xanthoprymna* and *Oe. chrysopygia*. *Ibis* (13)5, 198-201. Ivanov, A. I. (1941) *Oenanthe chrysopygia* de Fil. i *O. xanthoprymna* Hempr. & Ehrenb. *Izv. Akad. Nauk SSSR* 3, 381-384. Stepanyan, L. S. (1971) [On the taxonomic position of *Oenanthe xanthoprymna* Hemprich & Ehrenberg from the Transcaucasus and the Badakshan region.] *Nauch. Dokl. vyssh. Shk. Biol. Nauki* 1971(6), 26-31. Kumerloeve, H., M. Kasparek, & K.-O. Nagel (1984) Der Rostbürzel-Steinschmätzer, *Oenanthe xanthoprymna* (Hemprich & Ehrenberg 1833), als neuer Brutvogel im östlichen Anatolien (Türkei). *Bonner zool. Beitr.* 35, 97-101. Kasparek, M. (1986) New records of the Red-rumped Wheatear, *Oenanthe xanthoprymna*, in Turkey. *Zool. Middle East* 1, 54-56.

OENANTHE MOESTA

Red-rumped Wheatear

Habitat Flat dry steppe, fairly densely covered with low vegetation of *Artemisia*, Chenopodiaceae, or other plants and scrub, amply provided with rodent burrows which are needed for nesting.

Distribution No map. Individuals of this species, or probably of this species, were said to have been seen in April and May near Viransehir, between Gaziantep and Kilis, between Uzungeçit and Uludere, and north of Halfeti (Glimmerveen & Hols 1986), without evidence of breeding. These areas are now known to be within the range of breeding or migrating *O. xanthoprymna* (in fact, some of the *O. moesta* recorded had *O. xanthoprymna* nearby), and though the features described seem to point to *O. moesta*, the characteristics mentioned also appear to fit *O. xanthoprymna*, of which the extent of individual variation (especially in tail pattern) was not fully known at the time. It is not clear whether the specific habitat of *O. moesta* occurs in Turkey. The nearest known breeding site of *O. moesta* is in southern Syria.

Geographical Variation The only subspecies which may eventually occur in Turkey is *brooksbanki* Meinertzhagen, 1923, described from a bird wintering at Al Jid in westernmost Iraq, and which is scarcely different from *moesta* of North Africa (see Roselaar in Cramp 1988).

References Glimmerveen, U., & H. Hols (1986) Red-rumped Wheatear, *Oenanthe moesta*, in Turkey. *Zool. Middle East* 1, 26-28.

MONTICOLA SAXATILIS

Rock Thrush   Kaya Ardici

Habitat Barren rocky outcrops in steppe or in mountains, extensive limestone screes, eroded canyons, or alpine meadows with boulders, either bare, lightly wooded, or amidst forest. At (400-) 800-1500m in the west, in the interior of the Black Sea Coastlands, and at the fringe of the Central Plateau, at 1500-3000m on the Ulu Dag and in the Taurus, at 2000-3000m in the Van area, and at 1850-4000m in the East.

Distribution See map 43. Confined to mountainous regions, but everywhere scarce or only locally common. Absent from the steppe of the Central Plateau and of the South-East,

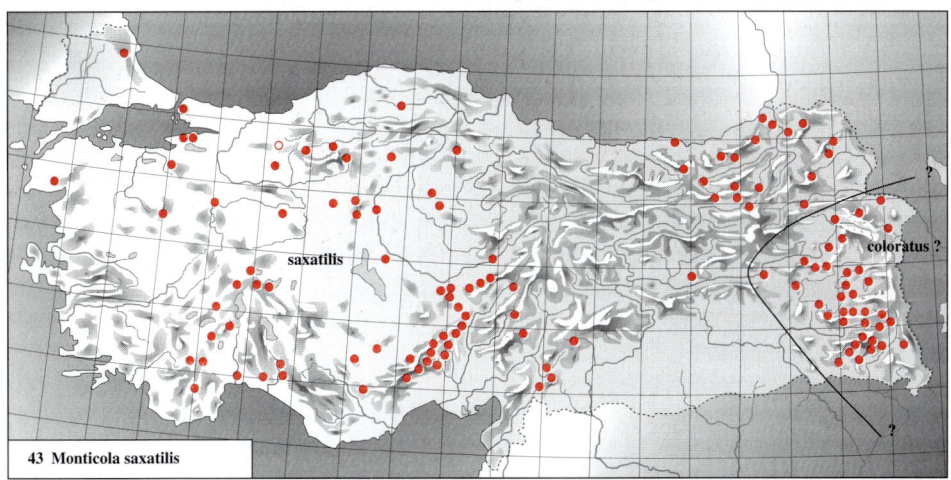

43 *Monticola saxatilis*

except on some isolated mountains. Mapping is difficult due to occurrence of spring migration until early June and autumn migration from early August onwards, making the period during which breeding can be assumed rather short.

Geographical Variation

Subspecies described or recorded in the region:

(W) *M. s. saxatilis* (Linnaeus), 1766, Switzerland. [Male in spring brightly coloured, breast to under tail-coverts rufous-cinnamon. Wing of ♂ from southern Europe 124.8 (119-129) (n=15), of ♀ 120.0 (115-126) (n=10) (CSR); wing of ♂ from the Levant, Armenia, and Iran 122.4 (118-127) (n=7), of ♀ 120.3 (118-125) (n=3) (Paludan 1940, Nicht 1961, CSR).

(T) *M. s. coloratus* Stepanyan, 1964, 'Chnys-Kala', Turkey (=Hinis Kale,on eastern slope of Bingöl Daglari). [Said to be similar to *saxatilis*, but colour of ♂ more saturated, breast to under tail-coverts deep rusty-red (Stepanyan 1964).]

Subspecies recognized in Turkey: birds from the western half of Turkey as well as those from Varsambeg (Verçenik) in the north-east are not different from *saxatilis* from southern Europe. A single male from Gürpinar (just south-east of Van Gölü) was not different either, and thus the subspecies *coloratus*, based on 4 males collected in Hinis Kale (Stepanyan 1964) is perhaps not valid, especially in view of the individual variation in depth of colour of breeding males elsewhere. However, males seen on the Nemrut Dag (just west of Van Gölü) by Dufourry (1990) appeared to be very intensely coloured, and research on a larger number of breeding males than currently available is required to settle the validity of *coloratus*. Wing of ♂ from Turkey 127.5 (122-130.5) (n=3), of ♀ 122.5, 123.5 (n=2) (CSR), wing of ♂ from Çankiri (north-east of Ankara) 125 (Kummerlöwe & Niethammer 1934-35).

References Stepanyan, L. S. (1964) [Geographical variation of the Rock Thrush *Monticola saxatilis* L.] *Sb. Trudy Zool. Mus. MGU* 9, 228-231.

MONTICOLA SOLITARIUS
Blue Rock Trush   Mavi Kayaardici

Habitat Largely confined to steep and broken limestone cliffs and slopes at any level (e.g., on cliffs at the coast), but generally below the altitudes inhabited by *M. saxatilis*; also, caves, stone walls, and rocky wadis; mainly at 0-1700m, but to 2500m on the Ulu Dag, in the

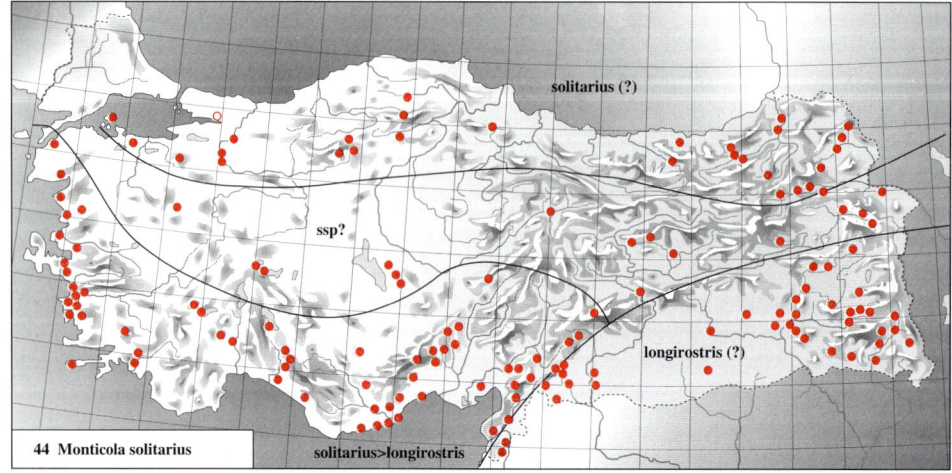

44 *Monticola solitarius*

Taurus, and in the east.

Distribution See map 44. Locally common in Western Anatolia, the Southern Coastlands, and on the slopes of the lower valleys of the East and South-East, scarce in the interior of the Black Sea Coastlands, absent from the Central Plateau and from the steppe of the South-East, except along its fringe and on some isolated mountains; apparently absent from Thrace.

Geographical Variation

Subspecies described or recorded in the region:

(W)   ***M. s. solitarius*** (Linnaeus), 1758, Italy. [Upperparts of ♀ dark brown with olive tinge; underparts heavily marked on buff ground-colour, with uniform buff chin, small uniform buff spots on throat and breast (c. 2-3 mm across), and fairly broad black-brown bars on belly and flanks. Wing of ♂ from northern Italy, the Alps, the Balkan countries, and northern Greece 127.0 (123-133) (n=16), of ♀ 122.6 (120.5-126.5) (n=4), tail of both sexes 84.0 (78.5-92) (n=14), bill of both sexes 29.8 (26.5-32.5) (n=20) (CSR). Breeds also in the Caucasus area.]

(W)   ***M. s. behnkei*** Niethammer, 1943, Samaria (Levka mountains, Crete, Greece). [Colour of ♀ intermediate between *solitarius* and *longirostris*. Wing of ♂ from Sicily, southern Italy, southern Greece (including Crete), and of migrants from Egypt 125.0 (122-130) (n=14), of ♀ 120.9 (116-124) (n=10), tail 80.0 (76-85) (n=19), bill 29.2 (27.5-31.5) (n=22) (CSR).]

(E)   ***M. s. longirostris*** (Blyth), 1847, between Sind (Pakistan) and Ferozepore (Punjab, India; in winter). [Upperparts of ♀ brown-grey, less dark than *solitarius*; pale spots on chest larger, c. 4 mm across; dark bars on belly and flanks narrower, browner; ground-colour of underparts paler, pink-buff to cream. Wing of ♂ from Iran and Afghanistan 123.0 (118-127) (n=14), of ♀ 117.1 (114-119) (n=9), bill 28.0 (26.6-30.1) (n=22) (Roselaar in Cramp 1988).]

Subspecies recognized in Turkey: birds from Western Anatolia (Izmir) and the Taurus area (Antalya, Anasha, Haruniye) examined are similar to *behnkei* in colour and size; wing of ♂ of these 124.2 (121-129) (n=4), of ♀ 120.5 (115.5-124) (n=8), bill of both sexes 29.6 (28.2-31.0) (n=12) (CSR). However, the characters of *behnkei* are rather poor, as its colour is close to Italian *solitarius* (birds of France and Iberia are larger and darker) and its size is near *longirostris* (tail of *longirostris* and '*behnkei*' mainly below 84 in ♂, below 82 in ♀; in *solitarius*, virtually always over these values). Thus, birds of Western Anatolia and the Taurus are better considered to be intermediate between *solitarius* and *longirostris*. True *longirostris* occurs in the Levant, Dihok (northern Iraq), and Iran, close to the Turkish border, and probably the birds breeding in the Antakya area and in the South-East belong to *longirostris*, too. The subspecies of the birds breeding locally in the Black Sea Coastlands is obscure: birds wintering in Istanbul and some birds wintering in the Levant are inseparable from *solitarius* in colour and size, and these wintering *solitarius* probably have their origin in northern Turkey, thus connecting the known breeding ranges of *M. s. solitarius* in the Balkan countries with that in Georgia, Armenia, and Azerbaijan.

TURDUS TORQUATUS
Ring Ouzel   Kolyeli Ardiç

Habitat Low shrubs at or above the tree line on rocky mountain slopes. At about 1300-1500m on the Ulu Dag and in the interior of the western Black Sea Coastlands, at 1800-2700m in the Karanfil and Ala Dagi, and from 2000 to over 3000m in the mountains of the east, but recorded as breeding hardly above sea level near Kesan (Thrace).

Distribution See map 45. Known from some widely separated mountain tops in the west, but the range is apparently more continuous in the mountains of the East. Elsewhere, the species is probably underrecorded. Birds observed between 16 May and 8 June in suitable habitat near Bergama, above Burdur, near Kizilcahamam, and north of Çorum may indicate breeding.

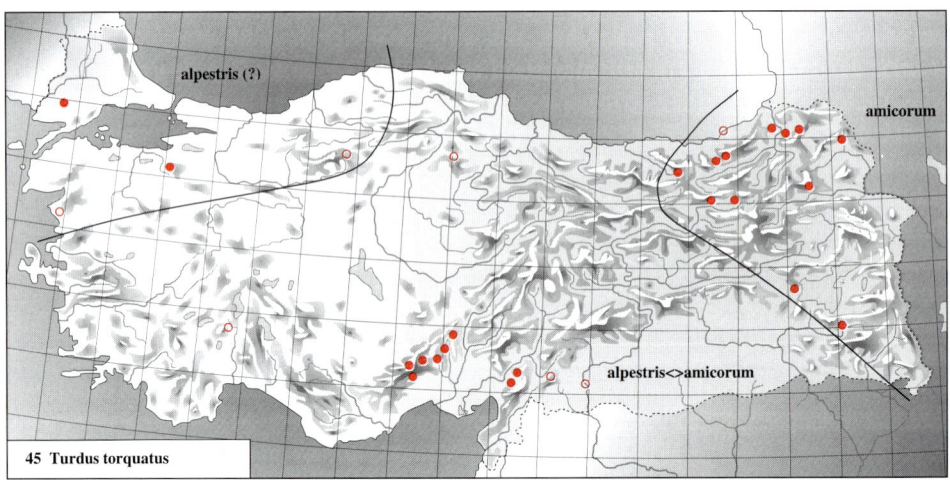

45 *Turdus torquatus*

Geographical Variation

Subspecies described or recorded in the region:

(W) *T. t. alpestris* (Brehm), 1831, Tirol (Austria or Italy). [Breast and belly of ♂ black, feathers with narrow and concealed white central marks and with white fringes 2-3 mm wide, appearing black with white scaling; greater upper wing-coverts and tertials with grey-white fringes of *c*. 2 mm wide, not joining into a large white patch. Breast and belly of ♀ black, each feather with easily visible large white centre and broad white fringe, breast and belly appearing heavily variegated black-and-white or white with broad black scaling; fringes of greater coverts and tertials grey, only a few central coverts white. Wing of ♂ from the Alps and the Balkan countries 144.4 (140-149) (n=8), of ♀ 138.5 (133.5-143.5) (n=11), bill of both sexes 25.3 (23.2-26.8) (n=19) (CSR).]

(E) *T. t. amicorum* Hartert, 1923, Kislovodsk (northern Caucasus, Russia). [Breast and belly of ♂ black, feathers with white fringes 1.5-2 mm wide when fresh (latter absent when worn), centres of feathers without white marks; most greater coverts and tertials with broad and conspicuous white fringes, together forming large white patch on inner wing. Underparts of ♀ as in ♂, thus without the large white feather centres of the ♀ of *alpestris*; white fringe along most greater coverts of ♀ *c*. 2 mm wide, generally not forming a confluent white patch, closely similar to extent of white on wing of the ♂ of *alpestris*, but with more white than in the ♀ of *alpestris*. Wing of ♂ from the Caucasus, Transcaucasia, and northern Iran 144.8 (141-150) (n=7), of ♀ 136.8 (134-139) (n=3), bill 27.0 (25.8-28.2) (n=6) (Stresemann 1928, CSR).]

Subspecies recognized in Turkey: 4 birds examined from the foot of the Taurus mountains (Haruniye and Adana) are neither *alpestris* nor *amicorum*, but intermediate between both: the amount of white on the wing is more or less close to *alpestris* (differing between sexes, as in the other subspecies), the amount of white on the underparts is intermediate (slightly closer to *amicorum*). Though collected in winter, these birds probably breed in the nearby Taurus as they do not resemble any population outside Turkey. No birds were seen from the north-west; birds breeding in Thrace are likely to be *alpestris*, as this subspecies breeds in the mountains of Bulgaria nearby; a bird collected Ankara (male, 17 April, wing 135) and probably originating from the mountains of the western Black Sea Coastlands has been assigned to *alpestris* (Kummerlöwe & Niethammer 1934-35). Of 3 birds from north-east Turkey (two from 'Zebatos'=Cayeli in September, probably from the mountains nearby, and one from Sarikamis in May), 2 are identical to *amicorum* of the Caucasus, but one of the Cayeli birds is more or less intermediate between *alpestris* and *amicorum*. The population from eastern Turkey is best included in *amicorum*. Wing of ♂ from the Taurus 147 (n=1), of ♀ 139.2 (137-142.5) (n=3), bill 25.7 (24.0-26.5) (n=4); wing of ♂ from north-east Turkey 143.2 (140-147.5) (n=3), bill 26.5 (26.2-27.1) (n=3) (CSR). As deduced from the increase in bill length eastward throughout Turkey, *alpestris* apparently grades into *amicorum*, and this is also reflected by the change in plumage.

TURDUS MERULA
Blackbird   Karatavuk
Habitat All types of open forest, plantations, parklands, and gardens, where scrub or trees border open ground, preferably near water or near humid sites, occurring mainly at 500-2000m, but in the west locally down to sea level and in the east locally to over 2000m (but not above the tree line). In the Taurus, in the bottoms of ravines and gorges with some cover. Not as closely associated with humans as in western and central Europe, but occurs locally in parks in the west (Kumerloeve 1958a).

Distribution See map 46. Common and widespread in Thrace, Western Anatolia, the Black Sea Coastlands, and the Southern Coastlands, very local in the mountain valleys of the East and South-East. Absent from the coastal strip of the Southern Coastlands, from the Central Plateau (except its fringes and some isolated forests), and from the bare arid parts of the South-East.

Geographical Variation
Subspecies described or recorded in the region:
(W/E) ***T. m. aterrimus*** (Madarász), 1903, Vladikavkaz (Ordzhonikidze, northern Caucasus, Russia). [Entire plumage of ♂ dull black with faint brown tinge (less deep and glossy black than in *merula* from western, central, and northern Europe), belly usually with slight olive-brown tinge and with faint olive-grey feather fringes, especially in first year birds. Upperparts of ♀ deep olive-brown to olive-black; chin and throat contrastingly streaked sooty and grey-white, upper breast dull rufous-brown with sooty spots, lower breast and belly medium grey with clear olive tinge and paler grey feather fringes; bill of ♀ in spring usually horn-brown with some yellow at tip. Wing of ♂ from Romania, Bulgaria, and Transcaucasia 129.3 (124.5-133) (n=19), of ♀ 123.6 (118.5-129) (n=10), bill of both sexes 25.9 (24.2-27.2) (n=29) (CSR); wing of ♂ from northern Iran (south of Elburz mountains) 128.1 (125-133) (n=7), of ♀ 120.3 (118-126) (n=9) (Stresemann 1928, Paludan 1940, Schüz 1959, Diesselhorst 1962).]

(S) ***T. m. syriacus*** (Hemprich & Ehrenberg), 1833, Lebanon. [Male slate-black, belly with faint brown tinge, slightly less brown than in *aterrimus*. Upperparts of ♀ dark sooty grey to olive-black, underparts entirely dark grey with slightly paler grey feather fringes, with a slight brown tinge on the breast, and with inconspicuous paler grey streaking on chin

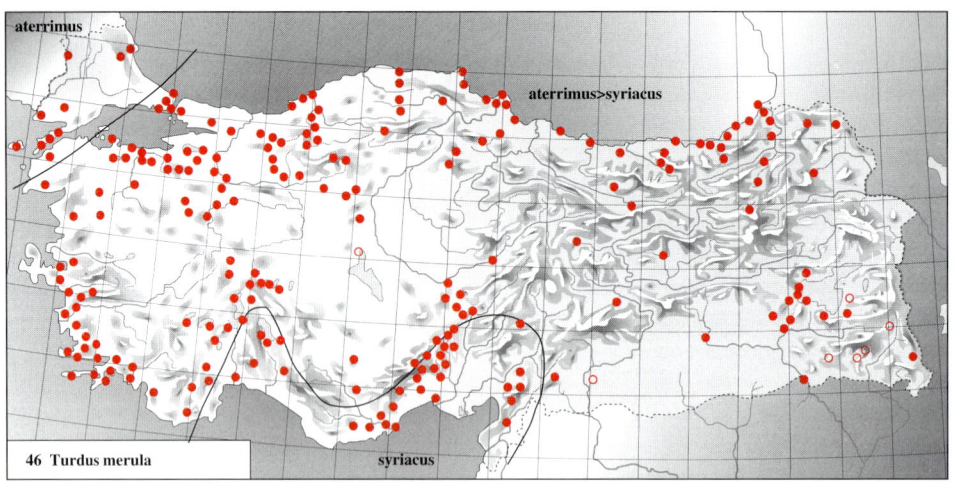

46 Turdus merula

and upper throat; bill of ♀ usually yellow except for the base of the upper mandible. Wing of ♂ from Israel, Syria, and western Iran 124.6 (123-126) (n=4), of ♀ 122.8 (120-125) (n=4), bill 26.8 (25.5-28.0) (n=6) (CSR).]

Subspecies recognized in Turkey: several birds examined, but some were juveniles which are probably less suitable for assessment of subspecies; some others were from winter and may have included birds from elsewhere. Ignoring a winter female from Maras (wing 125), which was *aterrimus*, and a very glossy black winter male from Gaziantep (wing 131.5), which may have been *merula*, and assuming that all remaining winter birds had been raised locally, the following pattern arises: all birds examined had a colour inseparable from that of *syriacus*, rather than *aterrimus*, even though the latter subspecies is usually presumed to occur in Turkey. However, the wing of most birds except for those of the Taurus and Amanus mountains is about as long as in *aterrimus*: wing of ♂ from north-west Asia Minor (Sogukpinar, Bolu, Devrek, Zonguldak, Kastamonu) 129.7 (122.5-133.5) (n=5), of ♀ 125.2 (123-128) (n=3), bill of both sexes 25.8 (25.0-26.3) (n=6); wing of ♂ from the east (Sebinkarahisar, Trabzon, Rize, Tatvan) 130.0 (126-133) (n=3), of ♀ 124, 124 (n=2), bill 26.1 (25.6-27.1) (n=5); wing of ♂ from the Taurus and Amanus area (Beysehir, Pozanti, Antakya) 122.3 (116-126.5) (n=3), of ♀ 117.7 (115-122) (n=3), bill 25.8 (25.0-26.8) (n=4) (CSR, with some additional wing data from Kummerlöwe & Niethammer 1934-35, Kumerloeve 1964a, and Vauk 1973). Thus, birds from the Taurus and Amanus mountains are inseparable in both colour and size from *syriacus*, and also the bill colour of the 2 spring females was as in *syriacus*. In the north and east of Asia Minor, size and bill colour are as in *aterrimus*, but body colour appears to be similar to *syriacus*; birds from here are probably best considered to be *aterrimus* with an influence of *syriacus*. A bird from Thrace (♂, wing 126) is *aterrimus* (Rokitansky & Schifter 1971). The subspecies breeding in Western Anatolia south of the Ulu Dag has yet to be established.

References Kumerloeve, H. (1958a) Brutverbreitung und Verstädterung der Amsel in der Türkei. *Vogelwelt* 79, 60-61.

TURDUS PHILOMELOS
Song Thrush   Sarkici Ardiç

Habitat Damp deciduous and mixed forest on hills and slopes at 500-1900m, but locally down to sea level in the north-west.

Distribution See map 47. Restricted to northern Thrace, the Black Sea Coastlands, and the northern part of Western Anatolia. Song heard in south-west Anatolia and in the Taurus in May is supposed to be produced by late migrants (e.g., Vauk 1973), but the species may occasionally breed there.

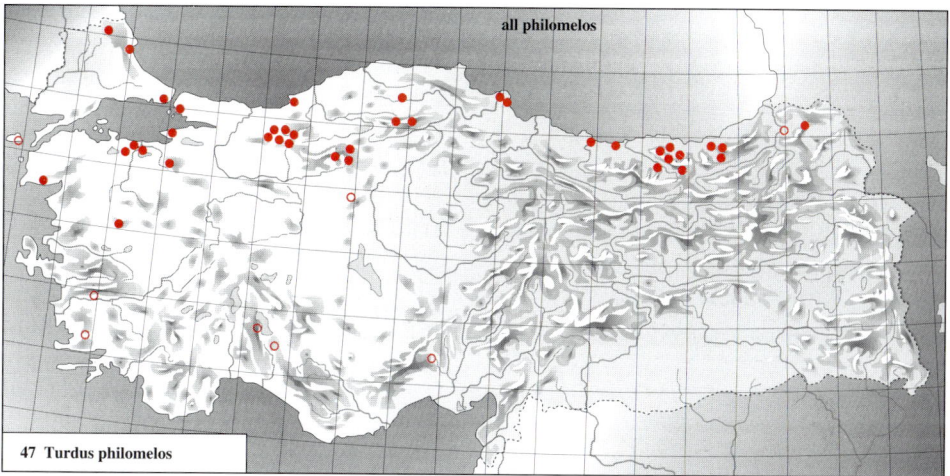

47 Turdus philomelos

Geographical Variation

Subspecies described or recorded in the region:

(W/E) **T. p. philomelos** C. L. Brehm, 1831, central Germany (on migration). [Upperparts olive-brown to olive-grey and ground-colour of underparts pale cream to white, colour depending on wear; less rufous-brown above and less buffish below than the subspecies living in western Europe. For geographical variation, see Voous (1959). Wing of Scandinavian ♂ migrating through the Netherlands 119.4 (113-125) (n=72), of ♀ 117.7 (113-123) (n=115) (Roselaar in Cramp 1988).]

Subspecies recognized in Turkey: quite a number of birds was examined, but these included only a single breeding bird (a male from Sogukpinar at the foot of the Ulu Dag, collected 1 July, wing 115). All birds were inseparable from *philomelos*. Wing of migrants from Turkey (sexes combined): 119.0 (115.5-123.5) (n=10) (CSR), or (in another sample from the literature) 116.5 (112-120.5) (n=11) (Weigold 1913-14, Rössner 1935, Kummerlöwe & Niethammer 1934-35, Kumerloeve 1961a). Birds paler and greyer than *philomelos* are said to occur in the Zagros mountains (south-west Iran) in winter (Vaurie 1959), and these may occur in eastern Turkey, too. The status of these birds is unknown, and it is not known whether the name *nataliae* Buturlin, 1929, based on birds from Krasnoyarsk in Siberia and used by Vaurie for these Zagros birds, is valid. Wing of ♂ from the Zagros 122 (118-128) (n=11) (Vaurie 1959).

References Voous, K. H. (1959) Individual and geographical variation in the Song-thrush, *Turdus philomelos* Brehm. *Ardea* 47, 28-41.

TURDUS VISCIVORUS

Mistle Trush   Ökseotu Ardici

Habitat Open coniferous or mixed forest, plantations, gardens, and orchards, mainly at 800-2000m, but lower in the extreme west and up to 2600m in the Taurus and in the mountains of the east.

Distribution See map 48; also, Kumerloeve (1957c). Local but widespread in Thrace, Western Anatolia, the Black Sea Coastlands, the Taurus, the Amanus mountains, and the Upper Aras Valley. Records in May and July in the Van and Hakkari areas may point to local breeding.

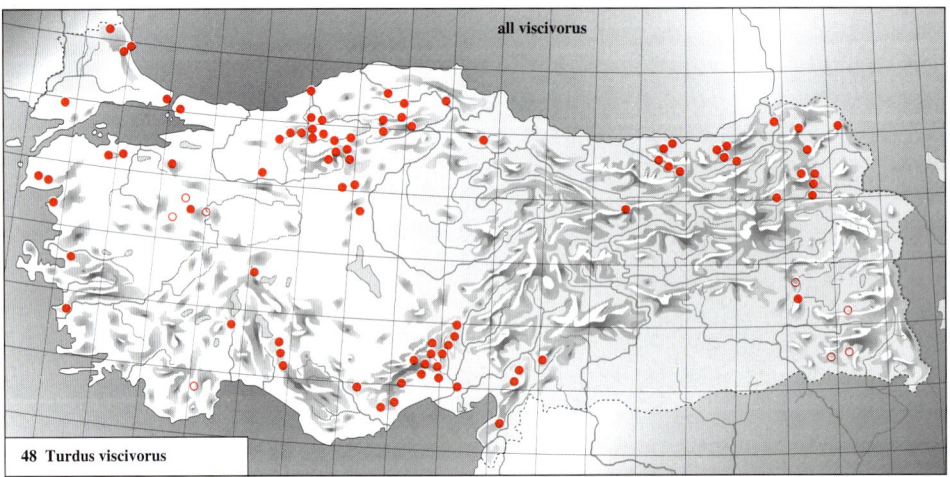

48 *Turdus viscivorus*

Geographical Variation

Subspecies described or recorded in the region:

(W/E) ***T. v. viscivorus*** (Linnaeus), 1758, Essex (England). [Upperparts light buffish olive-brown when plumage fresh, changing to olive-grey and cold grey when heavily worn; ground-colour of underparts white, in some birds strongly tinged tawny-yellow or cream when plumage fresh, purer white when worn, but much individual variation in tinge in fresh plumage (rather than geographical variation). Wing of birds of both sexes from Britain 154.4 (146-162) (n=19) (Vaurie 1955), from the Netherlands and Germany 154.9 (147.5-161, once 164.5) (n=47) (CSR), from former Yugoslavia, Bulgaria, and Greece 153.4 (149-162) (n=17) (Stresemann 1920, Makatsch 1950, CSR), from the Caucasus 156.8 (154-161) (n=4) (Stresemann 1920), and from south-west and north-west Iran 152.7 (150-156) (n=6) (Stresemann 1928, Paludan 1938, 1940, Schüz 1959). Many subspecies have been described from within the range of *viscivorus* (which occurs from western Europe east to western Siberia, the Caucasus area, and north-west and south-west Iran), based on slight variations in colour, but all variation is individual or due to wear and bleaching, and none is recognized here. These subspecies include, e.g., *bithynicus* Keve, 1943, described from Sogukpinar (Western Anatolia, Turkey), and *jubilaeus* Von Lucanus & Zedlitz, 1911, Slonim (western Belorussia).]

Subspecies recognized in Turkey: all birds examined are inseparable from *viscivorus* from England and the mainland of western Europe. Some birds from Sogukpinar and the Ulu Dag show some bright tawny-yellow on underparts in fresh plumage (*bithynicus* was based on such birds: see above), but others in equally fresh plumage from the same area are tinged cream below, and similar tawny-yellow birds occur occasionally in western Europe and as far east as Rzhev (near the Volga close to Moskva (Moscow), Russia). Apparently due to more intense bleaching in continental climates, birds from the Balkan countries and eastern Europe (*'jubilaeus'*) more rapidly fade to grey on the upperparts and to white on the underparts in spring and summer than in western Europe. Wing of birds from Turkey 153.7 (148-160) (n=13), bill 25.4 (24.3-26.8) (n=13) (CSR), in 3 other birds wing 153 in each (Kummerlöwe & Niethammer 1934-35, Rössner 1935, Jordans & Steinbacher 1948).

References Kumerloeve, H. (1957c) Ökologie der Misteldrossel im Bosporusraum. *Vogelwelt* 78, 194-195.

CETTIA CETTI
Cetti's Warbler   Setti Bülbülü

Habitat Dense bushes and thickets, mainly in wetlands, riverine vegetation, and scrub along water or in gardens in west, but also common to abundant on moist to fairly dry mountain slopes and gullies covered with oak scrub, brambles, or other dense vegetation in the east of the country, occurring up to at least 2500m.

Distribution See map 49. In all suitable habitat throughout the entire country.

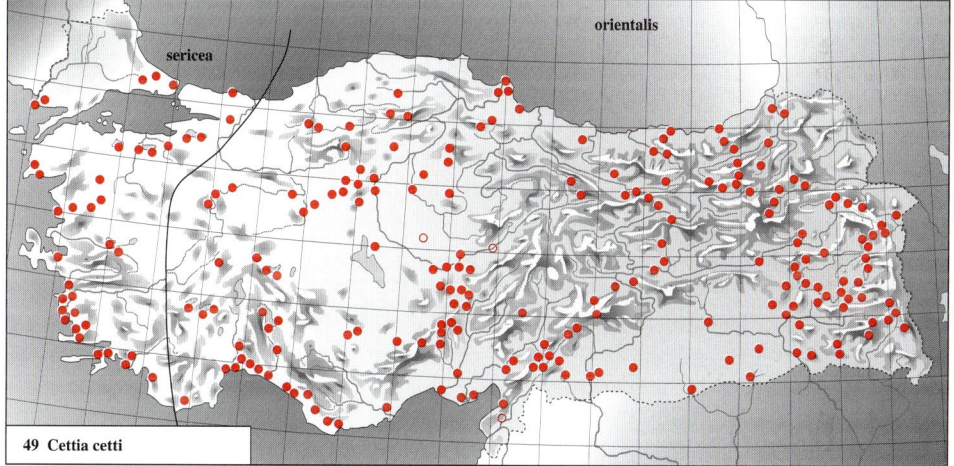

Geographical Variation
Subspecies described or recorded in the region:
- (W) ***C. c. cetti*** (Temminck), 1820, Sardinia (Italy). [Upperparts saturated rufous-brown; sides of head, breast, and flanks rather extensively olive- or rufous-brown. Smaller, wing of ♂ from Sardinia, Corsica, Sicily, and southern Italy 61.6 (59-65.5) (n=24), of ♀ 53.8 (52-56) (n=6); tail of ♂ 58.6 (55-61) (n=16), of ♀ 48, 51.5 (n=2) (CSR).]
- (W) ***C. c. sericea*** (Temminck), 1820, mouth of the Brenta River near Venezia (north-east Italy). [Rather like *cetti*, but the colour of the upperparts slightly more olive-brown, and size larger. Throat and breast washed grey, less white than in *orientalis*, flanks with rather extensive drab-brown wash. Wing of ♂ from Venezia (including the type specimens), former Yugoslavia, and Albania 64.5 (61-68) (n=59), of ♀ 58.1 (56-62) (n=17) (Stresemann 1920, CSR), tail of ♂ 60.7 (57.5-63) n=12), of ♀ 56.7 (54.5-59.5) (n=9) (CSR).]
- (S) ***C. c. orientalis*** Tristram, 1867, Palestine. [Paler; upperparts olive-brown (less saturated than in *sericea*), more greyish when worn; supercilium wider, whiter; breast and flanks paler grey-brown and this colour more restricted, white of belly more extensive. Rather large, size about similar to *sericea*. Wing of ♂ from Israel 66.1 (64.5-68) (n=5), of ♀ 59.2

(58-60) (n=3), tail of ♂ 63.0 (61-66.5) (n=5), of ♀ 57.0 (56.5-57.5) (n=3) (CSR); wing of ♂ from Turkey (eastward from the Central Plateau and the Southern Coastlands) 64.6 (62-68.5) (n=13), of ♀ 59.2 (55-61) (n=4), tail of ♂ 63.5 (57-69) (n=10), of ♀ 54 (n=1) (Weigold 1912-13, CSR).]

Subspecies recognized in Turkey: birds from central and eastern Turkey (Solak, Balikesir area, Karadirek, Ankara, Eregli, Gaziantep, Tatvan) are all inseparable from *orientalis*, which also occurs from the Levant east to Iraq and Qasr-e-Shirin in western Iran. In Thrace and in the coastal zone of Western Anatolia (e.g., Izmir, Tire, and Priene), birds are slightly smaller and darker, more or less intermediate between *orientalis* or *sericea* and *cetti*; wing of ♂ from here and from Crete 62.8 (60-65) (n=6), of ♀ 57.5 (56-58) (n=3), tail of ♂ 61.3 (59.5-63.5) (n=6), of ♀ 56.2 (54.5-58) (n=3) (Weigold 1913-14, CSR). For these birds, the names *reiseri* Parrot, 1910, described from Hercegovina and Greece, or *muelleri* Stresemann, 1919, described from Han Abdipassa in Makedonija (former Yugoslavia), are available, but the differences from *sericea* are too slight to warrant recognition of an intermediate subspecies, and birds from western Turkey are best included in *sericea*. The still larger and paler subspecies *albiventris*, occurring from Lorestan and Fars in Iran to Central Asia, is not found in Turkey; wing of ♂ of *albiventris* 67.6 (64-72) (n=34) (Vaurie 1954).

CISTICOLA JUNCIDIS
Zitting Cisticola   Yelpazekuyruk

Habitat Dense low vegetation (e.g., grasses, low reeds, or *Juncus*) in marshes near water or on damp sites in plains and valleys.

Distribution See map 50. Restricted to wetlands below c. 100m in Thrace, Western Anatolia, and the Southern Coastlands; breeding near Birecik and Ceylanpinar in the South-East is probably irregular. Records of birds along lakes at the foot of the western Taurus, in the east, and in the Kizilirmak Delta are probably due to post-breeding dispersal or to vagrant birds overshooting during migration, but the species may occasionally breed here.

Geographical Variation

Subspecies described or recorded in the region:

(W)   *C. j. juncidis* (Rafinesque), 1810, Sicily (Italy). [Upperparts rather contrastingly streaked black and cinnamon-buff (when plumage fresh) or yellow-buff (when worn); rump bright rufous; sides of breast and flanks extensively cinnamon-buff or yellow-buff. Wing of ♂ from Italy and Greece 51.1 (49-53) (n=11), of ♀ 47.5 (46-50) (n=14) (CSR). For more details of this and other species and subspecies of *Cisticola*, see Lynes (1930).]

(S)   *C. j. neurotica* Meinertzhagen, 1920, Saïda (Lebanon). [Ground-colour of upperparts paler, less warm buff, more pink-cinnamon (when fresh) to sandy-grey (when worn); rump paler, buff-cinnamon; underparts extensively white, buff on sides of breast and flanks rather pale and diluted. Size as in *juncidis*, wing of ♂ from Israel 50, 50.5 (n=2), of ♀ 49 (n=1) (CSR).]

Subspecies recognized in Turkey: birds from the Izmir area and probably elsewhere in Western Anatolia and Thrace are *juncidis*. Birds from the population breeding in the east of the Southern Coastlands (Göksu Delta to Amik Gölü) have apparently never been collected, but are likely to belong to *neurotica*, the subspecies which occurs from the Levant east to Iraq. Wing of ♂ of *juncidis* from Izmir 52.5 (n=1), of ♀ 47.5, 49 (n=2) (CSR).

References Lynes, H. (1930) Review of the genus *Cisticola*. *Ibis* (12)6, suppl., 1-673.

Warblers

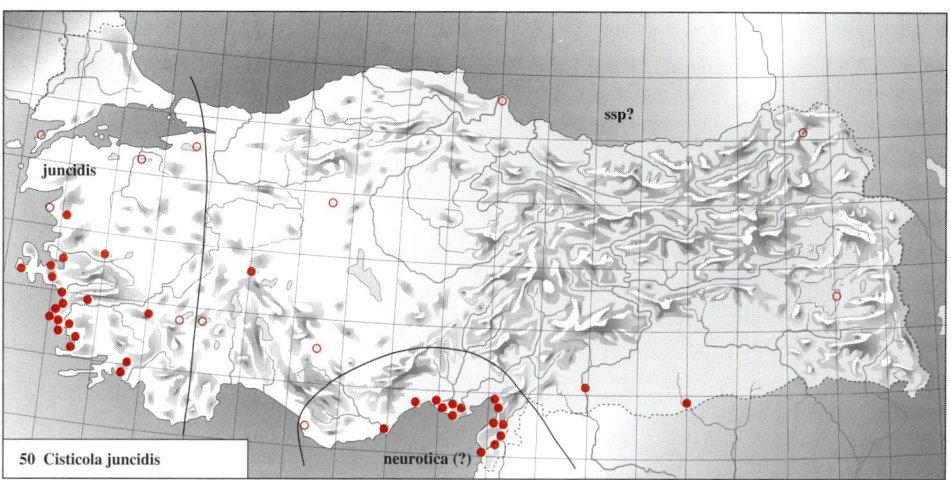

PRINIA GRACILIS
Graceful Warbler   Sürmeli Çalıkuşu
Habitat Flats covered with clumps of *Scirpus, Juncus,* or *Salicornia,* low bushes and scrub along fields and roads, or other low dense vegetation near open ground, mostly below *c.* 100m.

Distribution See map 51. Restricted to the Southern Coastlands east from Antalya, where numbers and distribution seem to fluctuate; also breeds in the Birecik area.

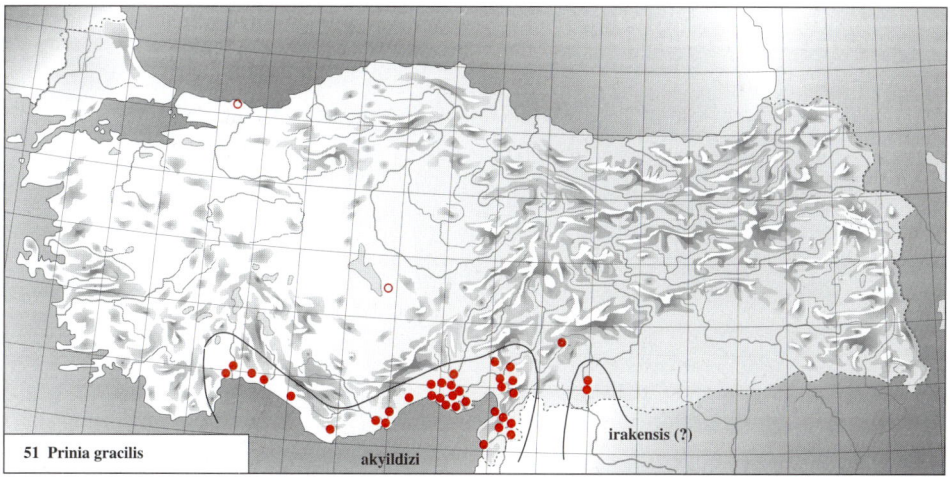

Geographical Variation
Subspecies described or recorded in the region:

(T)   *P. g. akyildizi* Watson, 1961, Antalya (Southern Coastlands, Turkey). [A bird of the *lepida*-group of subspecies, which occurs from Turkey, Iraq, and the north-east Arabian peninsula east throughout southern Asia. The *lepida*-group is characterized by relatively short bill and legs and by a narrow and ill-defined black subterminal tail-bar, in contrast to the *gracilis*-group of subspecies, which occurs from north-east Africa to the Levant and western Arabia. The upperparts of *akyildizi* show broad deep black shaft streaks on brown ground-colour; the underparts are white with an intense pink-buff suffusion and often with black-brown spots or streaks on the breast and flanks. Wing of ♂ of *akyildizi* from Turkey 43.3 (42-45.5) (n=7), of ♀ 39.8 (37.5-41) (n=5), bill of both sexes 11.1 (10.5-12.1) (n=11), tarsus 17.0 (16.6-17.4) (n=5) (Watson 1961, CSR).]

(S) **P. g. palaestinae** Zedlitz, 1911, El Mezra'a (El Lisan Peninsula, Dead Sea, Jordan). [Belongs to the *gracilis*-group. Upperparts sandy-olive or sandy-brown with rather narrow ill-defined black-brown streaks; breast and flank cream, virtually without dark streaks or spots or with traces only; tail with broad and contrasting black subterminal bar. Wing of ♂ from the Lebanon, Israel, Jordan, and northern Egypt 45.2 (43-47) (n=31), of ♀ 43.0 (41-45) (n=12), bill 12.2 (11.0-13.3) (n=32), tarsus 18.3 (16.9-19.2) (n=33) (Roselaar in Cramp 1992).]

(S) **P. g. irakensis** Meinertzhagen, 1923, Baghdad (Iraq). [Belongs to the *lepida*-group. Similar to *akyildizi*, but streaks on upperparts less heavy and ground-colour paler brown. Wing of ♂ 44, 44.5 (n=2), of ♀ 41.5 (41-42) (n=3) (Paludan 1938, Diesselhorst 1962).]

Subspecies recognized in Turkey: birds examined from Turkey (all from the Southern Coastlands) are *akyildizi*. In the Levant, *palaestinae* is restricted to the Dead Sea region, southern Syria, and the coastal area of southern Israel; the birds from the coastal strip further north in Israel, Lebanon, and Syria already tend to *akyildizi*. For distribution in this area, see Kumerloeve (1959). Probably, *irakensis* is the subspecies occurring along the Euphrates (Firat) in Syria, and either this subspecies or *akyildizi* breeds along the Firat near Birecik. Further data from Birecik are needed.

References Kumerloeve, H. (1959) Distribution de *Prinia gracilis palaestinae* Zedlitz dans la région côtière de Liban. *Alauda* 27, 30-32.

LOCUSTELLA NAEVIA
Grasshopper Warbler   Tarla Ardıçkusu

Habitat Dense willow bushes and thorny scrub within beds of reeds and rushes in moist floodland, up to *c*. 2700m.

Distribution See map 52. Restricted to the upper Murat valley above Agri, close to the breeding grounds in Armenia and Georgia. Song was heard at Amik Gölü on 15 July (Vielliard 1968), but this was probably uttered by birds during early post-breeding dispersal. Occasional song recorded elsewhere in May or August is likely to be produced by migrants.

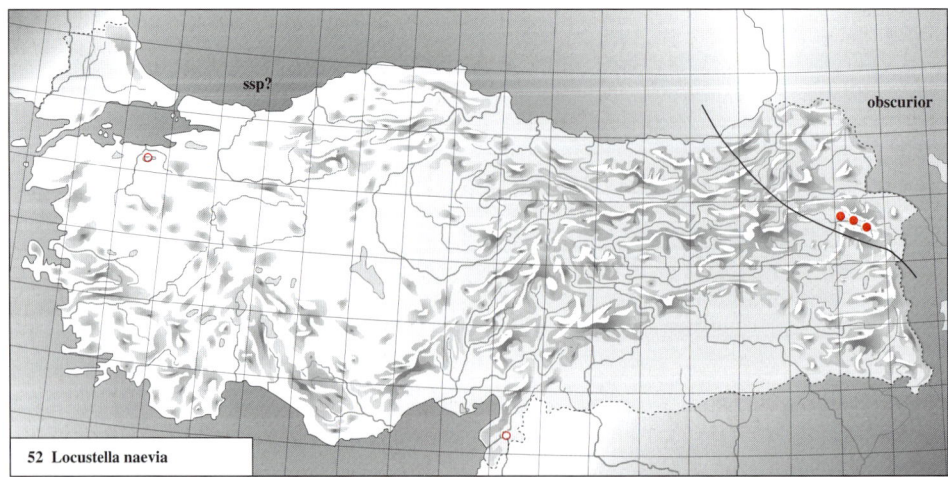

52 Locustella naevia

Geographical Variation
Subspecies described or recorded in the region:

(E) **L. n. obscurior** Buturlin, 1929, Mikhailovskaya Colony (northern Caucasus, Russia). [Rather like *naevia* from western and central Europe, but dark spots on upperparts heavier, blacker, and more contrasting; ground-colour of upperparts more olive (less brown), greyer than in *naevia* when plumage worn; flanks more rusty-cream, sometimes with black streaks; dark spots and streaks on the upperparts are as dark as in the subspecies *straminea* and *mongolica* from eastern Europe and Central Asia, but the ground-colour of the upperparts is less pale; also, both latter subspecies are much smaller. Wing of ♂ of *naevia* from western Europe 64.6 (61-68) (n=31), of ♀ 63.7 (60-66) (n=16) (Roselaar in Cramp 1992), wing of ♂ of *obscurior* from the northern Caucasus 65 (60-68) (n=10) (Vaurie 1959).]

Subspecies recognized in Turkey: only a single adult breeding ♂ from the upper Murat valley was examined, with wing 64, tail 54, and bill 13.7. This bird was inseparable in plumage and size from *obscurior* from the Caucasus and western Transcaucasia (CSR). Note: the song of the **River Warbler** *L. fluviatilis* (Irmak Ötlegeni) is frequently heard in spring and sometimes up to mid-June, but indication of breeding is absent.

LOCUSTELLA LUSCINIOIDES
Savi's Warbler   Savi'nin Dere Ardıçkusu

Habitat Large reedbeds, interspersed with *Carex, Juncus, Typha,* and sometimes a few willows, growing on damp ground or in a thin layer of water rather than in deeper water; occurs up to at least 2000m.

Distribution See map 53. Restricted to a number of wetlands scattered throughout the country; locally common, e.g., in the Kizilirmak delta.

Geographical Variation
Subspecies described or recorded in the region:

(W) **L. l. luscinioides** (Savi), 1824, Pisa (Italy). [Upperparts uniform dull brown; underparts extensively drab-brown or dull olive-brown with restricted white or buff-white on throat and belly. Wing of ♂ from the Netherlands 70.5 (69-73) (n=16), of ♀ 68.7 (66-71) (n=6); wing of ♂ from Romania 72.4 (70-77) (n=12), of ♀ 68.7 (67-71) (n=5); bill of ♂ from

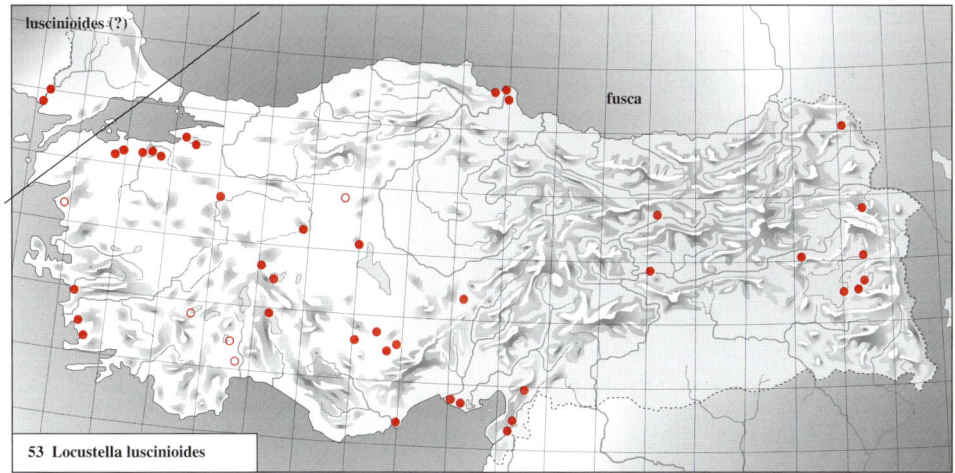

53 Locustella luscinioides

both areas combined 16.1 (15.6-17.1) (n=24), of ♀ 15.3 (14.8-15.9) (n=7) (Roselaar in Cramp 1992).]

(E) **L. l. fusca** (Severtzov), 1872, Arys' and Chimkent (southern Kazakhstan). [Upperparts grey-brown or dark olive-grey, less deep brown than in *luscinioides*; underparts more extensively white, breast and flanks with more restricted and paler buff-brown; tips of under tail-coverts sometimes whitish. Wing of ♂ from Asia Minor 69.0 (68-70) (n=4), bill 16.7 (16.3-17.2) (n=4) (CSR).]

Subspecies recognized in Turkey: only a few breeding birds examined, but these originated from widely separated localities (Manyas Gölü, Amik Gölü, and Van). All are closely similar, and they agree with *fusca*, which breeds from Iran and Transcaspia eastward. Birds breeding in the Meriç delta are likely to be *luscinioides*, as are those on the Greek side of the delta (Bauer *et al.* 1969).

ACROCEPHALUS MELANOPOGON
Moustached Warbler   **Bıyıklı Ardıçkuşu**

Habitat Dense and tall reeds along lake shores and in marshes, up to *c.* 2000m.

Distribution See map 54. Locally common in a number of wetlands, e. g., 1000-1500 pairs in the Kızılırmak delta (Hustings & Dijk 1994). The situation is obscured by widespread occurrence of migrants and wintering birds; only those areas where the species was recorded between late May and early August are shown in black, but it may breed elsewhere. Probably underrecorded due to the close resemblance of its pale subspecies *mimica* to *A. schoenobaenus*.

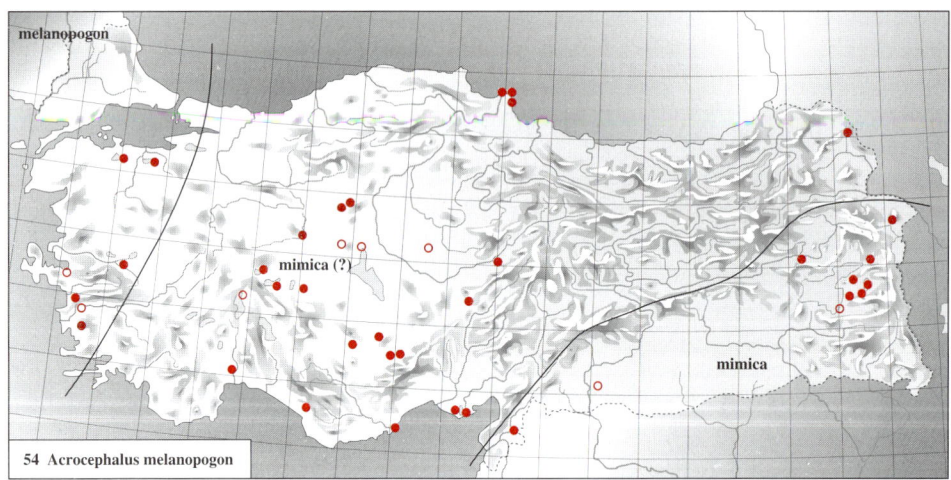

54 Acrocephalus melanopogon

Geographical Variation

Subspecies described or recorded in the region:

(W) **A. m. melanopogon** (Temminck), 1823, Campania (Italy). [Upperparts bright rufous-brown, heavily streaked black, cap almost uniform black; underparts buff- or tawny-brown with a restricted amount of white on throat and mid-belly. Wing of ♂ from Italy and Greece 59.3 (58-62) (n=19), of ♀ 57.3 (55-59) (n=12), bill of both sexes 15.2 (14.0-16.4) (n=30) (Roselaar in Cramp 1992).]

(E) **A. m. mimica** (Madarász), 1903, Tedzhen (Turkmenistan), Khorasan, and Seistan (eastern

Iran). [Upperparts dull olive-grey, centre of cap distinctly streaked olive-grey; underparts largely white with a restricted amount of pale pink-brown on flanks. Rather similar to *A. schoenobaenus*, but tail more rounded (difference in length between inner- and outermost feather 7-12 mm instead of 3-8 mm) and wing more rounded (see Roselaar in Cramp 1992). Wing of ♂ from the shores of the Caspian Sea east to Pakistan 62.2 (60.5-63.5) (n=10), of ♀ 57, 60.5 (n=2), bill 15.7 (15.3-16.3) (n=9) (CSR).]

Subspecies recognized in Turkey: only birds from Izmir, Amik Gölü, and Van were examined. Those of Amik Gölü and Van are inseparable in colour and size from *mimica* from the shores of the Caspian Sea; wing of ♂ from Amik and Van 61.9 (59-64) (n=4), of ♀ 61.5 (n=1), bill 15.6 (14.9-16.1) (n=5) (CSR). Birds from Izmir are best included in *melanopogon*, though birds examined from here differed slightly from *melanopogon* from Greece in showing a slightly paler ground-colour on the cap and slightly less heavy black streaks on the mantle and scapulars. Though these Izmir birds were collected in winter, they are probably birds breeding locally, in view of the difference from winter birds from Greece. Wing of ♂ from Izmir 59, 60 (n=2), bill 15.7, 16.1 (n=2) (CSR). Birds from the north of Western Anatolia are also probably *melanopogon*, those from the Central Plateau are likely to be *mimica*, but confirmation is required.

ACROCEPHALUS SCHOENOBAENUS
Sedge Warbler   Çit Ardiçkusu

Habitat Tangled vegetation and dense scrub amidst reedbeds along lake shores, ditches, irrigation canals, and in marshes; occurs up to at least 2500m.

Distribution See map 55. Locally common in a number of wetlands. Widespread occurrence of singing migrants up to late May makes assessment of breeding distribution difficult. Black dots refer either to proved breeding or to presence of several singing birds in June or July.

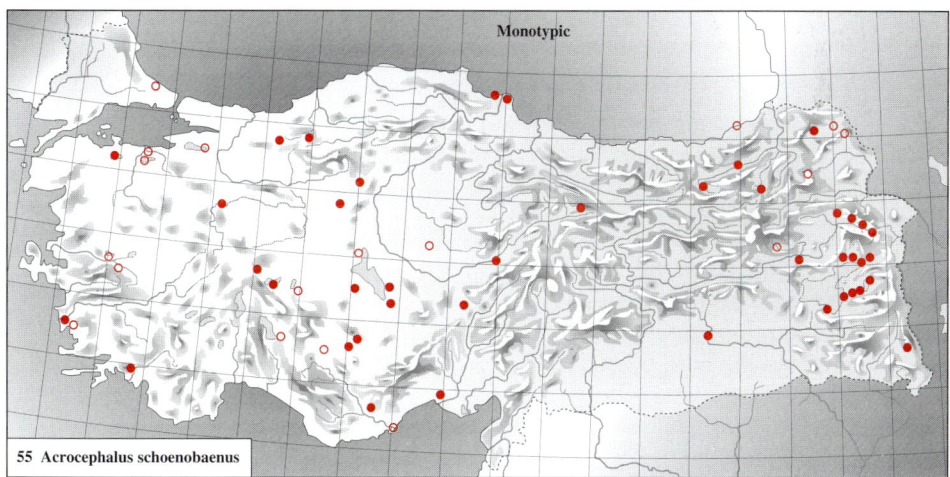

55 Acrocephalus schoenobaenus

Geographical Variation The species shows a slight variation in wing and bill length, but no subspecies are recognized. *A. schoenobaenus* (Linnaeus), 1758, was originally described from southern Sweden. Birds from western Europe (north to southern Sweden) tend to be smaller than the birds from further north and east: wing of ♂ from the Netherlands 66.8 (63-71) (n=100), of ♀ 64.7 (62-68) (n=61), bill of both sexes 14.7 (13.8-15.5) (n=99), against wing of ♂ from Poland, Romania, and European Russia 68.4 (66-72) (n=16) and bill of ♂ 15.2 (14.6-15.8) (n=14) (CSR). In Turkey, wing of breeding ♂ from Manyas Gölü, the Eskisehir area, Eber Gölü, Van, and Agri is 68.0 (65-70.5) (n=9), of ♀ 64.5, 66 (n=2), bill 15.1 (14.3-16.1) (n=11), close to the size of birds from eastern Europe (CSR).

ACROCEPHALUS AGRICOLA
Paddyfield Warbler
Habitat Dense vegetation of reeds and scrub in marshes and along lakes, up to at least 2000m.
Distribution See map 56. Recorded in the east only, where first found in Van in 1986 (Amcoff et al. 1986, Berg & Bosman 1988).

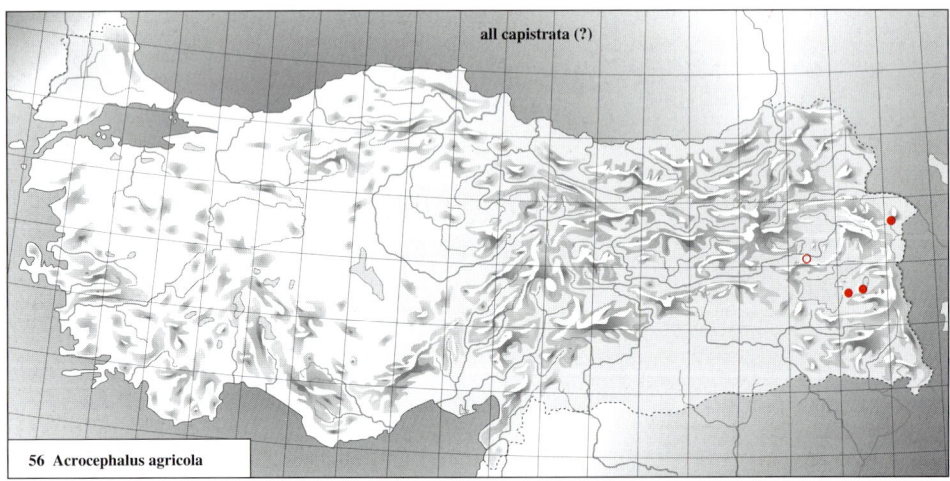

56 Acrocephalus agricola

Geographical Variation
Subspecies described or recorded in the region:
(W) **A. a. septima** Gavrilenko, 1954, Poltava area (eastern Ukraine). [Upperparts brown-grey in worn plumage; underparts in first autumn birds more sandy-yellow. For measurements, see Roselaar in Cramp (1992). Occurs along the western and northern shores of the Black Sea east to at least c. 35ºE, but the exact boundary with the next subspecies is not known. See Stepanyan & Matyukhin (1984).]
(E) **A. a. capistrata** Severtzov, 1873, Turkestan. [Upperparts brown-olive in worn plumage; underparts in first autumn more rufous-orange. Wing of both sexes from the lower Volga river 58.5 (56.5-60) (n=11), bill 14.8 (14.2-15.8) (n=11), wing of birds from Transcaspia eastward 58.5 (56-62) (n=19), bill 14.7 (14.1-15.7) (n=18) (CSR).]

Subspecies recognized in Turkey: according to a published colour photograph (Berg & Bosman 1988), Van birds are inseparable from birds of the lower Volga river and thus the subspecies capistrata is involved. Birds captured were apparently not measured (or the measurements were not published).

References Stepanyan, L. S., & A. V. Matyukhin (1984) [On the systematic situation of the European population of the Paddyfield Warbler (Acrocephalus agricola).] Ornitologiya 19, 212. Berg, A.B. van den, & C. A. W. Bosman (1988) Paddyfield Warbler, Acrocephalus agricola, at Van Gölü, eastern Turkey. Zool. Middle East 2, 16-18.

ACROCEPHALUS PALUSTRIS

Marsh Warbler   Bataklık Saz Ardıçkusu

Habitat Dense shrub and trees amidst lower vegetation on moist ground, along fields, brooks, irrigation canals, and rivers, in graveyards, and in luxuriant growth of abandoned or used gardens; occurs up to at least 2500m.

Distribution See map 57. Scattered breeding records in the north-west and east; locally common in the upper Kara and Aras valleys. Many heard singing at a great number of localities from mid-May to early June in the north, south-west, and south, but the species is a late spring migrant and is not likely to breed here. Breeding is presumed only for localities where several birds were singing between 10 June and late July.

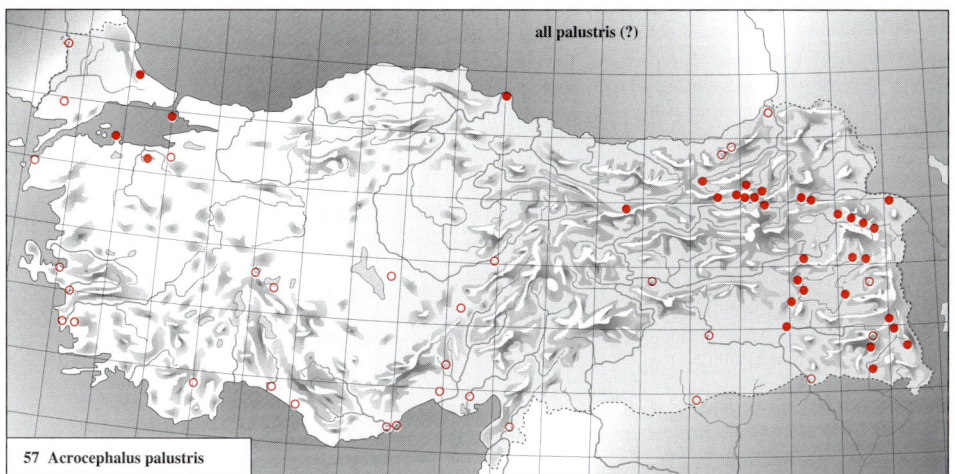

57 Acrocephalus palustris

Geographical Variation

Subspecies described or recorded in the region:

(W)   ***A. p. palustris*** Bechstein, 1798, Thüringen (eastern Germany). [Upperparts grey-olive with slight buff tinge when fresh, purer grey-olive when worn; underparts white, tinged cream-yellow on breast and flanks. Wing in ♂ from western and central Europe 69.3 (67-73) (n=42), of ♀ 68.0 (65-71) (n=47), bill of both sexes 16.0 (14.7-17.1) (n=47); wing of ♂ from the Balkan countries and south European Russia 67.8 (66.5-69) (n=4), of ♀ 67.3 (65.5-70) (n=3), bill 15.7 (14.9-16.4) (n=7) (CSR).]

(E)   ***A. p. laricus*** Portenko, 1955, Lar valley, Damavand (northern Iran). [Like *palustris*, but slightly paler, especially juvenile. Wing of ♂ from Iran in spring 68.2 (67-70) (n=6), of ♀ 67.0 (66-68) (n=3) (Paludan 1938, Schüz 1959, Érard & Etchécopar 1970, CSR; however, all these birds are probably migrants on their way to breeding areas further north and thus presumably are *palustris*.]

Subspecies recognized in Turkey: all birds examined are inseparable from *palustris*. *A. p. laricus* is apparently not different from *palustris* when the plumage is slightly to heavily worn, as in spring and summer, and perhaps this subspecies should not be separated at all. However, among birds wintering in South Africa, those in freshly moulted plumage show 2 colour-types: some are similar to *palustris* from central Europe, others have colder and more greenish upperparts as well as whiter underparts with more restricted buff on breast and flanks (Clancey 1975). These latter birds are considered to be *laricus* by Clancey, but there is no proof that these differently coloured birds came from Iran. They may have had their origin in the eastern part of the species' range, in which case *turcomana* Severtzov, 1873, may be the valid name. This latter subspecies was described from birds from Krasnovodsk (on the eastern shore of the Caspian Sea, Turkmenistan) and is generally not recognized. Wing of ♂ from Turkey 69.2 (66-71) (n=8), of ♀ 67.5, 68 (n=2), bill 16.0 (15.0-16.6, once 17.3) (n=9) (Jordans & Steinbacher 1948, CSR; this sample may contain a few migrants, but most birds were apparently breeding).

References Clancey, P. A. (1975) Miscellaneous taxonomic notes on African birds 42. *Durban Mus. Novit.* 10(19), 231-238.

ACROCEPHALUS SCIRPACEUS
Reed Warbler   Saz Ardıçkusu

Habitat Reedbeds in and along water and in marshes, up to *c.* 2000m.

Distribution See map 58. Probably occurs in all suitable reedbeds throughout Turkey. Often less numerous than *A. arundinaceus,* but locally more abundant, e.g., 1500-2000 pairs in the Kizilirmak delta, against 275-325 pairs of *A. arundinaceus* (Hustings & Dijk 1994).

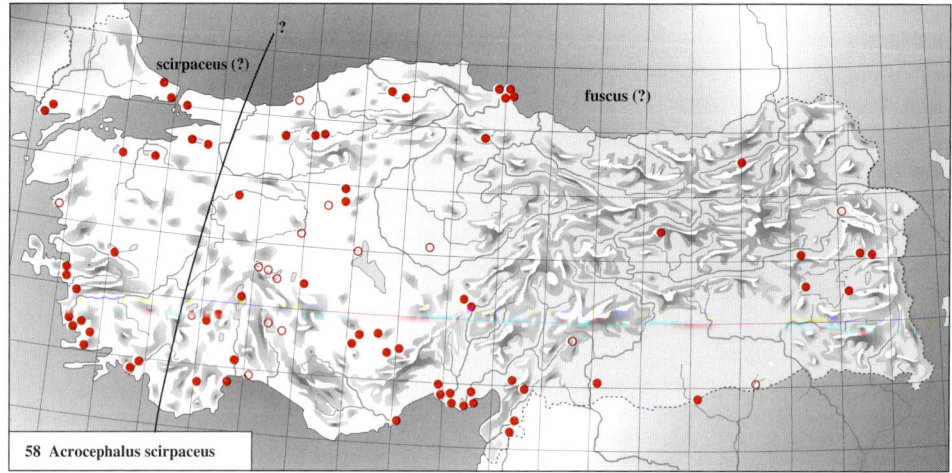

58 Acrocephalus scirpaceus

Geographical Variation

Subspecies described or recorded in the region:

(W)   ***A. s. scirpaceus*** (Hermann), 1804, Alsace (France). [Upperparts brownish-olive, rump clearly more rufous. Wing of ♂ from western and central Europe 66.2 (63-69) (n=56), of ♀ 65.3 (62-69) (n=49), bill of both sexes 17.0 (16.1-18.1) (n=42) (CSR).]

(E)   ***A. s. fuscus*** (Hemprich & Ehrenberg), 1833, northern Arabia (on migration). [Upperparts grey-olive to brown-grey, rump slightly paler or more buff, hardly contrasting with the remainder of the upperparts; supercilium, eye-ring, and underparts whiter; sides of breast and flanks with paler and more restricted greyish-cream. Wing of both sexes from southern Russia and central Asia 66.0 (62-72) (n=12) (CSR). Closely similar in colour to *A. palustris,* differing mainly in structure, especially in the position of the notch on the outer primaries (see Roselaar in Cramp 1992).]

Subspecies recognized in Turkey: both races occur, but their distribution is scarcely known. Birds from Thrace (20-22 May) examined by Rokitansky & Schifter (1971), as well as birds from Manyas Gölü (1 October), Amik Gölü (6 and 21 May), and Ceylanpinar (12 May) seen by the author were *scirpaceus*, but all these are likely to have been migrants. Birds examined during May at Beysehir Gölü (Vauk 1973) were migrants, too, as was a possible *fuscus* from Ankara on 29 April (Kummerlöwe & Niethammer 1934-35). A bird from Elmali, collected 11 June, was *fuscus*, and this is probably the only bird likely to be a local breeding bird. As birds breeding in Greece are *scirpaceus* (Bauer *et al.* 1969) and those of the Levant, Transcaucasia, and Iran are *fuscus* (Vaurie 1959), birds from Western Anatolia and Thrace can be assumed to be *scirpaceus*, those from further east in Turkey *fuscus*, but there is no proof. Wing of live migrants at Beysehir Gölü 66.8 (64-69) (n=16) (Vauk 1973); wing of all other Turkish birds cited above 65.7 (62-70) (n=9), bill 16.6 (15.9-17.2) (n=4).

ACROCEPHALUS ARUNDINACEUS
Great Reed Warbler   Büyük Saz Ardiçkusu
Habitat Mainly tall reeds in wetlands, especially those standing in water, but locally in stands of trees or dense scrub (e.g., tamarisk) at fringes of water and in rice fields; occurs up to *c.* 2000m.

Distribution See map 59. Probably in all suitable habitat throughout Turkey, where generally common to abundant.

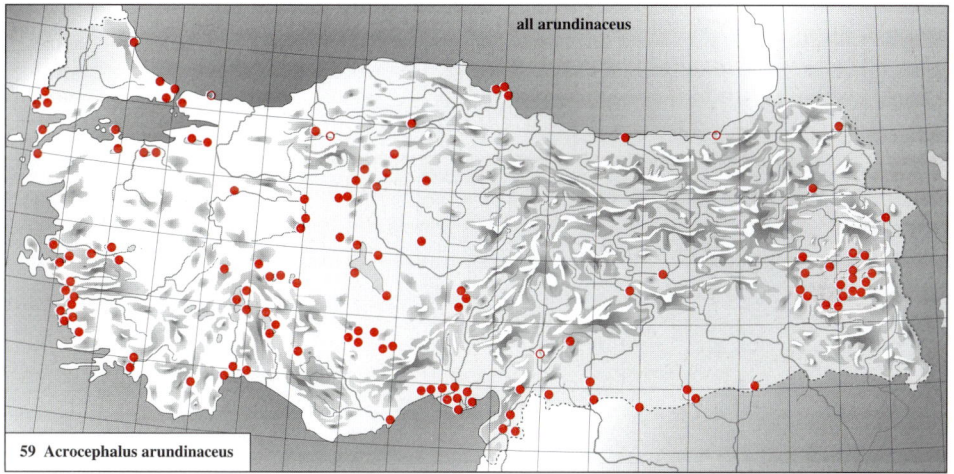

59 Acrocephalus arundinaceus

Geographical Variation
Subspecies described or recorded in the region:

(W)   ***A. a. arundinaceus*** (Linnaeus), 1758, Gdansk (Poland). [Upperparts rufous-brown, brightest on rump; underparts cream-white with warm buff tinge on flanks. Wing of ♂ from western Europe 96.8 (94-101) (n=48), of ♀ 92.3 (89-95) (n=14) (Roselaar in Cramp 1992).]

(E)   ***A. a. zarudnyi*** Hartert, 1907, Dzharkent (Panfilov, eastern Kazakhstan). [Upperparts grey-olive; supercilium and underparts whiter than in *arundinaceus*; size as in *arundinaceus*. Both subspecies apparently occur side-by-side at a number of localities in eastern Europe and western Asia, and they perhaps could be better considered as morphs of an otherwise monotypic species; in the west, *arundinaceus* predominates, in the east *zarudnyi*. Also, the difference in colour could in part be due to a difference in bleaching and abrasion as a consequence of different moult strategies between eastern and western birds.]

Subspecies recognized in Turkey: all birds examined as well as those recorded in the literature are *arundinaceus*, except for 4 birds collected 30 April to 19 May in Ankara, which were 'even paler and greyer than *zarudnyi*' (Kummerlöwe & Niethammer 1934-35) and which were probably migrants. Wing of Turkish ♂ (both *arundinaceus* and *zarudnyi*): 96.1 (91.5-100) (n=10), of ♀ 96.0 (91-99) (n=3) (Weigold 1912-13, Kummerlöwe & Niethammer 1934-35, Kumerloeve 1963a, Rokitansky & Schifter 1971, CSR; some birds are probably wrongly sexed). Wing of unsexed migrants at Beysehir 97.0 (93-101) (n=11) (Vauk 1973).

HIPPOLAIS PALLIDA
Olivaceous Warbler   Gri Mukallit

Habitat All types of groves, thickets, tall scrub, or stands of small or large trees on open ground or in open forest, in dry valleys or on dry hills, but sometimes on damp ground; also, parks, gardens, and tangled scrub near water. Avoids tall closed forest as well as trees amidst dense undergrowth. Mainly at 0-1000 (-1500) m, in the valleys of the east locally to *c.* 2000m.

Distribution See map 60; also, Kumerloeve (1961a). Common in the west and centre, more scarce and local in the higher valleys of the east.

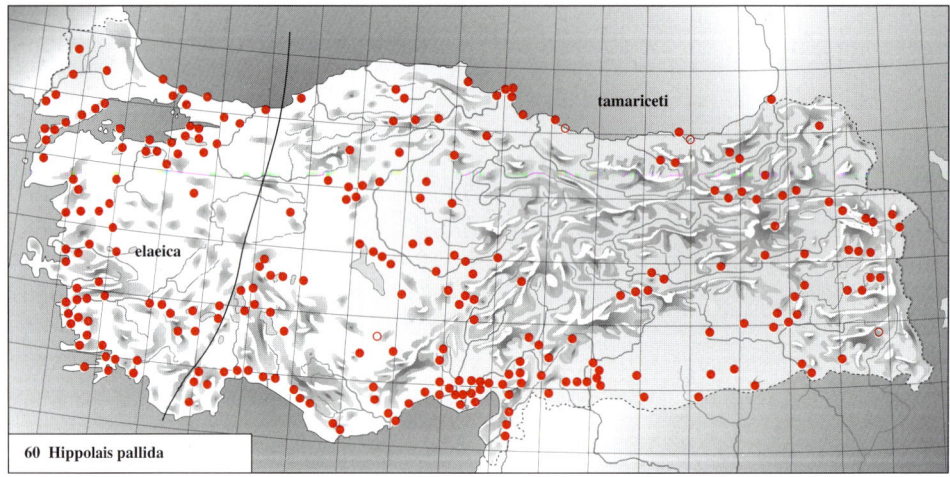

60 Hippolais pallida

Geographical Variation
Subspecies described or recorded in the region:

(W) **H. p. elaeica** (Lindermayer), 1843, Greece. [Upperparts light brown-grey, underparts white with slight pale brown-grey tinge on flanks; bill longer. Wing of ♂ from the Balkan countries and Greece (including Crete) 67.6 (66-71) (n=27), of ♀ 64.9 (62-66, once 68.5) (n=12), bill of both sexes 17.0 (16.4-18.0, once 16.0) (n=39) (CSR).]

(E) **H. p. tamariceti** Severtzov, 1873, Syr Darya river, 'Turkestan' (=southern Kazakhstan or eastern Uzbekistan). [Upperparts slightly paler and more sandy than in *elaeica*; bill shorter. Wing of ♂ from the Levant and Iran 66.9 (65.5-70) (n=13), of ♀ 64.5, 65 (n=2), bill 15.6 (15.4-15.8) (n=3) (Stresemann 1928, Paludan 1938, Schüz 1959, Diesselhorst

1962, Érard & Etchécopar 1970, CSR); wing of ♂ from Turkmenistan and Uzbekistan 69.2 (67-71) (n=8), bill of ♂ 16.0 (15.6-16.5) (n=7) (CSR).]

Subspecies recognized in Turkey: according to bill measurements, both subspecies occur in Turkey. Long-billed *elaeica* occurs in Thrace and Western Anatolia east to Burdur, short-billed *tamariceti* was examined from many localities further east (Solak, Elmali, Ankara, the Taurus mountains, Pozanti, Birecik, Ceylanpinar, Siirt, and Agri). No obvious difference in colour was detected. Wing of ♂ of *elaeica* from Western Anatolia east to Burdur 67.1 (65-71) (n=26), of ♀ 64.6 (62.5-66) (n=10), bill 16.9 (16.4-18.3, once 16.0) (n=33) (CSR). Wing of ♂ of *tamariceti* from Solak, Elmali, and Ankara eastward 67.5 (64.5-70) (n=8), of ♀ 65.0 (63.5-66.5) (n=3), bill 16.2 (15.6-16.4, once 17.1) (n=11) (CSR). Wing of ♂ from Thrace 68 (n=1), of ♀ 65.5, 66 (n=2), bill 16.4 (15.9-17.0) (n=3); these 3 birds (collected Terkos Gölü on 21/22 May 1967 by Rokitansky, perhaps still on migration) had bill rather short and laterally compressed, as if intermediate with *H. caligata*, but other measurements, wing formula, and colour were as in *H. pallida elaeica* (CSR).

HIPPOLAIS LANGUIDA
Upcher's Warbler   Beyaz Mukallit

Habitat Thorny bushes and trees (e.g., oak or acacia) in steppe country or on sunny rocky slopes at borders of cultivation, as well as in gardens, orchards, and vineyards; mainly at 500-1500m, but occasionally up to *c.* 1800m.

Distribution See map 61; also, Kumerloeve (1958b, 1969e). Restricted to the fringe of the steppe country in the South-East, from the eastern slopes of the Amanus and Anti-Taurus mountains east to the mountains of the Hakkari and Van areas; also, locally in the Çukurova area and some records west to Silifke. An isolated colony has been found near Delice (east of Ankara) since 1974 (Schubert 1979b). May occur in the Turkish part of the Aras valley, close to the breeding sites in Nakhichevan (Transcaucasia).

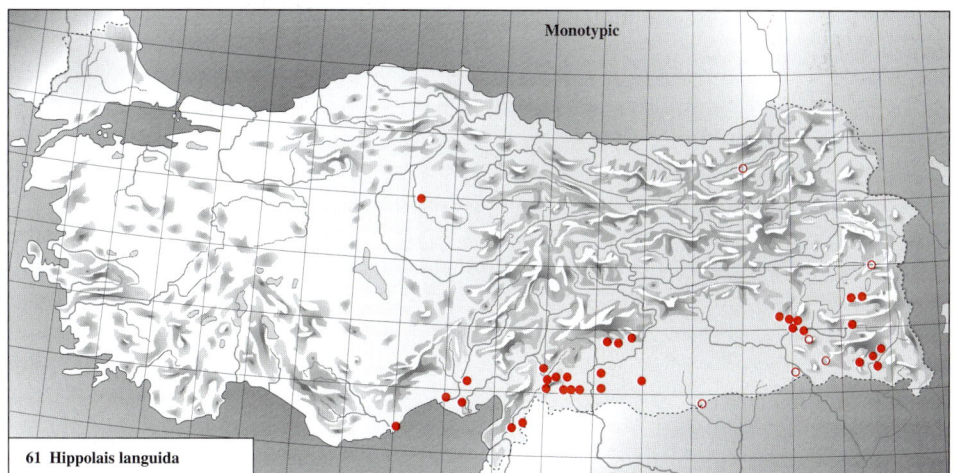

61 Hippolais languida

Geographical Variation Slight, if any; no subspecies recognized here. *H. languida* (Hemprich & Ehrenberg), 1833, was originally described from the Lebanon. A paler subspecies with more sandy-grey upperparts is sometimes recognized in the eastern part of the range; it has been named *magnirostris* Severtzov, 1873, and was described from the Karatau mountains and the western foothills of the Tien Shan mountains (southern Kazakhstan). No obvious difference in colour or size was found in specimens examined. Wing of ♂ from Turkey 76, 78.5 (n=2), of ♀ 75, 75.5 (n=2), bill of both sexes 17.5 (16.9-18.8) (n=3) (Kumerloeve 1970a, CSR), wing in Nakhichevan 75.4 (72.4-78) (n=8) (Worobiev 1934), wing of ♂ from elsewhere 76.9 (73-80) (n=21), of ♀ 73.8 (70-77) (n=21), bill 17.8 (16.6-18.6) (n=25) (Paludan 1938, 1940, Érard & Etchécopar 1970, CSR).

References Worobiev, C. A. (1934) Sur la distribution géographique des oiseaux en Transcaucasie. *Oiseau* 4, 155-159. Kumerloeve, H. (1958b) Vom Dornbuschspötter, *Hippolais languida* (H. & E.) im Hatay (türkisch-syrischer Grenzbereich). *Anz. orn. Ges. Bayern* 5, 137-141. Kumerloeve, H. (1969e) Zur Westausdehnung des Brutareals des Dornbuschspötters (*Hippolais languida*). *J. Orn.* 110, 500-501. Schubert, W. (1979b) Der Dornbuschspötter, *Hippolais languida*, als Brutvogel in Mittelanatolien (Türkei). *Bonner zool. Beitr.* 30, 158-159.

HIPPOLAIS OLIVETORUM
Olive-tree Warbler   Zeytinlik Mukallidi

Habitat Dry slopes covered with scrub and trees (e.g., pine, oak, almonds, and olives), including olive groves, gardens, and orchards, and sometimes in rather damp woodland; mainly at 0-500m.

Distribution See map 62; also, Kumerloeve (1961a). Largely confined to the coastal zone of the Mediterranean Sea and the Sea of Marmara as well as in Thrace, with some isolated localities elsewhere (Gaziantep, Kizilcahamam, Seyfe Gölü), which are perhaps not regularly occupied. Records in the South-East in May and July refer probably to late migrants or early post-breeding dispersal.

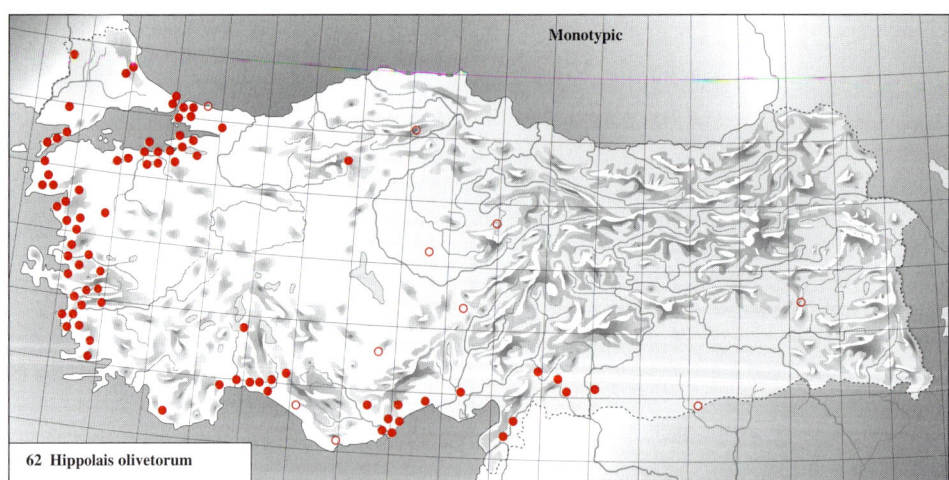

62 *Hippolais olivetorum*

Geographical Variation None. *H. olivetorum* (Strickland), 1837, was originally described from Zákinthos (Ionian Islands, Greece). Wing of ♂ from the Balkan countries and Greece 87.8 (85-91) (n=19), of ♀ 86.3 (81-90) (n=13), bill of both sexes 19.9 (18.7-21.2) (n=30); wing of ♂ from Turkey 87.0 (85-90) (n=10), of ♀ 85.5 (82.5-89) (n=4), bill 20.1 (19.2-20.7) (n=14) (CSR).

HIPPOLAIS ICTERINA

Icterine Warbler   Sari Mukallit

Habitat Bushy undergrowth of moist open forest as well as bushes and trees in riverine floodlands, gardens, parks, and cemeteries; in Turkey, only close to sea level.

Distribution See map 63. Some records of singing birds in June in the north-west may refer to local breeding, but those from May and early June and from late July onwards are very likely to be on migration (e.g., a bird singing in the Kizilirmak delta on 31 May and a bird at Çukurbag in the south on 28 July).

63 Hippolais icterina

Geographical Variation

Subspecies described or recorded in the region:

(W)   ***H. i. icterina*** (Vieillot), 1817, Nancy (France). [Upperparts grey- or green-olive; underparts yellow; outermost functional primary 1-4 mm shorter than the longest primary, tip of the outermost primary usually between the tips of the 3rd-4th outermost primaries. Wing of both sexes from central Europe and the Balkan countries 78.0 (72-83) (n=85), bill 16.6 (15.6-17.5) (n=70) (CSR). Occasionally the yellow pigment is absent, the upperparts then appearing grey and the underparts white; *borisi* Von Jordans, 1940, described from Bulgaria, is based on such a bird, but this aberrant plumage is not a valid subspecies (type specimen examined).]

(E)   ***H. i. alaris*** Stresemann, 1928, forest south of Khorramabad (northern Iran). [Upperparts slightly darker than in *icterina*, especially crown; wing slightly shorter; tip of the outermost functional primary slightly shorter than the tip of the 4th outermost when wing is closed (Stresemann 1928). Unfortunately, these characters were not found in other specimens from northern Iran, and *alaris* is probably not a valid subspecies (Paludan 1938, Vaurie 1954); also, there is much variation in length of the outermost primary elsewhere in the breeding range (see Roselaar in Cramp 1992). However, two forms are found on the wintering grounds in southern Africa, one with brighter citrine-green upperparts and deeper chrome-yellow underparts, which is presumed to be *icterina*, the other with colder more grey-olive upperparts and paler primrose-yellow underparts, which may correspond with *alaris* from northern Iran, as well as with possibly similar birds from Transcaspia (*'schuchowi'* Snigirewski, 1931): see Clancey (1989). However, the species is considered to be monotypic by Stepanyan (1990). Wing of ♂ from northern Iran 76.2 (75-78) (n=6) (Stresemann 1928, Paludan 1938, Vaurie 1954); wing of a ♂ from Lenkoran (south-east Azerbaijan), with dull olive-grey upperparts and with outermost primary between 4th and 5th outermost, 79.5 (CSR).]

Subspecies recognized in Turkey: none examined. A migrant male from Ankara (10 May; wing 78) is considered to be intermediate between *icterina* and *alaris* (Kummerlöwe & Niethammer 1934-35), but recognition of *alaris* is doubtful and preferably no subspecies should be recognized.

Note: a ringing record of the **Melodious Warbler** *H. polyglotta* in Kavak (Thrace) on 25 July (Eyckerman *et al.* 1992) is far beyond the known range of the species and on an unusual date. Perhaps a yellowish juvenile *H. pallida* was involved, as these are difficult to separate from *H. polyglotta*.

References Clancey, P. A. (1989) Taxonomic and distributional findings on some birds from Namibia. *Cimbebasia* 11, 111-133.

SYLVIA CANTILLANS
Subalpine Warbler   Aksakal Ötlegen

Habitat Dense bush and scrub on stony slopes as well as thick undergrowth of woodland, usually close to the coast. Generally, below *c.* 500m, but up to over 1000m in the hills of inland Western Anatolia

Distribution See map 64. Common only in southern Thrace and north-west Anatolia, very local in the western part of the Southern Coastlands. Records of this species further east probably all refer to misidentified *S. mystacea*, which is rather similar to *S. cantillans* in colour, especially *S. m. mystacea*.

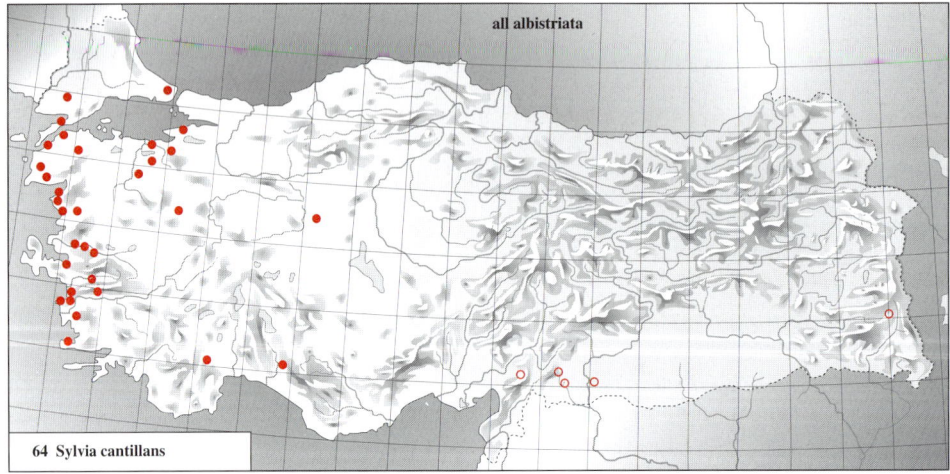

64 Sylvia cantillans

Geographical Variation
Subspecies described or recorded in the region:

(W) **S. c. albistriata** (C. L. Brehm), 1855, Egypt and south-east Europe. [Chin to breast of ♂ deep chestnut-red, contrasting with the grey-white flanks and the whitish belly; tip of outermost functional primary usually between tips of 3rd-5th outermost in the closed wing (in the subspecies of the western Mediterranean basin, chin to breast of ♂ are paler, more pinkish, merging into the pink of the flanks; tip of outer primary usually between the tips of the 4th-6th outermost; also, bill on average shorter). Wing of ♂ *albistriata* from the Balkan countries and Greece 62.7 (58-67) (n=36), of ♀ 61.9 (58-65) (n=14), bill of both sexes 13.4 (12.4-14.3) (n=45) (CSR).]

Subspecies recognized in Turkey: all Turkish birds are referable to *albistriata,* as could be expected, as the other races of the species do not occur further east than Italy and Tunisia. These extralimital subspecies, which are paler and more pinkish below than *albistriata,* are rather similar to *S. mystacea.* Wing of ♂ of *albistriata* from Thrace and Western Anatolia 62.3 (58-65) (n=5), bill 13.2 (13.1-13.5) (n=3) (Kumerloeve 1961a, CSR).

SYLVIA MYSTACEA
Ménétries's Warbler   Pembegögüs Ötlegen

Habitat Low scrub and thickets on steppe and on mountain slopes, especially tamarisk (see Konrad 1985); also, hedges and scrub along fields, brooks, and irrigation canals; occurs up to *c.* 2000m.

Distribution See map 65. Local in the South-East and East, mainly in steppe country but also in the mountains further north and east, where probably seriously underrecorded.

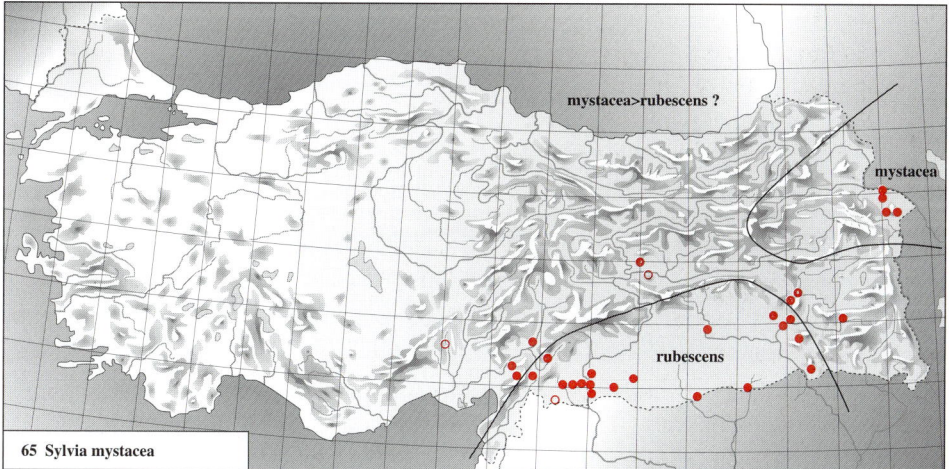

65 Sylvia mystacea

Geographical Variation
Subspecies described or recorded in the region:

(E)   ***S. m. rubescens*** Blanford, 1874, Shiraz and Esfahan (Iran). [A pale subspecies; frontal part of cap of ♂ plumbeous-black, merging into the medium grey of the remainder of the upperparts, the latter suffused sandy-brown when plumage fresh; underparts silky-white, often with a pale vinous-pink wash on the throat and a pale grey wash on the flanks when plumage fresh. This species is rather like the ♂ of *S. melanocephala momus,* but the cap of the latter is deep black and sharply defined at rear, not gradually merging into the grey of the remainder of the upperparts, and the underparts of *momus* are white with light grey flanks and without a pink flush on the throat. Separation of ♀ is more difficult; see Roselaar in Cramp (1992). Wing of both sexes of *rubescens* from western Iran 59.8 (58-62) (n=13), bill to skull 12.9 (12.6-13.7, once 12.4, twice 12.2) (n=13) (CSR); wing in Syria 55.7 (55-56) (n=3) (Baumgart & Stephan 1987), in Iraq 56.5, 57.5 (n=2) (Harrison 1959). For characters and measurements of this and other subspecies, see also Kazakov (1973).

S. m. semenowi Zarudny, 1904, described from Bakhtiari (Iran), is a synonym of *rubescens*.]

(E) **S. m. mystacea** (Ménétries), 1832, Sal'yana (Azerbaijan, south-east Transcaucasia). [A dark subspecies; front part of the cap of ♂ dull black, merging into the dark grey of the remainder of the upperparts, the latter suffused brown if the plumage is fresh; chin to throat deep vinous-pink, flanks pale ash-grey, the remainder of the underparts white. Upperparts closely similar to those of *S. cantillans albistriata*, but cap darker; underparts like those of *S. cantillans cantillans*. Differs from *S. cantillans* in blacker ground-colour of wings and tail. Bill shorter and finer than in *rubescens*. Wing of both sexes of *mystacea* from Transcaucasia and north-west Iran 59.8 (57-63) (n=10), bill to skull 12.1 (11.8-12.5) (n=10) (CSR); wing of ♂ from northern Iran 57.8 (55-61) (n=15) (Stresemann 1928, Schüz 1959, Diesselhorst 1962).]

(E) **S. m. turcmenica** Zarudny & Bilkevitsch, 1918, Murgab and Tedzhen rivers (Turkmenistan). [Upperparts pale, as in *rubescens*; underparts vinous-pink, extent as in *mystacea* but colour slightly paler; wing slightly longer than both these subspecies. Wing of ♂ of *turcmenica* from north-east Iran and Transcaspia 62.2 (60-65) (n=6), bill 13.2 (12.5-13.8) (n=6) (CSR). Apparently grades into *mystacea* in northern Iran (north of the Elburz) and into *rubescens* in central Iran; a common migrant or wintering bird in central and southern Iran and in the Arabian peninsula.]

Subspecies recognized in Turkey: birds examined from the Amik Gölü area, Birecik, and Gaziantep are all *rubescens*; wing of these 60.1 (58-62) (n=9), bill 12.9 (12.5-13.4) (n=9) (CSR). Wing of probable *rubescens* from the Urfa area 59.8 (59-61) (n=7) (Weigold 1912-13). *S. m. mystacea* occurs on the mountain slopes of the Aras area in eastern Turkey. The subspecies of the birds occurring in the Asvan area (near Elazig) and just south of Van Gölü (Bitlis, Siirt, Sirvan, Çatak, Eruh) is not known, but published measurements suggest that either *mystacea* breeds here or intergrades between *mystacea* and *rubescens*. Wing of birds captured at Asvan 59.8 (57-62) (n=15), bill 12.4 (12.0-13.0) (n=28) (Andrew et al. 1972; see also Harrison et al. 1972, 1973); wing of a ♂ from Siirt 63 (Kumerloeve 1968).

References Andrew, P., M. C. Harrison, & R. B. H. Smith (1972) Taxonomy of Ménétries' Warbler. *Bristol Orn.* 1, 207-208. Harrison, M. C., R. B. H. Smith, & P. Andrew (1972) Rep. Asvan Exped. 1971. Oxford. Harrison, M. C., P. C. Lack, & W. J. A. Dick (1973) Rep. Asvan Exped. 1972. Oxford. Kazakov, B. A. (1973) [Taxonomic status and geographical variation of *Sylvia mystacea* Ménétr.] *Vestnik Zool.* 1973(2), 66-69. Konrad, V. (1985) Samtkopf-Grasmücke (*Sylvia melanocephala*) und Tamarisken-Grasmücke (*Sylvia mystacea*) doch zwei 'gute' Arten? *Orn. Mitteil.* 37, 81.

SYLVIA MELANOCEPHALA
Sardinian Warbler **Karabas Küçük Ötlegen**

Habitat Maquis, dense scrub, and tangled vegetation on dry rocky slopes, mainly in the coastal zone, at (0-) 100-500 (-1000) m. In the overlap area with *S. mystacea* (e.g., in the Halfeti area: Konrad 1985), *S. melanocephala* inhabits orchards and gardens near human habitation, *S. mystacea* low tamarisk scrub along water.

Distribution See map 66. Occurs Thrace, Western Anatolia, and the Black Sea Coastlands east to Ünye (Steiner 1970), and very local in the Southern Coastlands, perhaps east to near Nizip (Vielliard 1968). Isolated records in the northern part of the Central Plateau during May-July may refer to stragglers of this species or perhaps to *S. mystacea*.

Warblers

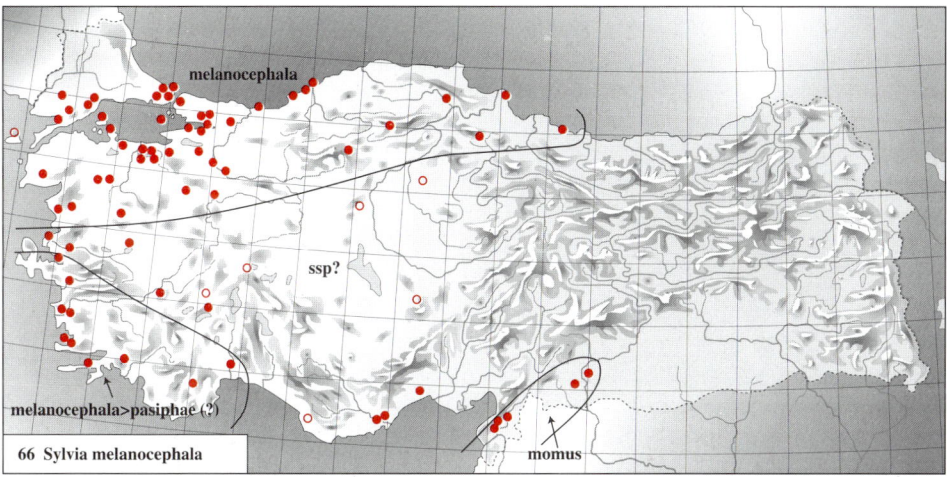

66 Sylvia melanocephala

Geographical Variation

Subspecies described or recorded in the region:

(W) ***S. m melanocephala*** (Gmelin), 1789, Sardinia. [Upperparts (except black cap of ♂) plumbeous-grey, washed dark olive-brown when plumage fresh; sides of breast and flanks extensively grey; large, tail relatively long. Wing of both sexes from the western Mediterranean basin 59.9 (57-64) (n=64), tail 59.7 (55-68) (n=51), bill 14.2 (13.3-15.4) (n=60) (Roselaar in Cramp 1992).]

(W) ***S. m. pasiphae*** Stresemann & Schiebel, 1925, Crete (Greece). [As *melanocephala*, but grey on sides of breast more extensive, sometimes forming a full band across the upper breast; smaller than *melanocephala*. Wing on Crete 56.6 (55-59) (n=19), on Rodhos 56.1 (55-57.5) (n=4), tail on both islands 54.2 (52-56) (n=10), bill on both 13.5 (12.9-14.2) (n=22) (CSR); occurs east to Karpathos and Rodhos and perhaps this subspecies elsewhere in the southern Sporadhes and Kikladhes; grades into *melanocephala* in the southern half of mainland Greece.]

(S) ***S. m. momus*** (Hemprich & Ehrenberg), 1833, Egypt (in winter). [Upperparts medium grey (apart from deep black cap of ♂), paler than in *melanocephala* and *pasiphae*, tinged paler grey-brown when plumage fresh; underparts extensively white, grey on side of breast and flanks pale and restricted; smaller, tail relatively shorter. Wing of both sexes from the Levant 57.3 (55-60) (n=34), tail 53.6 (50-57) (n=34), bill 13.1 (12.7-14.4) (n=30) (CSR).]

Subspecies recognized in Turkey: birds from the western Black Sea Coastlands, Thrace, and Western Anatolia are *melanocephala*, although those of Izmir (and perhaps others in the centre and south of Western Anatolia) show slightly more extensive grey on the side of the breast, tending somewhat to *pasiphae*. Wing in western Turkey 58.8 (56.5-61) (n=5), tail 57.6 (56-60) (n=4), bill 13.9 (13.5-14.4) (n=4) (Eyckerman *et al.* 1976, CSR). Birds from the Amanus area north-east to Halfeti (Vielliard 1968, Konrad 1985) are *momus*.

References Steiner, H. M. (1970) Ein Samtkopfgrasmücken-Vorkommen in degradiertem Buchen-Buschwald bei Ünye (Vilayet Ordu, Türkei). *Egretta* 13, 48-49. Eyckerman, R., M. Louette, & M. Becuwe (1976) Neuer fernfund von Samtkopfgrasmücke (*Sylvia melanocephala* Gmelin). *Vogelwarte* 28, 232-233.

SYLVIA RUEPPELLI

Rüppell's Warbler   Maskeli Ötlegen

Habitat Dense scrub on dry grassy and stony slopes and on wastelands, e.g., maquis of oak, beech, or juniper scrub; mainly at 0-800m, but up to *c*. 1500m in the Taurus and Amanus mountains.

Distribution See map 67. Common in the south of Western Anatolia and in the Southern Coastlands, local in the coastal zone of the Sea of Marmara. Some isolated sites, perhaps not permanently occupied, are reported from near Nallihan and Ankara along the northern fringe of the Central Plateau.

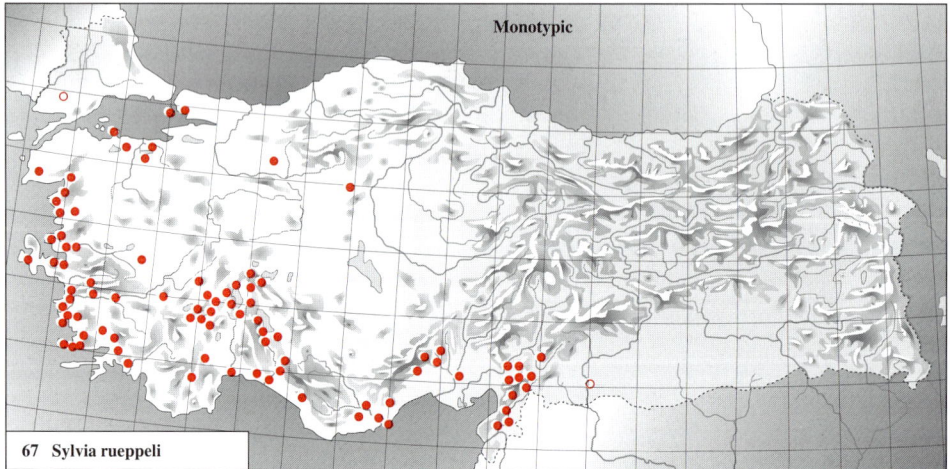

67 *Sylvia rueppeli*

Geographical Variation None reported. *S. rueppelli* Temminck, 1823, was originally described from Kandia (Crete, Greece). Wing of both sexes from Greece 69.3 (66-72.5) (n=18), wing from Western Anatolia (mainly Izmir, Aydin, and Mugla) 70.0 (67-74) (n=51), wing from Korkuteli, Keciborlu, Burdur, and the Taurus area 70.0 (65.5-74) (n=14), and wing from the Levant 67, 68 (n=2) (CSR). None examined from the Amanus area.

SYLVIA HORTENSIS

Orphean Warbler   Orfe Ötlegeni

Habitat Tall scrub and densely-crowned trees, situated in open woods with dense ground cover, as well as in plantations, in gardens, or along streams; also, tangled vineyards, scrub-clad side valleys, and bushes of *Berberis* mixed with trees on rocky slopes; up to 1700m in the Taurus mountains and to 2000m in the east.

Distribution See map 68. Widespread but generally sparse in Thrace, the Black Sea Coastlands, Western Anatolia, and the Southern Coastlands; very local in mountain valleys in the East and South-East.

Geographical Variation

Subspecies described or recorded in the region:

(W/E)   **S. h. crassirostris** (Cretzschmar), 1826, Nubia (northern Sudan, on migration). [Cap plumbeous-grey to dull black, gradually merging into the brown-grey of the remainder of the upperparts; underparts white with some cream-grey wash on the sides of breast, flanks, and under tail-coverts; bill rather long. Wing of both sexes from former Yugoslavia,

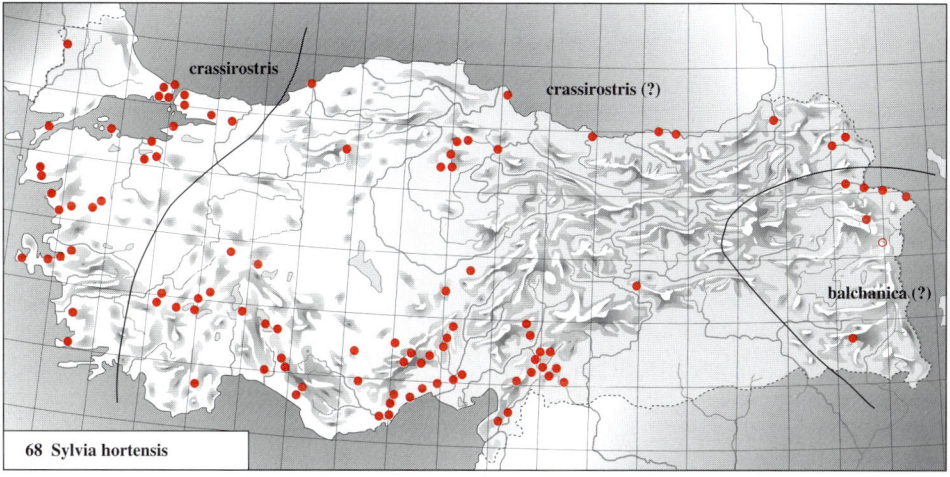

68 Sylvia hortensis

Greece, and the Levant 79.7 (77.5-82.5) (n=13), tail 64.9 (62-69) (n=13), bill 18.8 (17.6-20.0) (n=13) (CSR). Also, breeds in the Caucasus area, while *hortensis* (Gmelin), 1789, described from France, is restricted to the western Mediterranean basin; the latter is similar to *crassirostris*, but sides of breast, flanks, and under tail-coverts are extensively washed drab-grey, breast and belly are tinged pink-buff, and the bill is on average shorter.]

(E) **S. h. balchanica** Zarudny & Bilkevitsch, 1918, Bolshoy Balkhan mountains (western Turkmenistan). [Cap black, sharply defined from the medium grey of the remainder of the upperparts, which are slightly tinged brown when the plumage is fresh; underparts as white as in *crassirostris*, wing and tail slightly longer. Wing in Iran and Turkmenistan 82.6 (80-85) (n=14), tail 65.2 (62-69) (n=14), bill 19.0 (18-20.5) (n=14) (Vaurie 1954). In central Asia, replaced by *jerdoni*, which is similar in plumage, but has a longer bill and a higher wing/tail ratio.]

Subspecies recognized in Turkey: only birds from Izmir examined, and these are *crassirostris* according to plumage, though the bill is rather short: wing 80.5 (78-83) (n=4), tail 66.5 (63.5-69.5) (n=4), bill 18.4 (17.6-19.1) (n=4) (CSR). A relatively short bill was also apparent in a bird examined by Stresemann (1928) from Eregli. Wing of a probable migrant from Beysehir 77 (Vauk 1973). No data available on birds from the Taurus and Amanus mountains, nor on those from the entire Black Sea Coastlands and the east; as *balchanicus* occurs in Iran and *crassirostris* in the Caucasus area, *balchanicus* can be expected to occur in at least south-east Turkey and *crassirostris* in north-eastern Turkey. More information is needed.

SYLVIA NISORIA
Barred Warbler   Çizgili Ötlegen

Habitat Extensive dense impenetrable scrub, preferably thorny (e.g., bramble), up to c. 1650m.
Distribution See map 69; also, Kumerloeve (1961a). Locally common in Thrace, the Black Sea Coastlands, and the East. Many observations in the southern half of the country, mainly in May but occasionally in June-July; most of these probably refer to migrants (in spring) or to early post-breeding dispersal (in summer), but breeding was proved in Silifke in 1990 (Colin 1990).

Songbirds of Turkey

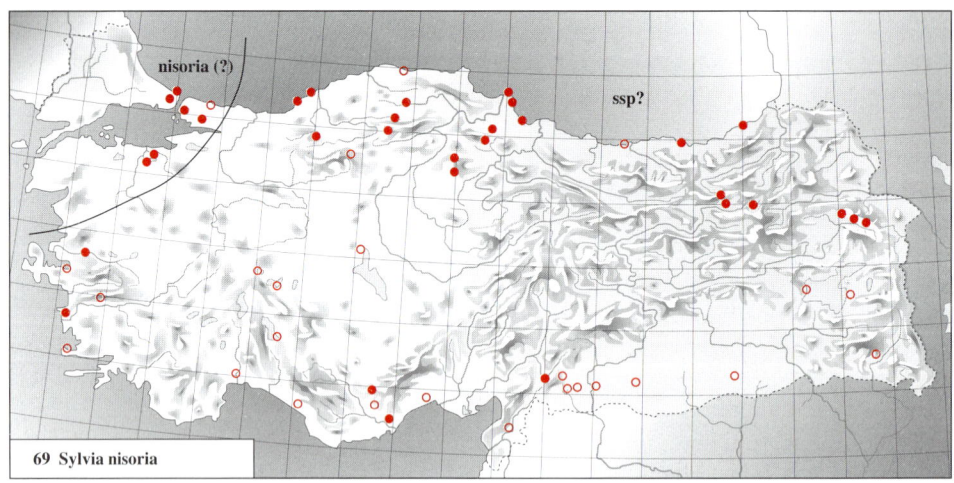

69 *Sylvia nisoria*

Geographical Variation
Subspecies described or recorded in the region:
- (W) ***S. n. nisoria*** (Bechstein), 1795, central and northern Germany. [Upperparts slightly duller grey in adult and duller brown in juvenile. Wing in Europe (sexes and ages combined) 88.6 (84-94) (n=36), bill 17.8 (16.7-18.8) (n=35) (CSR).]
- (E) ***S. n. merzbacheri*** Schalow, 1907, Kashkasu (Tien Shan mountains, Kyrgyzstan). [Upperparts of adult perhaps slightly paler and purer grey, of juvenile slightly paler and more sandy grey-brown. Wing in central Asia 88.5 (85-93) (n=10), bill 18.0 (17.5-18.7) (n=10) (CSR); perhaps better not recognized, as difference in colour and size is negligible.]

Subspecies recognized in Turkey: adult birds proved breeding in Turkey as well as birds collected after 14 May and presumed to be breeding locally have markedly shorter wings than the birds from Europe and central Asia cited above: wing in Turkey 83.8 (82-86) (n=6), bill 17.4 (17.0-17.7) (n=6) (CSR). These adults do not overlap in size with *nisoria* and *merzbacheri* from the samples above, which have wing 87 and over. Similarly short-winged birds are reported from Crimea (83-87, n=8: Johansen 1944-57), Iran (2 birds from 12-13 May with wing 85 each, but presumed to be migrants: Paludan 1938, Schüz 1959), and the Caucasus (Snigirewski 1928); a ♂ from Tbilisi (Georgia) had wing 85 and bill 17.3 (CSR). Perhaps the birds from the zoogeographical region formed by Asia Minor, south-west and northern Iran, Transcaucasia, the Caucasus, and the Crimea form a separable small subspecies, but additional data are needed. Birds from Turkey which are presumed to be migrants, all *nisoria* and all collected before mid-May, have wing 89.4 (86-95) (n=9), bill 17.4 (16.4-18.0) (n=6) (Hartert 1903-10, Kumerloeve 1961a, CSR); wing of spring migrants from Beysehir 88.5 (85-95) (n=12) (Vauk 1973).

References Snigirewski, S. I. (1928) Beiträge zur Avifauna der Wüste Kara-Kum (Turkmenistan). *J. Orn.* 76, 587-607.

SYLVIA CURRUCA
Lesser Whitethroat   Akgerdan Ötlegen
Habitat Open pine forest or plantations with patches of tangled undergrowth on hill tops, edges of coniferous forest at upper border of tree line, as well as mountain slopes with scattered dense scrub and stunted coniferous trees, from 900m up to at least 2600m.

Warblers

Distribution See map 70. Widespread in hills and mountain valleys, but often local, and apparently virtually absent (or underrecorded) in Thrace and Western Anatolia; common in the Taurus mountains and in the Yesilce area (near Gaziantep).

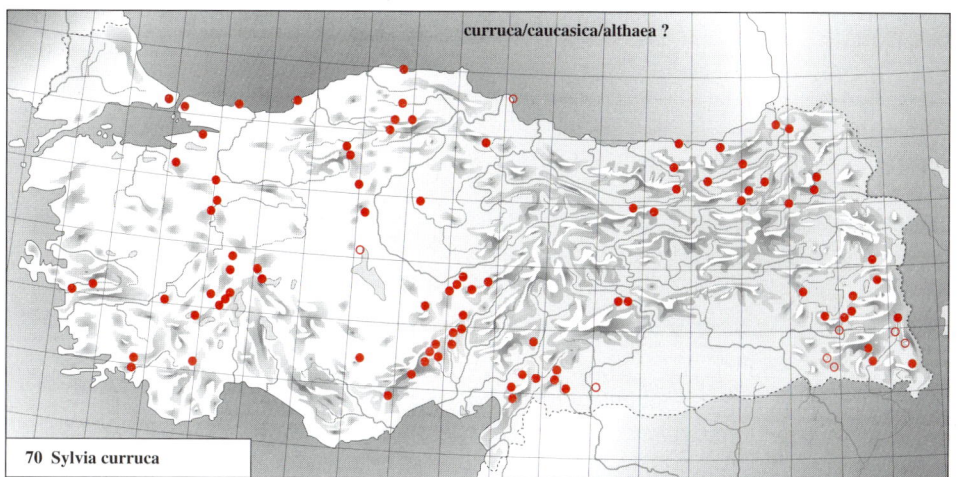

70 *Sylvia curruca*

Geographical Variation

Subspecies described or recorded in the region:

(W) ***S. c. curruca*** Linnaeus, 1758, Sweden. [Cap medium ash-grey, bordered below by a dark grey or dull black mask through the eye; remainder of upperparts dull drab-grey; underparts white with pale drab-grey or buff-grey wash on sides of breast and flanks; tip of outermost primary mainly between the tips of the 4th-5th outermost in the closed wing. Wing of both sexes from west and central Europe 65.7 (62-69) (n=70), bill 13.2 (12.4-14.0) (n=32) (Roselaar in Cramp 1992).]

(E) ***S. c. caucasica*** Ognev & Banjkovski, 1910, Mtskheta (near Tbilisi, Georgia, Transcaucasia). [Cap slightly darker grey than in *curruca*, less contrasting with the dark mask; upperparts darker grey, less drab; white of mid-underparts more washed with buff; tip of outer primary between the tips of the 5th-7th outermost. Wing 64.3 (62-66) (n=19); bill as in *curruca* (Snigirewski 1929). According to Snigirewski (1929) and Stepanyan (1990), this subspecies occurs throughout Transcaucasia west to eastern [and western?] Asia Minor and east to Iran; according to Matvejev & Vasic (1973), it even breeds west to the mountains of Greece, southern Bulgaria, and the south and west of former Yugoslavia. *S. c. zagrossiensis* Zarudny, 1911, described from Zagros mountains (south-west Iran) is either a synonym of *caucasica*, a synonym of *althaea*, or a separate form close to *althaea*. *S. c. caucasica* is intermediate in characters between *curruca* and *althaea*, and hence it is placed by some in the *curruca*-group of subspecies, but by others in the *althaea*-group of subspecies.]

(E) ***S. c. althaea*** Hume, 1878, Kashmir. [Upperparts still darker than in *caucasica*, cap slate-grey, hardly contrasting with the black mask, remainder of upperparts dusky brown-grey; sides of breast and flanks tinged grey, flanks and mid-breast sometimes with pink-buff suffusion; wing and tail blacker but outermost tail feather with more white; outer primary as in *caucasica*; large, bill rather thick. Wing of both sexes from Iran, Afghanistan, and Kashmir 68.4 (66-72) (n=7), bill 14.2 (13.6-14.5) (n=6) (CSR); wing of birds from Afghanistan 69.0 (65-71) (n=13) (Paludan 1959). Stated to show local overlap with some subspecies of the *curruca*-group without apparent interbreeding (e.g., Vaurie 1959), and therefore sometimes considered to form a separate species, which may include *caucasica*. However, evidence of overlap is slight or non-existent (e.g., see Desfayes & Praz 1978).

Subspecies recognized in Turkey: very difficult to decide, as the taxonomic situation is confusing. *S. c. curruca* is a common migrant in Turkey, and some authors consider breeding birds to be inseparable from *curruca*, though others consider *caucasica* to be the breeding form (see above). Recognition of *caucasica* as a valid taxon is often disputed, and some authors include it in *curruca* (e.g., Vaurie 1959), others in *althaea* (e.g., Williamson 1968). If *caucasica* is not considered to be a valid race and if *S. althaea* is specifically separated from *S. curruca*, it is not clear to which of the two species the birds of Turkey belong. Most birds examined from Turkey are inseparable from *curruca*, but the majority were probably migrating rather than breeding locally, except probably for a male from Uluborlu (western Taurus) collected 25 June; wing of all these *curruca* 64.6 (62-66.5) (n=8), bill 13.2 (12.5-13.9) (n=8) (CSR). Wing of migrant *curruca* from the Urfa area 64.9 (60-67) (n=8) (Weigold 1912-13). A single heavily worn ♂ from Sogukpinar (Ulu Dag), collected 4 July, appeared to agree with the description of *caucasica*; wing of this bird 62, bill 12.1 (CSR). The song of *curruca* and *caucasica* (or *althaea*, or whatever the Turkish birds may be) are said to differ; *curruca*-like songs were heard in Kizilcahamam, Osmancik, and Köse (Black Sea Coastlands) as well as above Silifke in the Taurus, while different songs (sometimes seen to be uttered by dark birds) were noted in Gümüshane, Bayburt (both in the north-east), south of Van Gölü, in the Gaziantep-Yesilce-Birecik-Halfeti area, and in the Ala Dag in the Taurus (P. J. Dubois, P. S. Hansen, S. Harrap, and R. Martins, per E. Dunn and M. Wilson). The *curruca*-like songs could have been produced by birds still on migration. A migrant *althaea* has been captured on 14 April in the Çukurova delta (Berk *et al.* 1988). Perhaps true *althaea* breeds in Turkey, or some of the local *caucasica* are very similar to *althaea*. Whether more forms breed in Turkey, or whether birds from south-east Turkey are *althaea* and grade through an intermediate *caucasica* in central Turkey into *curruca* in Thrace, or whether Turkish breeding birds form a uniform subspecies with distinct characteristics worthy of recognition (e.g., to be named *caucasica*), or whether all birds from Turkey and the Caucasus area should be included in *S. c. curruca* has yet to be established.

References Snigirewski, S. (1929) Uebersicht der Formen von *Sylvia curruca* (Linn.). *J. Orn.* 77, 252-261. Williamson, K. (1968) Identification for ringers 3. The Genus *Sylvia*. BTO Field Guide 9, Tring.

SYLVIA COMMUNIS
Common Whitethroat　　　　　　　　　　　　　　　　　　　　　　Çali Ötlegeni

Habitat Thickets of brambles, oak, or other tangled vegetation and scrub, situated in open forest, forest edges, at the tree line, in alpine meadows, in gardens, or along water; sometimes in olive groves, vineyards, or orchards. Mainly at 200-1900m in the western and central part of Turkey, up to *c.* 3000m in the east.

Distribution See map 71. Everywhere, and generally common, but absent from the coastal strip along the Mediterranean and from the more barren steppe country in the South-East.

Geographical Variation
Subspecies described or recorded in the region:

Warblers

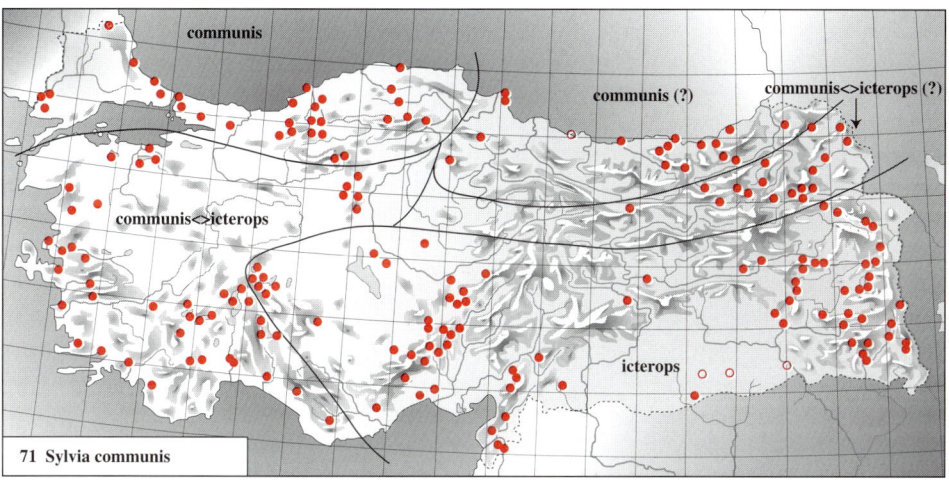

71 Sylvia communis

(W) ***S. c. communis*** Latham, 1787, Kent (England). [Cap of ♂ ash-grey, mantle and scapulars tawny-brown or dull olive-brown; underparts white, tinged vinous-pink on throat and breast, (pink-)buff on flanks, and cream-pink on belly and vent; outer fringes of inner wing feathers rufous-cinnamon. Wing of both sexes from the Netherlands 71.9 (67-77) (n=123), bill 14.0 (13.0-15.1) (n=39) (CSR).]

(T) ***S. c. traudeli*** Kumerloeve, 1969, Van (eastern Turkey). [Mantle and scapulars dark brownish sooty-grey, darker than *icterops* (below), much darker than *communis*, more contrasting with the grey of the cap (Kumerloeve 1969a). Wing of the type series from eastern Turkey 73.4 (71-75.5, once 78.5) (n=10), bill 14.0 (13.2-14.7) (n=10) (CSR).]

(E) ***S. c. icterops*** Ménétries, 1832, Talysh (Azerbaijan, south-east Transcaucasia). [Cap darker grey than in *communis*, mantle and scapulars dark grey-brown or dark olive-brown, less bright as in *communis*; underparts paler, throat and breast paler vinous-pink, flanks less extensive and paler buff; outer fringes of inner wing feathers paler pink-cinnamon. Wing of birds breeding in Iran 75.4 (71-76.5) (n=18) (Stresemann 1928, Paludan 1938, Schüz 1959, Diesselhorst 1962, CSR).]

Subspecies recognized in Turkey: comparison of Turkish birds with *communis* from western, northern, and central Europe and *icterops* from the Caucasus area and northern Iran shows that breeding birds from Haruniye and Eregli (in the Taurus) and from Tatvan, Van, Hakkari, Yüksekova, and Agri (in the east), as well as migrants from Ceylanpinar and Urfa are inseparable from *icterops*, and that *traudeli* is not a valid race. The variation in tinge and contrast of the dark upperparts of Turkish and Caucasian birds is due to differences in wear and abrasion, and the underparts of birds from the Taurus and eastern Turkey are pale, as in typical *icterops*. Further west, breeding birds from Beysehir, Uluborlu, Tire, Beynam, and Ankara combine upperparts rather similar to *communis* with pale underparts like those of *icterops*; this plumage is similar to that of birds from Crete and probably to that of birds from the Sporadhes and Kikladhes. All these birds, from western Asia Minor, the Aegean islands, and Crete, can be considered to form an intermediate population between *communis* and *icterops*. Birds from the Peloponnisos were closer to *communis*, but not yet completely like west European birds. Wing of intermediates from western Asia Minor 71.9 (68-73.5) (n=12), bill 13.9 (13.5-14.3) (n=4) (Stresemann 1928, Vauk 1973, CSR). Birds from Thrace, Gebze, and Sapanca Gölü in the north-west as well as birds from Rize and Verçenik in the north-east are inseparable from *communis* (Jordans

& Steinbacher 1948, Rokitansky & Schifter 1971, CSR); however, the birds from the north-east were collected in late August and early September and may have been migrants. Wing of the birds from the north-west 69.8 (65.5-74) (n=5) (CSR), and from the north-east 71.2 (69-73) (n=4) (Jordans & Steinbacher 1948). Wing of migrant *icterops* from the Urfa area 71.0 (68-77) (n=9) (Weigold 1912-13).

SYLVIA BORIN
Garden Warbler   Bahçe Ötlegeni

Habitat Rather open deciduous and mixed woodland, usually with some undergrowth, as well as forest edges and mature hedgerows in cultivation, mainly below 2000m.

Distribution See map 72; also, Kumerloeve (1962b). Restricted to northern Thrace and the coastal zone of the Black Sea Coastlands. Several records shown here as black refer to birds singing in the first half of June, and some of these may concern late migrants rather than local breeding birds.

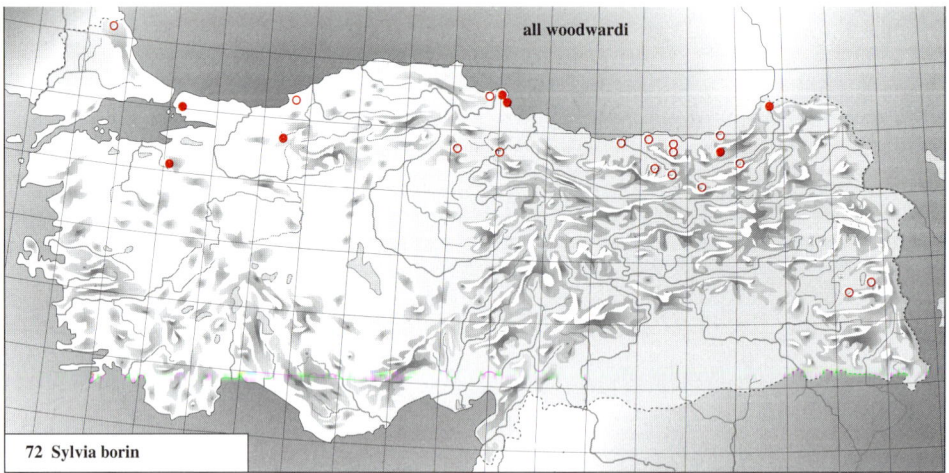

72 Sylvia borin

Geographical Variation

Subspecies described or recorded in the region:

(W) **S. b. borin** (Boddaert), 1783, France. [Upperparts light brown-olive; breast and flanks rather extensively washed buff-olive. Wing in western Europe 77.8 (73-82) (n=123), bill 14.5 (13.6-15.3) (n=106) (Roselaar in Cramp 1992).]

(W) **S. b. pateffi** Von Jordans, 1940, Bansko (south-west Bulgaria). [Upperparts grey; underparts more extensively white than in *borin*, breast and flanks paler and greyer. Wing in Bulgaria 80, 84.5, bill 14.3, 14.8 (n=2; the larger bird is the type specimen of *pateffi*) (CSR); wing in Greece 78.5 (77.5-80) (n=3), bill 14.2 (14.0-14.4) (n=3) (Watson 1962b).]

(E) **S. b. woodwardi** (Sharpe), 1877, Berea Hills (near Durban, Natal, South Africa; in winter). [Upperparts colder grey-olive than in *borin*, less brown; underparts whiter, breast and flanks with paler and more restricted cream-grey tinge. Wing in Kenya 80.4 (78-85) (n=12), bill 14.5 (13.7-15.2) (n=8) (CSR); wing in eastern Poland and European Russia 81.1 (79-85) (n=23) (Dunajewski 1938).]

Subspecies recognized in Turkey: *pateffi* from Bulgaria is sometimes recognized as an intermediate subspecies between the browner *borin* from northern, central, and western Europe and the paler and greyer *woodwardi* from eastern Europe and Asia. Examination of the type specimen and another bird from Bulgaria show that *pateffi* is inseparable from *woodwardi*. Thus, the intergradation zone between both subspecies is probably situated further west in the Balkan countries. Hence, the breeding birds from Turkey are likely to be *woodwardi*. All but 2 of the Turkish birds examined were indeed *woodwardi*, but, they were probably all migrants rather than local breeding birds. These birds, collected 5-23 May in Birecik, Ceylanpinar, and Agri (CSR), as well as a bird from Ankara collected 6 May (Kummerlöwe & Niethammer 1934-35) had wing 81.0 (78-84) (n=6), bill 14.9 (14.3-15.3) (n=4). Of 4 birds collected Rize (eastern Black Sea Coastlands), 17-21 August, probably also migrants, 2 were *borin*, 2 were *woodwardi*; wing of these birds 82.1 (80.5-84) (n=4), bill 15.0 (14.8-15.2) (n=4) (CSR). The bill of all these Turkish migrants appears to be rather long, but due to lack of samples of birds from most of the breeding grounds, nothing can be said about their probable origin.

References Dunajewski, A (1938) Zwei neue Vogelformen. *Acta Orn. Mus. Zool. Polonici* 2, 157-160. Kumerloeve, H. (1962b) A propos de la Fauvette des jardins en Asie-mineure. *Alauda* 30, 214-216. Watson, G. E. (1962b) La Fauvette des jardins *Sylvia borin* migratrice et nidificatrice sur les iles de la mer Egée. *Alauda* 30, 210-213.

SYLVIA ATRICAPILLA
Blackcap   **Karabas Ötlegen**

Habitat Open woodland and bushes, usually with ample undergrowth, at 0-2000m.

Distribution See map 73; also, Kumerloeve (1961a) and Kasparek (1990b, 1992). Largely restricted to areas north of 40ºN. Common further south during late April and May, in full song and showing territorial behaviour, but most of these observations apparently refer to migrants. Local breeding in areas other than shown here cannot be excluded.

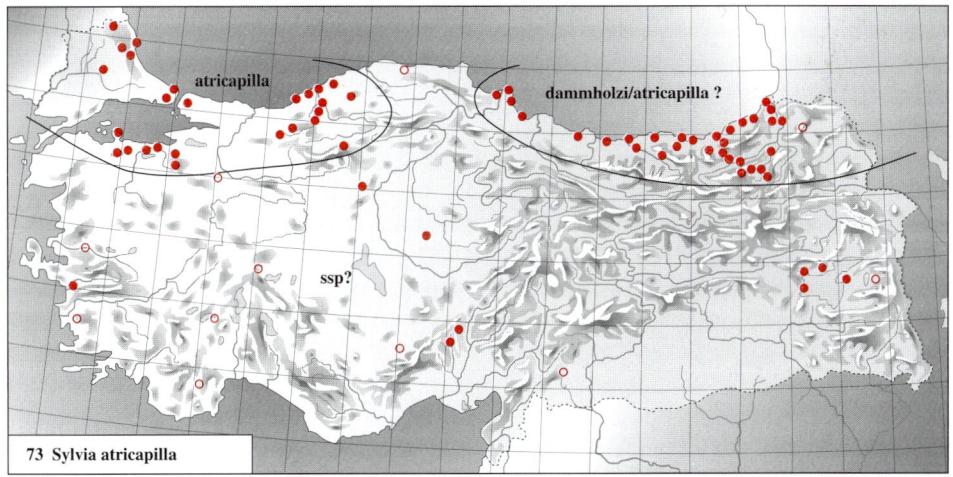

73 *Sylvia atricapilla*

Geographical Variation
Subspecies described or recorded in the region:

(W)   **S. a. atricapilla** (Linnaeus), 1758, Sweden. [Upperparts olive-brown, breast and flanks grey; larger and wing-tip more pointed than in the subspecies from the western Mediterranean basin and the Atlantic islands. Wing of both sexes in western and central Europe

74.3 (70-80) (n=107) (Roselaar in Cramp 1992).]

(E) **S. a. dammholzi** Stresemann, 1928, Pish Kuh (Gilan, northern Iran). [Like *atricapilla*, but upperparts colder olive-grey and breast and flanks with paler and more restricted grey wash. Wing in northern Iran 74.0 (70-78) (n=11) (Stresemann 1928, Paludan 1938, Schüz 1959, Érard & Etchécopar 1970). Also, breeds in the Caucasus and Transcaucasia (Stepanyan 1990).]

Subspecies recognized in Turkey: birds examined and those recorded in the literature, from Thrace east to the Amik Gölü, the Urfa area, and Ankara (collected late March to late May as well as September-October), and birds from Rize in the north-east (collected 1-16 August), are all inseparable from *atricapilla*; wing of these birds 75.1 (71-78.5) (n=27) (Weigold 1912-13, Kummerlöwe & Niethammer 1934-35, Rössner 1935, Jordans & Steinbacher 1948, Rokitansky & Schifter 1971, CSR). However, most of these were undoubtedly migrants. It seems probable that birds breeding in the western half of Turkey are *atricapilla*, but the subspecies in the eastern half is not known. As *dammholzi* breeds in Transcaucasia, one may expect this subspecies in the east, and it is supposed to occur here according to Kasparek (1990b), but according to Vaurie (1954) all Turkish birds are *atricapilla*, although they are slightly duller and paler than birds from the British Isles and northern Europe. It is not clear, however, whether Kasparek and Vaurie actually examined breeding birds from Turkey and further research is needed.

References Kasparek, M. (1990b) Zur Brutverbreitung und zum Zug der Mönchsgrasmücke (*Sylvia atricapilla*) im Nahen Osten. *Vogelwarte* 35, 169-176.

PHYLLOSCOPUS TROCHILOIDES
Greenish Warbler   Ardıç Bülbülü

Habitat Mature deciduous, coniferous, or mixed forest (with pine, fir, spruce, juniper, beech, oak, chestnut, alder, poplar, or a mixture of these), alternating with glades with ample low rhododendron scrub and other vegetation; occurs at about 800-1500m in the northwest but up to at least 2100m in the east. See Albrecht (1987).

Distribution See map 74; also, Steiner (1962), Kumerloeve (1967b), and Albrecht (1987). Restricted to several localities in the Black Sea Coastlands. Probably more widely distributed between Bolu and Ordu than known at present.

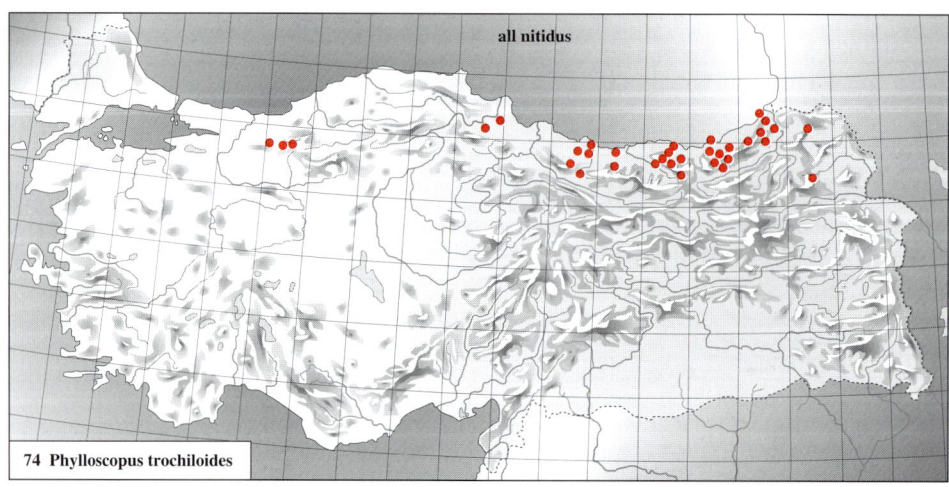

74 *Phylloscopus trochiloides*

Geographical Variation
Subspecies described or recorded in the region:
(E) **P. t. nitidus** Blyth, 1843, Calcutta (India, in winter). [Upperparts yellow-green, underparts tinged yellow; upperparts not as pure green and underparts less pure white than in *viridanus* from eastern Europe and west and west-central Asia (see Albrecht 1987). Wing of ♂ from Iran 65.0 (63-67) (n=8), of ♀ 61 (n=1) (Stresemann 1928, Schüz 1959, Érard & Etchécopar 1970); the size of *viridanus* is generally smaller: see A. J. van Loon in Cramp (1992).]

Subspecies recognized in Turkey: all birds breeding in Turkey are referable to *nitidus*, the subspecies breeding from Turkey and the Caucasus east through Iran and Afghanistan to south-west Tajikistan. Wing of ♂ from Tire (Western Anatolia) and Rize (eastern Black Sea Coastlands), all from August and not certain to have been bred locally, 65.0 (63-67) (n=6), of ♀ 61 (n=1), bill of both sexes 13.2 (13.0-13.6) (n=4) (Jordans & Steinbacher 1948, CSR). *P. t. nitidus* was formerly often considered to be a separate monotypic species rather than a subspecies of *P. trochiloides*, as a consequence of reported overlap between *nitidus* and *P. trochiloides viridanus* without apparent interbreeding in west-central Asia. According to I. A. Neufeldt (in Cramp 1992), no overlap exists, and thus both are better united under *P. trochiloides*, in view of the close relationship between both. Quite a number of Turkish *nitidus* do not show much yellow and are hard to separate from *viridanus*. See Albrecht (1987) for details on identification and taxonomy.

References Steiner, H. M. (1962) 'Zilpzalp' und Grüner Laubsänger in NO-Kleinasien. *Egretta* 5, 57-60. Kumerloeve, H. (1967b) Zum Vorkommen von Laubsängern (*Phylloscopus*) im östlichen und südlichen Kleinasien. *Vogelwarte* 24, 143-145. Albrecht, J. S. M. (1987) Some notes on the identification, song and habitat of the Green Warbler in the western Black Sea Coastlands of Turkey. *Sandgrouse* 6, 69-75.

PHYLLOSCOPUS BONELLI
Bonelli's Warbler   **Dag Sögütbülbülü**

Habitat Slopes and hills covered with maquis and scattered trees, or in open pine or mixed forest, at 250-1700m.

Distribution See map 75; also, Kumerloeve (1967b). Restricted as a breeding bird to Western Anatolia (but apparently rare or absent in the south-west), the western part of the Black Sea Coastlands, and the Taurus Mountains; apparently a scarce breeding bird for which confirmation of breeding is required in Thrace and the Amanus Mountains. Scattered occurrences further east probably refer to migrants, but recorded breeding in the extreme South-East.

Geographical Variation
Subspecies described or recorded in the region:
(W) **P. b. orientalis** (Brehm), 1855, Wadi Halfa (northern Sudan, on migration). [Upperparts greyer than in *bonelli* from central Europe and from the countries surrounding the west Mediterranean basin, under wing-coverts and axillaries paler yellow; wing longer but bill shorter; tip of outermost functional primary mainly between the tips of the 4th-5th outermost instead of mainly between the 5th and 6th. Wing of ♂ 66.9 (63-72) (n=45), of ♀ 64.2 (61.5-69) (n=35), bill 12.2 (10.7-13.6) (n=28) (Williamson 1967, A. J. van Loon in Cramp 1992, CSR). Breeds from the Balkan eastwards.]

Subspecies recognized in Turkey: all Turkish birds are of the eastern subspecies *orientalis*. Wing of ♂ from Turkey 67.6 (65-69.5) (n=10), of ♀ 63.9 (60-65) (n=7), bill 12.1 (11.2-13.0)

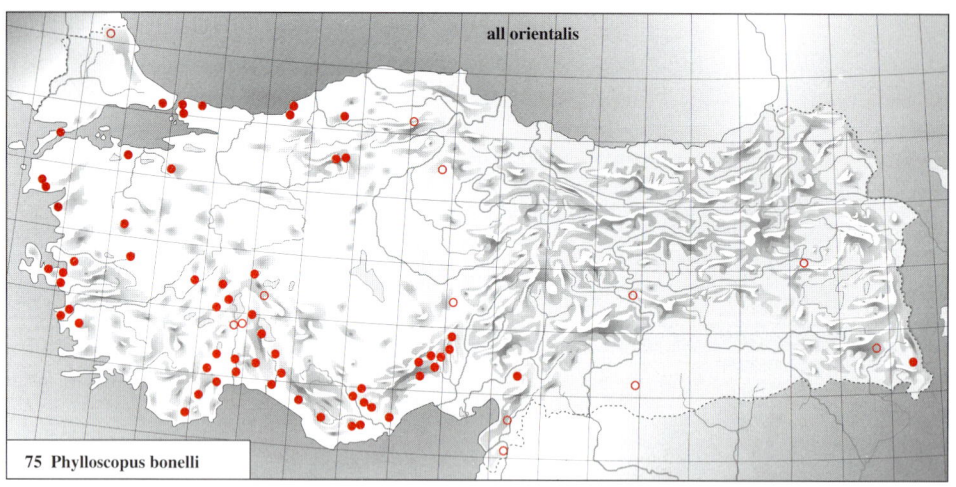

75 Phylloscopus bonelli

(n=17) (Weigold 1912-13, A. J. van Loon, CSR). The voice of both subspecies of *P. bonelli* is rather different in structure, and perhaps they are separate species (see Helb *et al.* 1982 and Kitson *et al.* 1983).

References Williamson, K. (1967) Identification for ringers 2. The Genus *Phylloscopus*. BTO Field Guide 8, Tring. Helb, H.-W., H.-H. Bergmann, & J. Martens (1982) Acoustic differences between populations of western and eastern Bonelli's Warblers (*Phylloscopus bonelli*, Sylviidae). *Experientia* 38, 356-357. Kitson, A. R., R. F. Porter, & R. A. Hume (1983) Call of Bonelli's Warbler. *Brit. Birds* 76, 537.

PHYLLOSCOPUS SIBILATRIX
Wood Warbler   Orman Söğütbülbülü

Habitat Mixed stands of mature and younger trees, either deciduous or mixed with conifers, with dense but not fully closed canopy, occurring from sea level to the upper limit of deciduous trees.

Distribution No map. Only recorded in the Istranca Daglari near Dereköy (northern Thrace) on 27 July, where perhaps breeding; all other records refer to migrants (see, e.g., Kumerloeve 1970c).

Geographical Variation None. Some subspecies have been described in the past, but none is now considered to be valid. *P. sibilatrix* (Bechstein), 1793, was originally described from the mountains of Thüringen (Germany). Wing of ♂ from Europe 76.7 (74-79) (n=17), of ♀ 73.4 (72-76) (n=12) (A. J. van Loon in Cramp 1992); wing of migrant ♂ from Turkey 76, 77 (n=2), of ♀ 71, 75, and 76.5 (n=3), of an unsexed bird 69 (n=1) (Weigold 1912-13, Kummerlöwe & Niethammer 1934-35, Jordans & Steinbacher 1948, Kumerloeve 1967b, 1970b).

References Kumerloeve, H (1970c) Remarques sur le statut du Pouillot siffleur (*Phylloscopus sibilatrix*) en Proche-Orient. *Gerfaut* 60, 226-227.

PHYLLOSCOPUS LORENZII
Caucasian Chiffchaff   Lorenz Bülbülü

Habitat Mountain forest, mainly of pine and juniper, at 1700m and over, generally at higher altitude than *P. collybita*.

Distribution See map 76; also, Kumerloeve (1967b). As far as known, restricted to the eastern part of the Black Sea Coastlands and the northern part of the East, east of the Zigana pass and north of the Aras valley. Recorded from Çatak near Van Gölü on 5 August (Helbig 1984), but perhaps a migrant here; a bird showing the characters of *lorenzii* was recorded on 29 June near Kizilcahamam (Dorèl 1991). The distribution is not well known due to confusion with *P. collybita*. All records of 'Chiffchaffs - *Phylloscopus collybita/lorenzii/sindianus*' are shown on map 77; these may include some *P. lorenzii*.

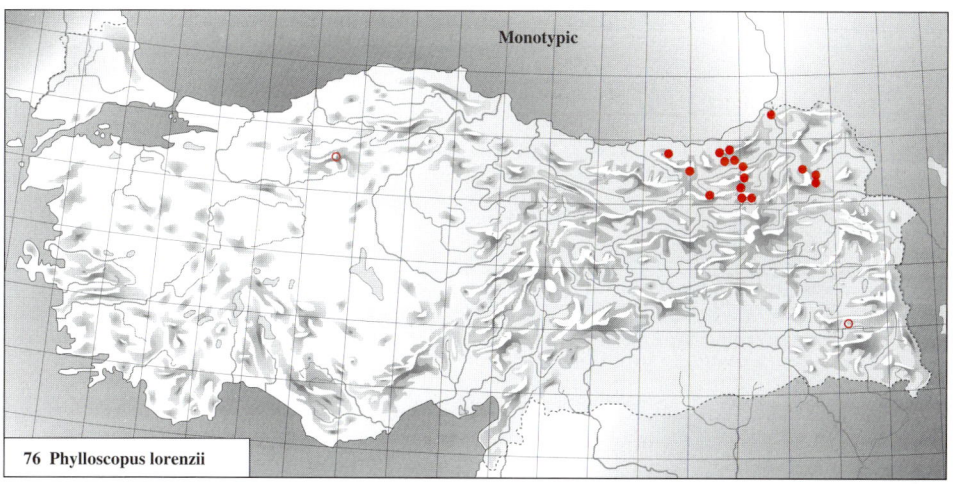

76 *Phylloscopus lorenzii*

Geographical Variation None. *P. lorenzii* (Lorenz), 1887, was originally described from the northern Caucasus. It is restricted to the Caucasus area, Turkey, and northern Iran. It is often considered to be a subspecies of *P. sindianus* from the mountains of west-central Asia (e.g., in Cramp 1992). Both *lorenzii* and *sindianus* are probably offshoots of *P. (collybita) tristis* from north Siberia (or *tristis* is an offshoot of *sindianus* or *lorenzii*) (see, e.g., Martens & Hänel 1981). As it is not certain that *lorenzii* and *sindianus* had a common ancestor rather than being independent offshoots, both can better be regarded as full species. Both are closely similar to *tristis* in plumage and could easily be considered to be subspecies of *collybita*, but *lorenzii* overlaps with *collybita* in the Caucasus area without apparent interbreeding and has thus achieved the status of a full species. The form *sindianus* does nòt overlap with a form of *collybita* and is therefore frequently considered to be a subspecies of *P. collybita* (e.g., by Stepanyan 1990), but the uniting of *P. sindianus* and *P. collybita* has an equally weak base as the uniting of *P. sindianus* and *P. lorenzii*, and, in fact, the uniting of *P. collybita* (including the races *brevirostris* and *abietinus*) and *P. tristis*. See also Stegmann (1934), Vaurie (1954), Watson (1962c), Martens & Hänel (1981), Martens (1982), and Marova & Leonovich (1993). The voice of *P. sindianus* is rather similar to that of *P. (c.) tristis* and differs from that of *P. collybita* (including *abietinus*) (Martens & Hänel 1981). *P. lorenzii* differs from Turkish *P. collybita brevirostris* in greyish-brown upperparts (in *brevirostris*, green-grey), dull white underparts with grey-drab or fulvous wash on breast and flanks (in *brevirostris*, white with some yellow streaking), white under tail-coverts (in *brevirostris*, cream and yellow), and cream-white under wing-coverts and axillaries, virtually without yellow (in *brevirostris*, mainly light yellow) (Watson 1962c); see also Shirihai (1986). From other subspecies of *P. collybita*, it differs mainly in relatively longer tail and relatively rounder wing with shorter outer primaries. Wing of a single

Turkish ♂ examined (Sarikamis, 20 May; see Kumerloeve 1967b) 60.5, tail 51.5, bill 12.0; wing of ♂ *lorenzii* from the Caucasus 62.5 (61-62) (n=3), of ♀ 55 (CSR).

References Stegmann, B. (1934) Ueber die systematische Stellung von *Phylloscopus lorenzii* (Lorenz). *Orn. Monatsber.* 42, 76-77. Watson, G. E. (1962c) A re-evaluation and redescription of a difficult Asia Minor *Phylloscopus*. *Ibis* 104, 347-352. Martens, J., & S. Hänel (1981) Gesangsformen und Verwandtschaft der asiatischen Zilpzalpe *Phylloscopus collybita abietinus* und *Ph. c. sindianus*. *J. Orn.* 122, 403-427. Martens, J. (1982) Ringförmige Arealüberschneidung und Artbildung beim Zilpzalp, *Phylloscopus collybita*. *Zeitschr. zool. Syst. Evol. Forschung* 20, 82-100. Shirihai, H. (1986) Field characters of Mountain Chiffchaff. *Proc. 4th Intl. Identif. Meet. Eilat*, Nov. 1986, 60-63. Elat. Marova, I. M., & V. V. Leonovich (1983) [On hybridization between Siberian (*Phylloscopus collybita tristis*) and east European (*Ph. collybita abietinus*) Chiffchaffs in the zone of sympatry], in O. L. Rossolimo (ed) Gibridizatsiya i problema vida u pozvonochnykh. *Sb. Trud. Zool. Muz. MGU* 30, 147-163.

PHYLLOSCOPUS COLLYBITA
Chiffchaff   Cif Caf

Habitat Mainly in open deciduous, coniferous, or mixed forest on hills and mountain slopes. Usually at 500-2000m, but locally to sea level in the north-west and north and even higher in the east (but the latter perhaps partly or all *P. lorenzii*).

Distribution See map 77. Widespread north of c. 40°N (including Thrace) and in Western Anatolia, with some isolated occurrences in the mountains further south. The map includes data of unidentified *collybita/lorenzii* birds (e.g., from Steiner 1962); most records in the north-east probably refer to *P. lorenzii*, but overlap of both species, as in the Caucasus, cannot be dismissed.

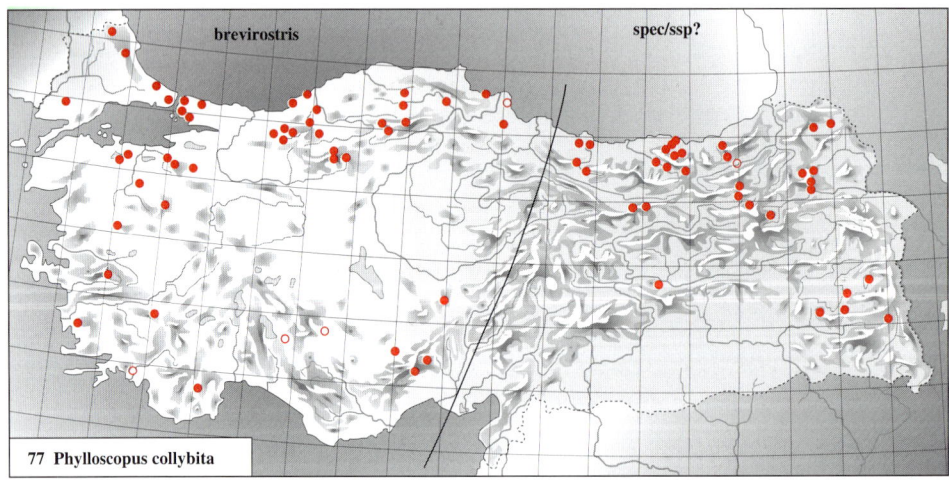

77 *Phylloscopus collybita*

Geographical Variation
Subspecies described or recorded in the region:

(W)   ***P. c. collybita*** (Vieillot), 1817, Normandy (France). [Upperparts in spring dark brownish-olive, rump brighter olive-green; underparts pale yellow with a restricted amount of white on the belly and with much pale drab-grey on throat, breast, and flanks shining through;

under tail-coverts mixed white and pale yellow; under wing-coverts and axillaries bright pale yellow; tip of the outermost functional primary usually equal to the tip of the 6th or 7th outermost primary or between these in the closed wing. Wing of ♂ from western and central Europe 60.8 (58-64.5) (n=34), of ♀ 55.5 (53-56.5) (n=12), bill of both sexes 12.2 (11.3-13.1) (n=25) (A. J. van Loon in Cramp 1992; CSR). Breeds western, central, and southern Europe east to Greece and the Balkan countries.]

(W) *P. c. abietinus* (Nilsson), 1819, Sweden. [Upperparts in spring much paler and greyer than *collybita*, greyish olive-green, rump brighter yellow-green; underparts including under tail-coverts white with light yellow wash on breast, flanks, and undertail, much less tinged grey than *collybita* and with the belly more extensively white; underwing and axillaries bright light yellow; outermost functional primary relatively longer, the tip usually equal to the tip of the 6th outermost or between the tips of the 6th and 7th outermost in the closed wing. Wing longer: wing of migrant ♂ from south European Russia 64.6 (63.5-65) (n=6), of ♀ 59.1 (58-62) (n=4), bill 12.3 (11.5-12.9) (n=10) (A. J. van Loon in Cramp 1992; CSR). Breeds in northern and eastern Europe, except for the extreme north-east.]

(T) *P. c. brevirostris* (Strickland), 1836, Izmir (Western Anatolia, Turkey; in winter). [According to Watson (1962c), like *abietinus*, but upperparts green-grey, breast with a buff or fulvous wash, yellow of underparts and underwing on average paler and less extensive, and tip of outermost primary usually between the tips of the 6th and 7th outermost. However, in breeding birds examined from the Ulu Dag (Western Anatolia) and Beycuma (western Black Sea Coastlands), the entire plumage, size, and wing structure are very similar to *collybita*, the only difference perhaps being a slightly paler yellow underwing and whiter under tail-coverts. Recognition of *brevirostris* is provisionally maintained, but examination of a larger number of breeding specimens may show that *brevirostris* is not separable from *collybita* (CSR). Stresemann (1928) considered a bird from Eregli on 28 May to be inseparable from *collybita*. Wing of breeding ♂ from the Ulu Dag, Sogukpinar, Beycuma, and Eregli and of wintering birds from Mytilene (Greece), Izmir (the type of *brevirostris*), Manavgat, Adana, Haruniye, and Osmaniye 60.5 (59-62) (n=10), of ♀ 55.5 (54.5-57) (n=4), bill 11.8 (11.4-12.5) (n=10) (Watson 1962c, CSR).]

Subspecies recognized in Turkey: breeding birds examined are *brevirostris*, but see above. In winter and on migration, *brevirostris*, *abietinus*, and perhaps *tristis* from north-east Europe and north Siberia occur (the latter close in size and structure to *abietinus*, but brown-grey and white instead of olive and yellow, and thus close in colour to *P. lorenzii*) (Kumerloeve 1967b, 1970b, CSR). According to Watson (1962c), *brevirostris* combines characters of *abietinus* and *P. lorenzii* and is perhaps an intermediate between both, which may grade into *lorenzii* in eastern Turkey. In specimens examined (but only a small number), no intermediate characters were found and evidence of intergradation was lacking. Further study is needed.

REGULUS REGULUS
Goldcrest   Altintavukcuk

Habitat Coniferous forests on hills and mountain slopes, including stunted pines and junipers at the upper tree limit; at 350-600m in the west (but to 1900m on the Ulu Dag), at (500-)1000-2400m in the Taurus and in the mountains of the Black Sea Coastlands.

Distribution See map 78; also, Kumerloeve (1964d). Common in suitable habitat in southern Thrace, the Black Sea Coastlands, the upper Aras Valley, Western Anatolia, and the Southern Coastlands.

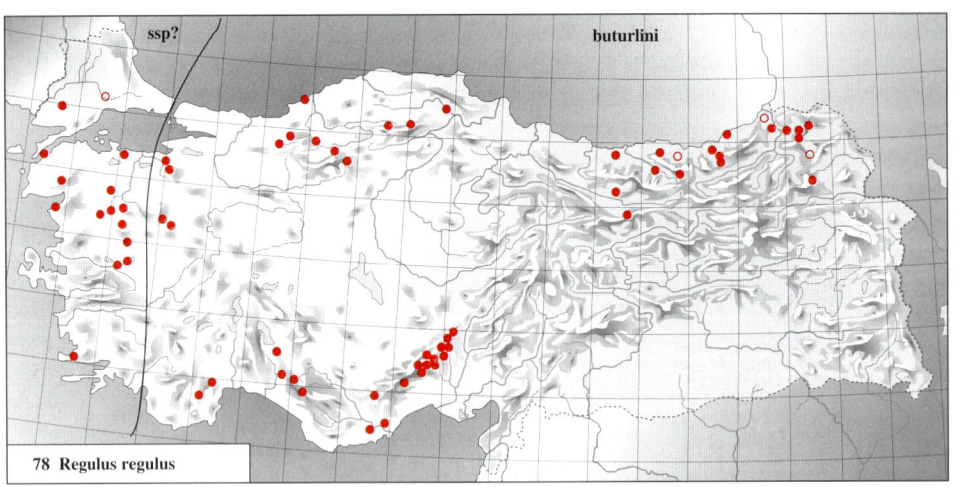

78 Regulus regulus

Geographical Variation

Subspecies described or recorded in the region:

(W) **R. r. regulus** (Linnaeus), 1758, Sweden. [Upperparts yellowish olive-green, underparts white, tinged yellowish-ochre on the breast and (especially) flanks. Wing in ♂ from the Netherlands 54.4 (51-57) (n=42), of ♀ 52.4 (50-54) (n=32) (A. J. van Loon in Cramp 1992). Occurs from western and northern Europe east to the Balkan countries, Greece, and west Siberia.]

(E) **R. r. buturlini** Loudon, 1911, Talysh (Azerbaijan, south-east Transcaucasia). [Stripe at each side of crown duller black than in *regulus*, upperparts colder and more greyish-green, especially hindneck which is greyish, less olive-green; breast and flanks pale drab-grey, belly whitish, almost without yellow-ochre. Breeds in the Caucasus, Transcaucasia, and Iranian Azarbaijan.]

Subspecies recognized in Turkey: all birds examined are clearly referable to *buturlini*. This subspecies was not recognized by Stepanyan (1990), but specimens of Goldcrests from Asia Minor are easily picked out from a series of *regulus* from Europe at all times of year, with help of the characters mentioned above for *buturlini*. Perhaps the birds from the Caucasus which Stepanyan probably used for checking the validity of the subspecies are not as distinct as those from Talysh, Iranian Azarbaijan, and Turkey. Wing of ♂ from Abant Gölü (near Bolu) and Kurayiseba (Sebinkarahisar) 56.7 (55-58) (n=3), of ♀ from Kurayiseba, Sarikamis, and Hacin Dagi (Taurus) 54.5 (53-56) (n=4) (CSR). It is not known to what subspecies birds from Thrace belong (perhaps *regulus*, as in Bulgaria). In winter, *regulus* is recorded from Bolu, Ankara, and Erzurum; wing of ♂ of these 55.5 (54-57) (n=4), of ♀ 50 (n=1) (Kummerlöwe & Niethammer 1934-35, Rössner 1935, Kumerloeve 1968).

References Kumerloeve, H. (1964d) Zur Brutverbreitung der beiden Goldhähnchen- (*Regulus*) Arten in Kleinasien. *Vogelwelt* 85, 120-122.

REGULUS IGNICAPILLUS

Firecrest   Sürmeli Altintavukcuk

Habitat The upper zone of the coniferous forest near the tree limit, in dense crowns of spruce or stunted pines. At c. 1800-1900m on the Egrigöz and Ulu Dagi (Western Anatolia), at c. 1000-1400m near Bolu and Kizilcahamam (western Black Sea Coastlands), and at c. 1800-2300m in the north-east.

Flycatchers

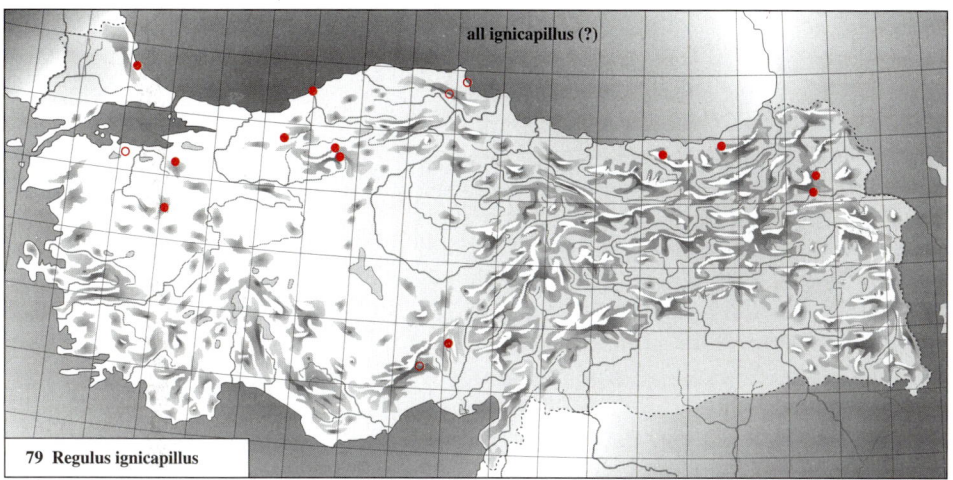

79 Regulus ignicapillus

Distribution See map 79; also, Kumerloeve (1964d). Restricted to a few scattered localities in the north of the country. Confirmation required for supposed breeding in the Istranca Daglari (Thrace), the Manyas-Bandirma Reserve (Marmara sub-region), and the Toros and Karanfil Dagi (Southern Coastlands).

Geographical Variation

Subspecies described or recorded in the region:

(W) ***R. i. ignicapillus*** (Temminck), 1820, 'France, Germany, etc.' [Upperparts yellowish-green, side of neck golden-yellow, underparts white, tinged pale brown-grey on throat and pale olive-buff on breast and flanks. Wing of ♂ from western and central Europe 53.3 (51-56) (n=34), in ♀ 50.2 (48-53) (n=14) (A. J. van Loon in Cramp 1992).]

(W) ***R. i. minimus*** Matvejev, 1947, Kapaonik (Serbia). [Not seen, but considered to be only slightly different from *ignicapillus* by Matvejev (1976); said to breed in the Balkan countries and western Turkey.]

Subspecies recognized in Turkey: breeding birds have apparently never been collected in Turkey, and hence nothing can be said about the subspecies occurring here. According to Stepanyan (1990), the birds breeding in the western Caucasus are *ignicapillus* (for occurrence here, see Fischer & Fischer 1976 and Schubert 1986). Vaurie (1959) also considers Asia Minor birds to be *ignicapillus*. Hence one may call Turkish birds *ignicapillus*, but as it is not at all certain whether Stepanyan and Vaurie based their opinion on examination of breeding specimens, the identification as *ignicapillus* is provisional.

References Fischer, W., & M. Fischer (1976) Ornithologische Beibachtungsergebnisse aus zwei reisen in den Kaukasus und nach Transkaukasien. *Beitr. Vogelk.* 22, 137-160. Schubert, P. (1986) Einige bemerkenswerte Beobachtungen an der kaukasischen Schwarzmeerküste bei Sotschi. *Beitr. Vogelk.* 32, 186-187.

MUSCICAPA STRIATA

Spotted Flycatcher   Gri Sinekkapan

Habitat Open pine forest, forest edges, and fringes of glades; mainly at 500-1000m, occasionally 0-1700m.

Distribution See map 80; also, Kumerloeve (1958c). Mainly confined to the area north of 40°N. Further south, many observations have been made in the second half of May, in early June and in July, and these may in part refer to local breeding, but only a few cases of probable breeding are known (e.g., nest-building as far south as Karatas in the Çukurova deltas: Groh 1968).

Geographical Variation

Subspecies described or recorded in the region:

(W) **M. s. striata** (Pallas), 1764, Holland. [Upperparts grey-brown, forehead and forecrown finely streaked paler grey-brown and black; underparts white, lower throat, breast, and flanks extensively marked with rather sharply defined grey-brown streaks. Wing of ♂ from Holland 85.8 (83-89) (n=25), of ♀ 86.0 (82.5-90) (n=21) (A. J. van Loon in Cramp & Perrins 1993); wing of ♂ from European Russia 87.6 (83-90) (n=44), of ♀ 87.0 (84-90), (n=24), bill of both sexes 11.7 (10.6-13.6) (n=68) (Peklo 1987).]

(W) **M. s. cretica** Stresemann & Schiebel, 1925, Crete. [Upperparts as in *striata*, but less brown; streaks of underparts ill-defined (Stresemann & Schiebel 1925, CSR). Wing of ♂ from Crete 87, 90.5 (n=2), of ♀ 90.5 (n=1) (CSR).]

(E) **M. s. neumanni** Poche, 1904, Masailand (north-east Tanzania; in winter). [Upperparts paler and greyer than in *striata*, less brown, forehead whitish, dark streaks more contrasting; streaks on lower throat narrower, sharply defined, those of breast and flanks more restricted in extent (especially on flanks), paler, more washed-out and less sharply defined than in *striata*. Wing of ♂ from central Asia 86.5 (82-92) (n=33), of ♀ 85.1 (81-90.5) (n=27), bill 12.2 (11.2-13.6) (n=60) (Peklo 1987).]

Subspecies recognized in Turkey: birds from western Turkey (Thrace and Western Anatolia south to Yatagan) are *striata*, showing dark upperparts and dark streaks on underparts, but the streaks on the breast are rather washed out and the streaking on the flanks is rather restricted, showing some influence of *neumanni*. Though paler and greyer birds occur west of Turkey (viz., *cretica*, which does not appear separable from *neumanni*), such birds were not found as breeding birds. Wing of ♂ from western Turkey 88.5, 89 (n=2), of ♀ 84.5, 85 (n=2) (CSR). The subspecies occurring further east in Turkey is not yet known; quite a number have been examined, but all were from late April and early May as well as August-September, and these may have been, at least in part, migrants from elsewhere. Some of these birds were similar to breeding birds described from Thrace and Western Anatolia, and thus were near to *striata* (examined from Zonguldak, Hacin Dagi in the Taurus, Maras, Hazer Gölü, and Rize); these may in part have involved local breeding birds. Others were (similar to) *neumanni* (examined from Hacin Dagi, Tarsus, Bolu,

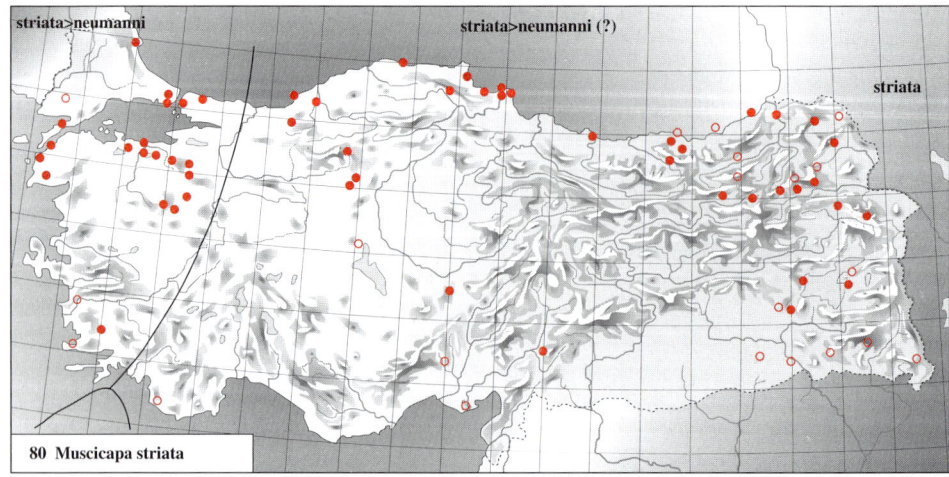

80 Muscicapa striata

Ankara, and Rize). Wing of probable migrant ♂ of *striata* from Turkey 88.4 (87-90.5) (n=6), of ♀ of *striata* 88.5 (n=1), of migrant ♂ of *neumanni* 88.1 (85-91) (n=8), of ♀ of *neumanni* 86.9 (85-90) (n=5) (Weigold 1912-13, Kummerlöwe & Niethammer 1934-35, CSR). As Transcaucasia is inhabited by *striata* (Stepanyan 1990), the subspecies breeding in the Black Sea Coastlands may be *striata* or near to *striata*, too.

References Stresemann, E., & G. Schiebel (1925) Neue Formen aus Kreta. *J. Orn.* 73, 658-659. Kumerloeve, H, (1958c) Der Grauschnäpper, *Muscicapa striata* Pallas, als Brutvogel in Anatolien. *Vogelwelt* 79, 111-112. Peklo, A. M. (1987) [*Flycatchers of the fauna of the USSR.*] Kiev.

FICEDULA PARVA
Red-breasted Flycatcher   Cüce Sinekkapan

Habitat Moist mature deciduous or mixed forest, mainly oak and beech; occurs at c. 1000m in the western Black Sea Coastlands (Albrecht 1981).

Distribution See map 81. Restricted to a few localities scattered throughout the Black Sea Coastlands. Breeding is difficult to prove, as the species is a common migrant during May and from August onwards, at a time that inconspicuous local birds may breed. Recorded 'between Maras and Göksun' as early as 24 July (Vielliard 1968).

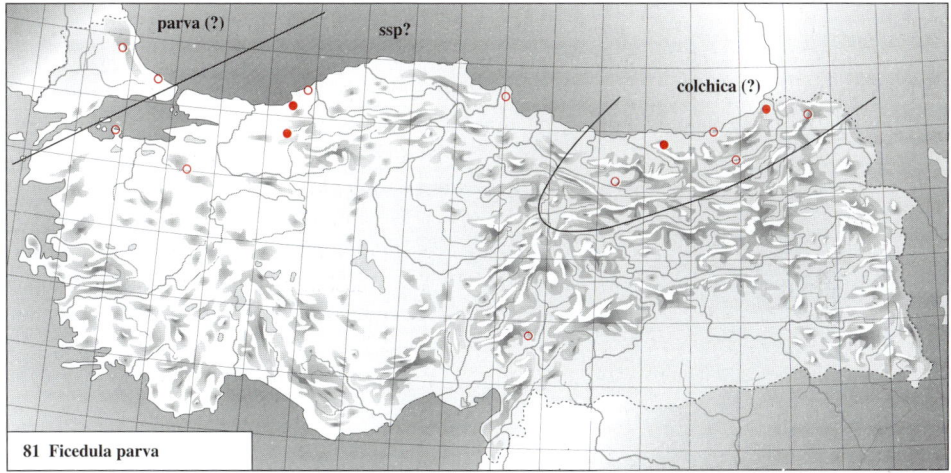

81 Ficedula parva

Geographical Variation
Subspecies described or recorded in the region:

(W)   ***F. p. parva*** (Bechstein), 1794, Thüringerwald (Germany). [Mantle and side of head of adult ♂ ash-grey; throat and upper breast orange-red; flanks tinged cream-buff. Wing of ♂ from European Russia 67.2 (65-70) (n=66), of ♀ 65.5 (63-70) (n=15); bill to the nostril of both sexes 6.7 (6.1-7.5) (n=81) (Peklo 1987).]

(E)   ***F. p. colchica*** (Von Dombrowski), 1911, Western Transcaucasia. [Mantle and side of head of ♂ darker and more extensively grey; rufous of breast more intense, extending further down on lower breast, cream-buff wash on flanks more extensive, and bill on average perhaps longer. Wing of ♂ from the Caucasus area 66.7 (64-68.5) (n=18), of ♀ 64.6 (63-67) (n=7); bill to nostril 7.0 (6.1-7.5) (n=25) (Peklo 1987); wing of ♂ from Transcaucasia and northern Iran 67.5 (64-70) (n=14), of ♀ 64.8 (63.5-67) (n=5) (Stresemann

1928, Paludan 1940, Schüz 1959, Érard & Etchécopar 1970).]

Subspecies recognized in Turkey: only females and 1st-year males from Turkey have been examined, which are apparently not suitable for identifying subspecies; moreover, as all have been collected May and August-October, they may have been migrants, even though the gonads were sometimes enlarged (e.g., a male from Zonguldak collected in suitable habitat in May) or the plumage was heavily abraded as early as May, which may indicate local breeding (e.g., a ♀ from Terkos Gölü, Thrace). Wing of ♂ from Turkey 66.4 (64-68) (n=5), of ♀ 65.7 (63.5-67) (n=6) (Kummerlöwe & Niethammer 1934-35, A. J. van Loon, CSR). *F. p. colchica* is considered to be inseparable from *parva* by Vaurie (1959) and Stepanyan (1990), even though the characters ascribed to *colchica* were found to be valid in specimens from Iran (Stresemann 1928, Paludan 1940, Vaurie 1954, Schüz 1959, Érard & Etchécopar 1970). The supposed breeding area of *colchica* (the Caucasus area to northern Iran and Turkey) is not in contact with that of *parva* (central and eastern Europe to western Siberia) and the occurrence of two genetically isolated subspecies is thus not unlikely. Further study is needed. If any Red-breasted Flycatcher breeds in Thrace (e.g., in the Istranca Daglari), it would probably be *parva*, the subspecies breeding in Bulgaria.

References Albrecht, J. S. M. (1981) Red-breasted Flycatcher - a new breeding species for Turkey. *Sandgrouse* 3, 91-92.

FICEDULA SEMITORQUATA
Semi-collared Flycatcher   Yarimband Sinekkapan

Habitat Deciduous or mixed humid forest, parkland with large trees, riverine forest, plantations, and orchards, mainly at 500-2000m. Has occurred in the Taurus at an altitude of 2150m, along a river far from trees, on 19-23 July (Beaudoin 1976).

Distribution See map 82. A scarce and local breeding bird in the north-east, above Samsun, near Abant Gölü, on the Ulu Dag, in Thrace, and near the Van Gölü. As in *F. parva*, breeding is often difficult to establish, as local breeding birds are outnumbered by more numerous *Ficedula* migrants (of various species) in May, early June, and from late July onwards. Field identification of birds of the *F. hypoleuca-albicollis-semitorquata*-complex was usually not attempted before the publication of Curio (1959), but those recorded bree-

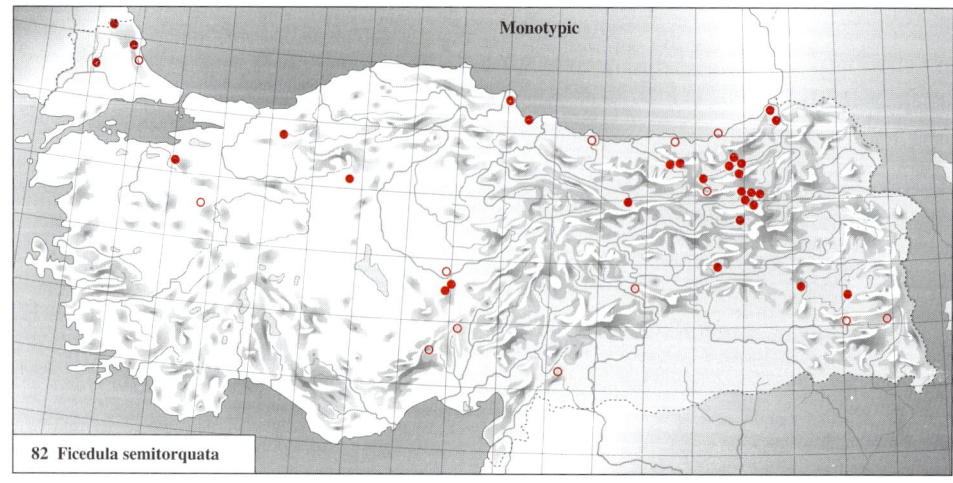
82 *Ficedula semitorquata*

ding are all presumed to be *F. semitorquata*. Records of birds in the south in late July (e.g., at Avanos, above Ürgüp, near Elazig, and on the Ala Dag) may either refer to local breeding or to early postbreeding dispersal, but the status of the birds observed by Beaudoin (1976), see above, is totally unclear.

Geographical Variation Apparently none, but the status of *transcaspica* Zarudny & Bilkevitch, 1918, described from Kopet Dag (where not certain to breed) has not been investigated. Otherwise, no subspecies have been descibed. *F. semitorquata* Von Homeyer, 1885, was originally described from the Caucasus. Wing of both sexes from the Caucasus area 80.5 (78-83) (n=17), from Iran 80.9 (75-85) (n=30), from Turkey 80.8 (77-85.5) (n=19), and from the Balkan countries and Greece 81.5 (80-82) (n=9) (Weigold 1912-13, Stresemann 1926, Paludan 1940, Schüz 1959, Érard & Etchécopar 1970, A. J. van Loon, CSR). The species has long been considered as a subspecies of Collared Flycatcher *F. collaris* from central Europe, following Stresemann (1926), but a study by Curio (1959) showed that separation as a full species was warranted. For identification, see Mild (1993, 1994a,b).

References Stresemann, E. (1926) Die Systematische Stellung von *Muscicapa semitorquata* E. von Homeyer. *Orn. Monatsber.* 34, 4-9. Curio, E. (1959) Beobachtungen am Halbringschnäpper *Ficedula semitorquata* im mazedonischen Brutgebiet. *J. Orn.* 100, 176-209. Mild, K. (1993) Die Bestimmung der europäischen schwarzweißen Fliegenschnäpper *Ficedula*. *Limicola* 7, 222-276. Mild, K (1994a) Fältbestämning av 'svartvita' flugsnappare. *Vår Fågelvärld* 53, 28-36. Mild, K (1994b) Field identification of Pied, Collared and Semi-collared Flycatchers. *Birding World* 7: 134-151, 231-240, 325-334.

PANURUS BIARMICUS
Bearded Tit   Biyikli Bastankara

Habitat Large reedbeds in marshes or along open water.

Distribution See map 83. Restricted to a few wetlands throughout the country, mainly on the Central Plateau. Presumed to breed in areas where recorded May-August. Numbers apparently show annual fluctuations. At the Amik Gölü, the endemic subspecies *kosswigi* was present up to at least 1956, but it was apparently extinct by 1962 (Kumerloeve 1963a, 1964b).

Geographical Variation

Subspecies described or recorded in the region:

(W)   **P. b. biarmicus** (Linnaeus), 1758, Holstein (northern Germany). [Head of ♂ medium blue-grey, upperparts rufous- or tawny-brown, breast and sides of belly pale vinous-pink, flanks pale tawny-brown; upperparts of ♀ tawny-brown with black spots or streaks on mantle and scapulars. Wing of adult ♂ from England and the Netherlands 60.8 (58-65) (n=49), of ♀ 59.2 (57-61) (n=45) (R. Sluys in Cramp & Perrins 1993; see also Sluys 1983). Occurs from the Baltic area and western and southern Europe east to the coast of former Yugoslavia, as well as in Albania and Greece.]

(T)   **P. b. kosswigi** Kumerloeve, 1958, Amik Gölü (southern Turkey). [Similar to *biarmicus*, but upperparts and flanks even darker rusty-brown (though mantle and scapulars of ♀ hardly streaked black); head of ♂ rather dark grey, breast and sides of belly intense vinous-pink. Wing of ♂ 61.5, 63.5, of ♀ 61, 61.5 (n=2 in both) (CSR). See also Kumerloeve (1958d). Never collected outside the area round Amik Gölü, where now extinct.]

(E)   **P. b. russicus** (C. L. Brehm), 1831, 'Russia, in winter to Hungary'. [Head of ♂ pale ash-grey, paler than in *biarmicus*, upperparts and flanks of both sexes paler tawny or greyish-buff, less rufous, mantle and scapulars of ♀ lightly streaked black; pink of breast and belly

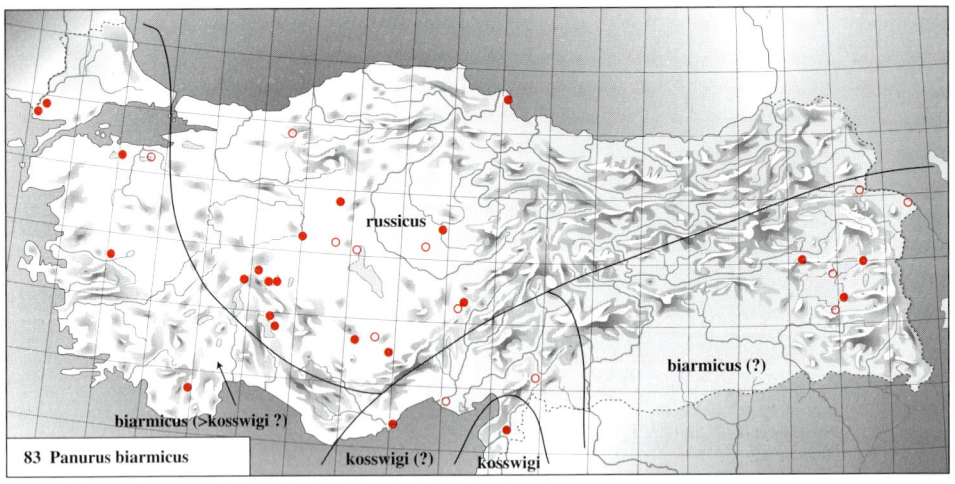

83 Panurus biarmicus

paler, mid-belly more extensively white. Wing of ♂ from Austria, Hungary, Romania, and southern European Russia 63.0 (60-65) (n=45), of ♀ 61.5 (59-64) (n=32) (R. Sluys in Cramp & Perrins 1993). Occurs from central Europe eastward, north to Belorussia, west to southern Slovakia, eastern Austria, Hungary, and northern Serbia, south to northern and eastern Bulgaria and the area north of the Caucasus.]

Subspecies recognized in Turkey: though a bird from Saloniki (northern Greece) was attributed to *kosswigi* by Spitzer (1973), those of northern Greece (apparently including the Meriç delta) are *biarmicus* according to Sluys (1983), and probably the birds of Manyas and Apolyont Gölü in north-west Anatolia are also this subspecies. The extinct birds of Amik Gölü are *kosswigi*. Birds from the Central Plateau are *russicus*. The subspecies of other Turkish populations is less clear. Transcaucasia is inhabited by a form similar to *biarmicus* but with virtually unstreaked upperparts in the female (Sluys 1983), and one may also expect this form to occur in the Van Gölü and Aras river area. Sluys (1983) includes this Transcaucasian form in *biarmicus*, but Stepanyan (1990) includes it in *russicus*. The few birds apparently breeding in the Çukurova and Goksü deltas may belong to *kosswigi*, in which case the subspecies may survive, but confirmation of breeding and of the subspecies is required. Birds collected in the Elmali area are either *biarmicus* (Jordans & Steinbacher 1948, Wolters 1968), *russicus* (Vaurie 1959, Sluys 1983), or intermediates between these (Kumerloeve 1958b, 1969f); they are rather dark (like *biarmicus*) but the female shows hardly any streaking on the upperparts. Apparently, western and southern Turkey as well as the neighbouring parts of Greece and Transcaucasia are inhabited by rather dark birds, with ground-colours similar to *biarmicus* but with the streaking of the female close to that of *russicus*; within this range, the darkest population (*kosswigi*) inhabited the area round Amik Gölü.

References Kumerloeve, H. (1958d) Eine neue Bartmeisenform vom Amik Gölü (See von Antiochia). *Bonner zool. Beitr.* 9, 194-199. Wolters, H. E. (1968) Aus der ornithologischen Sammlung des Museums Alexander Koenig. I. *Bonner zool. Beitr.* 19, 157-164. Kumerloeve, H. (1969f) Sur la sitation subspecifique des Mésanges à moustache (*Panurus biarmicus*) en Asie Mineure et ses alentours. *Aves* 6, 61. Spitzer, G. (1973) Zur Verbreitung der Formen von *Panurus biarmicus* in der Westpaläarktis. *Bonner zool. Beitr.* 24, 291-301. Sluys, R. (1983) Geographical variation and distribution of the Bearded Tit *Panurus biarmicus* (Linnaeus, 1758) (Aves). *Bijdr. Dierk.* 53, 13-32.

AEGITHALOS CAUDATUS

Long-tailed Tit   **Uzunkuyruk Bastankara**

Habitat Open coniferous, deciduous, and mixed forest and scrub, especially of oak, beech, and pine; mainly at 100-1300m, but to 2000m in the Ilgaz Daglari and to at least 1500m in the Taurus.

Distribution See map 84. Widespread and generally common in Thrace, the Black Sea Coastlands, Western Anatolia, and the Southern Coastlands, apparently local and scarce in the East, absent from the Central Plateau (except its fringes) and apparently from much of the South-East.

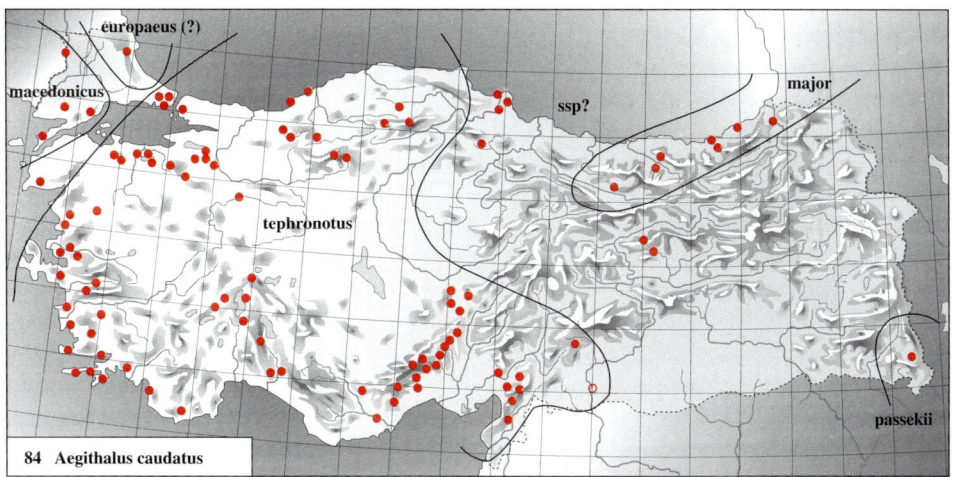

84 Aegithalus caudatus

Geographical Variation

Subspecies described or recorded in the region:

(W)   ***A. c. europaeus*** (Hermann), 1804, Basel (Switzerland). [Forehead, mid-crown, and sides of head dirty (buff-)white; each side of crown with rather narrow black stripe, not reaching bill, sometimes reduced in width or extent and head then largely white; upper mantle black; lower mantle to upper tail-coverts black, mixed vinous-pink; underparts white, sometimes with dark spots on upper breast, flanks washed pale pink, throat without dark patch; tail long. Wing of both sexes from central Europe 62.7 (56-67) (n=57), tail 88.7 (80-99) (n=37) (A. J. van Loon in Cramp & Perrins 1993). In this and other subspecies, plumages and measurements of juveniles are ignored in the subspecies accounts, as they differ considerably from adults.]

(W)   ***A. c. macedonicus*** (Dresser), 1892, Mount Olympus (Greece). [Like *europaeus*, but black of stripe at each side of crown broader and longer, reaching bill-base; sides of head narrowly and indistinctly streaked pink, grey, and white; vinous-pink on flanks slightly more extensive; occasionally, some dark grey on rump and upper tail-coverts; tail long. Wing in southern Bulgaria and northern Greece 63.2 (61-65, once 67) (n=18), tail 84.5 (78-88, once 92) (n=18) (CSR).]

(T)   ***A. c. tephronotus*** (Günther), 1865, Havancore (just east of the Bosporus, Turkey). [Forehead and mid-crown pale buff-brown, mixed with some drab-grey and off-white, bordered at each side by a broad black stripe, latter not reaching bill; upper mantle to upper tail-coverts uniform dark ash-grey, virtually without black or pink (some pink of the flank feathers is sometimes visible on rump); sides of head streaked buff and pale drab-brown; a large sooty patch on throat; white of underparts tinged buff, flanks rather extensively washed buff-pink; tail short. Wing in Turkey (Fethiye, Izmir area, Taurus mountains, Haruniye, Bolu, Ankara, Ilgaz Daglari, Kastamonu) 59.3 (56-64) (n=32), tail 69.5 (63-75, one each 59 and 77) (n=28) (A. J. van Loon, CSR; wing includes some data of Kummerlöwe & Niethammer 1934-35 and Kumerloeve 1961a).]

(E) ***A. c. major*** (Radde), 1881, Tbilisi (Georgia, western Trancaucasia). [Forehead fulvous-brown, mid-crown white with ill-defined drab-brown streaking, not sharply defined from a dark brown stripe at each side of the crown; upper mantle with a broad black band across, remainder of upperparts light ash-grey with a slight pink wash on outer scapulars; sides of head streaked drab-brown and dirty-white; throat dirty white, often with a mottled dark grey patch in the middle, remainder of underparts white with a slight cream wash, flanks rather extensively pink; tail long. Wing in the Caucasus 63.0 (59.5-65) (n=14), tail 80.9 (76-86) (n=13) (A. J. van Loon, CSR).]

(E) ***A. c. passekii*** (Zarudny), 1904, oak forest of the Zagros mountains (south-west Iran). [Like *tephronotus*, but central crown paler, buff-white, more contrasting with the black stripes at the sides of the crown; grey of upperparts paler; no black patch on throat; white of belly tinged cream; pink of flanks pale and restricted; tail short. Wing in north-west Iran (Lake Rezaiyeh and Qasr-e-Shirin to Fars) 60.2 (56-62.5) (n=51), tail 66.6 (59-72) (n=44) (Paludan 1938, Vaurie 1950, CSR).]

Subspecies recognized in Turkey: birds from south-west Thrace are *macedonicus*, but birds from the Istranca Daglari in northern Thrace appear to be inseparable from *europaeus*. As both subspecies intergrade in southern Bulgaria, the occurrence of intermediate birds in Thrace can be expected. *A. c. tephronotus* occurs just west of the Bosporus, as well as in western Asia Minor east in the north to at least the Ilgaz Daglari and in the south to the Anti-Taurus. The long-tailed subspecies *major* from the Caucasus area and western Transcaucasia ranges throughout the eastern part of the Black Sea Coastlands west to at least Sebinkarahisar. Wing of birds from Sebinkarahisar (Kurayiseba) 61.9 (61-63) (n=6), tail 79.4 (75-85) (n=5) (CSR). The birds from the isolated site south of Yüksekova in the extreme South-East are undoubtedly *passekii*. It is unknown whether *tephronotus* and *major* grade into each other in northern Turkey or whether they replace each other abruptly. The birds from western Asia Minor are sometimes called *alpinus*, but this subspecies, described from the Samamisian Alps in Gilan (northern Iran) by Hablizl in 1783, does not occur further west than eastern Transcaucasia; it differs from *tephronotus* in darker buff-brown forehead, mid-crown, and sides of head, broader and blacker stripes at sides of crown, darker slate-grey upperparts, and more uniform pinkish buff-brown underparts with a larger black throat patch.

PARUS PALUSTRIS

Marsh Tit Bataklik Bastankarasi

Habitat Open mixed and deciduous forest at 0-1000m.

Distribution See map 85. Restricted to northern Thrace, the Black Sea Coastlands, and the north of the Marmara region, with apparently isolated occurrences along the northern fringe of the Central Plateau. The species is perhaps more widespread in Western Anatolia, as it was recorded in May from Foça (Titcombe 1989) and in winter from Bafa Gölü and near Izmir (Weigold 1913-14, Kasparek 1988); a bird 'probably of this species' was encountered in July near Degirmen Bükü (Versluys 1992).

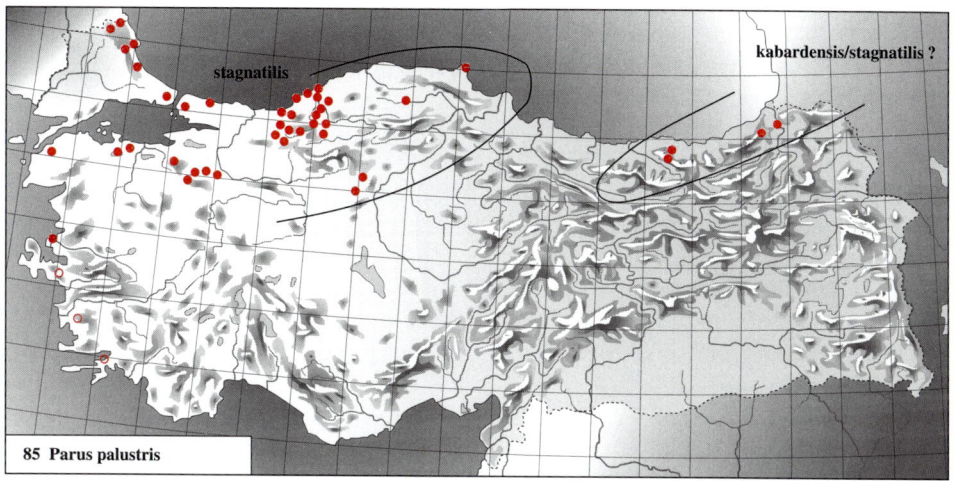

85 Parus palustris

Geographical Variation

Subspecies described or recorded in the region:

(W) ***P. p. stagnatilis*** C. L. Brehm, 1855, Galicia (south-east Poland or western Ukraine). [Upperparts pallid drab-grey with a slight buff tinge; underparts white, flanks with pale and restricted cream-grey tinge. Occurs in eastern Europe south to the plains of eastern Bulgaria; replaced by *palustris* in the mountains of Greece and the Balkan countries, and through central Europe to Scandinavia; this latter subspecies has upperparts darker drab-brown and flanks more extensively and darker pale drab-grey. Wing of ♂ of *stagnatilis* from eastern Romania and Bulgaria 66.8 (64-70) (n=23), of ♀ 64.0 (62-67) (n=10), tail of both sexes 53.2 (49-56) (n=15), bill 10.2 (9.7-10.7) (n=11) (CSR).]

(E) ***P. p. kabardensis*** (Buturlin), 1929, near Vladikavkaz (Ordzhonikidze, northern Caucasus, Russia). [None examined. According to the original description, upperparts paler and greyer than in *palustris*, underparts more extensively white; thus, apparently very similar to *stagnatilis*. Wing perhaps shorter, bill thicker, and tail relatively longer, but confirmation of these characters is required. Wing of ♂ from the Caucasus 63.6 (62-66) (n=6), of ♀ 60 (59-61) (n=4), tail 54.9 (53.5-57) (n=10) (Buturlin in Molineux 1930).]

Subspecies recognized in Turkey: according to Rössner (1935), birds from Bolu in western Black Sea Coastlands are 'similar to *kabardensis*, but bill slightly thicker'. However, these birds, as well as those from nearby Zonguldak and Devrek are inseparable from *stagnatilis* on plumage, though the bill is on average longer. The birds from north-west Turkey are provisionally included in *stagnatilis*. but direct comparison of Turkish birds with Caucasus specimens is required. Wing of ♂ from Bolu, Devrek, and Zonguldak 67.3 (65.5-69) (n=8), of ♀ 64, 66 (n=2), tail 54.4 (52-56) (n=10), bill 10.7 (10.3-11.3) (n=10) (CSR). No birds from north-east Turkey were examined.

PARUS LUGUBRIS

Sombre Tit   Mahzun Bastankara

Habitat Open deciduous or pine forest with scrubby undergrowth, scrub and small trees on stony slopes and in ravines, and (sometimes) olive groves; at 0-800m in the west of the country, at 100-1300 (-1500)m in the centre (but to 2200m in the Taurus), and at 100-1500 (-2200)m in the east.

Distribution See map 86. Widespread but only locally common, and absent from the steppe country of the Central Plateau, Thrace, and the South-East, as well as from the coastal zone of the Black Sea Coastlands and from Hatay. Rare in the mountain valleys of the South-East.

Geographical Variation

Subspecies described or recorded in the region:

(W) **P. l. lugubris** Temminck, 1820, Dalmatia and Hungary. [Cap of ♂ black-brown or dark sooty-brown, of ♀ dark greyish- or chocolate-brown, remainder of upperparts drab-brown; flanks pale cream-grey, breast and belly cream-white. Wing of both sexes from former Yugoslavia and north-west Greece 74.1 (71-78) (n=15) (CSR).]

(W) **P. l. lugens** C. L. Brehm, 1855, Attiki (Greece). [Cap more greyish-brown, less dark as in *lugubris*, upperparts paler drab-grey, breast and belly purer and more extensively white; slightly smaller, wing in mainland Greece (south from mount Olimbos) 72.3 (68-75.5) (n=29) (CSR). Includes *splendens* Gengler, 1920, described from Nevsa (eastern Bulgaria), which more or less combines the darker cap and larger size of *lugubris* with the paler body of *lugens*. Wing of *'splendens'* from eastern Romania 75.4 (71-79) (n=10) (CSR). For distribution and variation in the Balkan peninsula, see Catuneanu (1975).]

(T) **P. l. anatoliae** Hartert, 1905, 'Ahoory', Asia Minor. [Cap and throat patch of ♂ generally deep black, of ♀ black-brown; in both, cap and throat patch smaller and more sharply defined than in *lugubris* and *lugens*; upperparts darker and more olive-brown, breast and belly whiter but flanks more drab-brown or slightly rufous; size as in *lugens* or slightly smaller, bill 12.1 (11.3-12.8) (n=17), against 12.6 (11.9-13.5) (n=39) in *lugubris* and *lugens* (Roselaar in Cramp & Perrins 1993). *P. l. derjugini* (Nesterov), 1911, described from 'Gurjany', in the Çoruh valley near Artvin (eastern Black Sea Coastlands, Turkey), is an invalid synonym.]

(E) **P. l. dubius** Hellmayr, 1901, 'Palestine, Syria, and Persia'. [Cap deeper black and more sharply defined than in the previous subspecies, tail relatively shorter, bill longer and more slender; upperparts pallid cream-grey, paler than in others. Wing in western Iran 73.0 (69-76) (n=35) (Vaurie 1950, Eck 1980).]

Subspecies recognized in Turkey: birds from Thrace are *lugens*; wing of ♂ 75, 75.5 (n=2) (CSR). The remainder of Turkey is inhabited by *anatoliae*. Wing of *anatoliae* in the Ankara area 73.2 (72-74) (n=4), in Western Anatolia 71.0 (68-74) (n=17), in the western Taurus (Yesilova, Burdur, Korkuteli, Beyşehir) 72.2 (69-75.5) (n=11), in the eastern Taurus (Zebil,

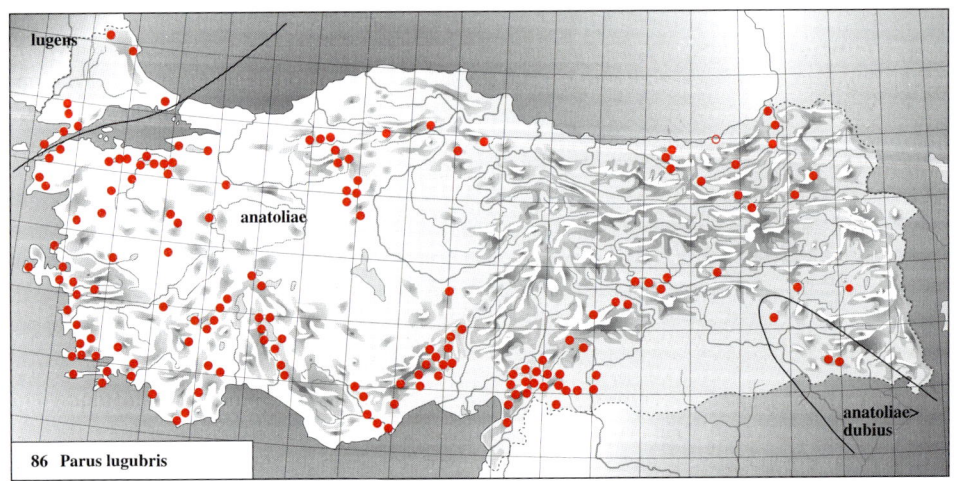

86 Parus lugubris

Anasha, Gozna, Pozanti, Hacin Dagi, Elazig) 72.6 (68.5-74.5) (n=10), in eastern Turkey and in Dihok (northern Iraq near the Turkish border) 71.7 (70-73) (n=5) (CSR, with some additional data from Weigold 1913-14, Hartert 1921-22, Kummerlöwe & Niethammer 1934-35, and Vauk 1973). Dihok birds (and probably those of the nearby Hakkari area) average slightly paler than birds from northern and western Asia Minor, tending somewhat to *dubius* from Iran, but they are still nearer to *anatoliae*.

References Catuneanu, I. (1975) [Verbreitungsgebiet und nisten von *Parus lugubris* in Rumänien.] *Stud. Comunic. stiinte nat. Muz. Brukenthal* 19, 273-295. Eck, S. (1980) Intraspezifische Evolution bei Graumeisen (Aves, Paridae: *Parus*, Subgenus *Poecile*). *Zool. Abh. Staatl. Mus. Tierkde. Dresden* 36, 135-219.

PARUS CRISTATUS
Crested Tit   Tepeli Bastankara

Habitat Near the Abant Gölü, recorded in copses of *Pinus* as well as in the transition zone between mixed and coniferous forest; in the Taurus mountains above Pozanti, recorded in coniferous forest.

Distribution No map. Recorded from 2 sites: abundant above the Abant Gölü on 6-9 July 1966 (Ganso & Spitzer 1967), and 5 birds seen in the Taurus above Pozanti on 31 August 1986 (Bijlsma & De Roder 1986). Both areas are much frequented by ornithologists, but no other records have been made, and confirmation of the occurrence is required. Similar isolated records exist in the Crimea and the Caucasus (Hartert 1903-10, Engel 1952, Mauersberger & Stephan 1967).

Geographical Variation

Subspecies described or recorded in the region:

(W)   ***P. c. bureschi*** Von Jordans, 1940, Pirin mountain above Bansko (south-west Bulgaria). [Upperparts dark olive-brown, underparts dirty white with restricted grey-buff tinge on flanks; close to *cristatus* from northern and eastern Europe, but slightly paler and greyer, though less so than in *baschkirikus* from the southern Ural mountains; also, bill longer than in *cristatus* and *baschkirikus*. Wing of *bureschi* from southern Bulgaria 66.1 (63-68) (n=7), bill to nostril 8.3 (7.8-8.8) (n=7) (CSR), against bill to nostril 7.6 (7.2-8.0) (n=29) in *cristatus* (CSR) and 7.6 (7.4-7.8) (n=3) in *baschkirikus* (Kohl 1967, CSR).]

Subspecies recognized in Turkey: none collected. If any occurs, the subspecies involved is perhaps *bureschi*, which is the one breeding nearest to Turkey.

References Engel, H. (1952) Die Verbreitung der Haubenmeise *Parus cristatus* L. *Bonner zool. Beitr.* 3, 41-74. Kohl, I. ('1965'=1967) Über die taxonomische Stellung der karpathischen Haubenmeisen, *Parus cristatus*. *Larus* 19, 158-178. Mauersberger, G., & B. Stephan (1967) *Parus cristatus*, in: E. Stresemann, L. A. Portenko, & G. Mauersberger (eds) *Atlas der Verbreitung palaearktischer Vögel* 2. Berlin.

PARUS ATER
Coal Tit   Çam Bastankarasi

Habitat Restricted to open or dense coniferous forest, sometimes mixed with oak, rarely purely deciduous (e.g., recorded in beech on the Çatal Dag in Western Anatolia), rarely in stunted junipers above the tree line. Altitudinal distribution thus closely related to that of coniferous forest: at 100-1800m in the west, at 0-2200m in the mountains of the central part of the country (e.g., Taurus and Ilgaz Daglari), and up to at least 2200m in the east.

Distribution See map 87. Common to abundant and widespread north of 40°N (but apperently very local in Thrace), in Western Anatolia, and in the Southern Coastlands, with some isolated occurrences in relict forest patches on the fringe of the northern Central Plateau (e.g., Beynam forest).

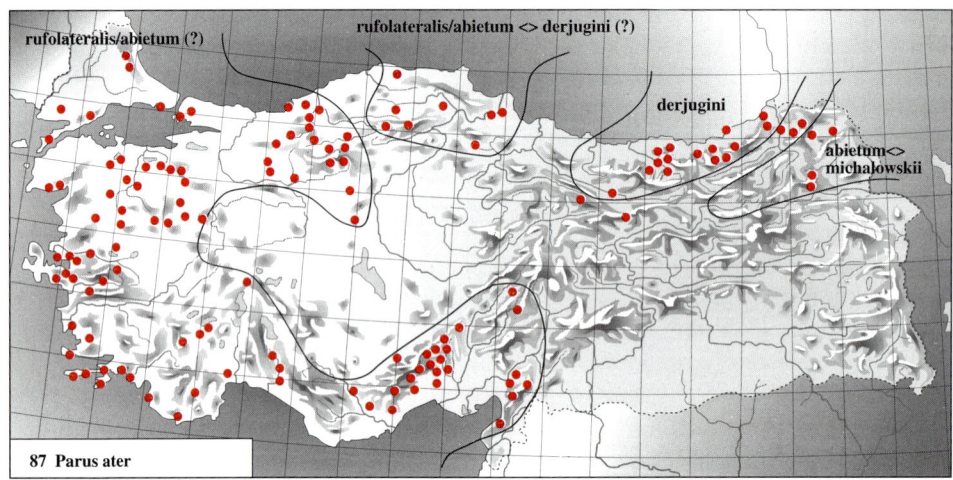

87 Parus ater

Geographical Variation

Subspecies described or recorded in the region:

(W)   ***P. a. ater*** Linnaeus, 1758, Sweden. [Upperparts medium bluish-grey, slightly washed olive when plumage fresh; flanks washed with a rather restricted amount of cream, underparts extensively white. Rather small, bill rather short and slender: wing of both sexes in Fenno-Scandinavia and western Russia 62.1 (58-66) (n=21), bill 10.5 (10.0-11.5) (n=21), bill depth at base 3.7 (3.4-4.0) (n=10) (CSR).]

(W)   ***P. a. abietum*** C. L. Brehm, 1831, the pine forests of Germany. [Upperparts darker blue-grey, more intensely washed olive than in *ater* in a comparable stage of plumage wear; flanks darker and more extensively greyish-buff. Slightly larger, wing in Central Europe 62.8 (58-67) (n=117), bill 11.1 (10.5-12.1) (n=64), bill depth 4.0 (3.6-4.2) (n=10); wing in former Yugoslavia, Romania, and mainland Greece 64.3 (61-67) (n=10), bill 11.6 (11.1-12.1) (n=10), bill depth 3.9 (3.8-4.3) (n=6) (CSR). Breeds in central and south-east Europe from the Netherlands to Romania, Bulgaria, and mainland Greece. Often included in *ater*, but distinguishable when series of specimens in a similar stage of wear are compared (Wolters 1968, CSR).]

(T)   ***P. a. derjugini*** (Zarudny & Loudon), 1903, Çoruh (Artvin, eastern Black Sea Coastlands, Turkey). [Upperparts rather dark bluish-grey with an extensive dark olive-brown or rufous-brown tinge, the latter largely concealing the grey unless the plumage is heavily worn; white on underparts rather restricted, flanks and lower belly extensively dark rufous-drab or dark tawny-grey. Rather large: wing in the eastern Black Sea Coastlands of Turkey (Sebinkarahisar, Cayeli, Borçka) 65.5 (62-68) (n=8), bill 11.6 (11.2-12.5) (n=8), bill depth 4.1 (3.8-4.4) (n=8) (CSR). Breeds in the coastal zone along the eastern Black Sea, from north-east Turkey to the southern slope of the western Caucasus. According to Vaurie (1957), the bill of birds from north-east Turkey is long, 13.2-14.0 (n=2), but this is not apparent in the series from near the type locality mentioned above; in size, *derjugini* is hardly different from *abietum* from the Balkan countries, but the colour difference is marked. Wing of ♂ from the western Caucasus area 68 (65-70) (Stepanyan 1990).]

(E)   ***P. a. michalowskii*** Bogdanov, 1879, Kirshalevi, Surami Pass (western Transcaucasia, Georgia). [Upperparts drab-grey or dark olive-grey, less dark and brown as in *derjugini*; belly extensively white, flanks with a rather restricted amount of cream-ochre. Large, bill rather thick: wing in the Caucasus and Transcaucasia 69.1 (67-72.5) (n=8), bill 11.7 (11.0-12.8) (n=8), bill depth 4.3 (4.2-4.5) (n=8) (CSR).]

Subspecies recognized in Turkey: birds examined from north-west Asia Minor (east to Beycuma and Gerede), Western Anatolia, and the Taurus east to Haruniye in the Anti-Taurus all have upperparts bluish-grey with a slight brown tinge, like *abietum*, less brown than in *derjugini* and *michalowskii*. However, the flanks of these birds are extensively buff-brown or rufous-buff, not as cream to pale drab-grey as in *ater* or as grey-buff or buffish-drab as in *abietum*. Wing of these birds 64.3 (61-68.5) (n=22), bill 11.8 (11.0-12.5) (n=14), bill depth 3.8 ((3.6-4.2) (n=14) (CSR). Perhaps the birds from western Asia Minor form a separable subspecies, for which the name *rufolateralis* Keve may be available (type specimen: a ♂ in Wien, collected Bolu Dagh on 24 August 1934 by Rössner, examined). Birds from the north-east are *derjugini* (see above). No birds from the Ilgaz Daglari in the Black Sea Coastlands, between the ranges of *abietum/rufolateralis* and *derjugini*, were examined; according to Kummerlöwe & Niethammer (1934-35), these birds show a thicker bill than the birds from further west and have a wing of 64.6 (61-67) (n=5). Possibly, the Ilgaz birds are intermediates between *abietum/rufolateralis* and *derjugini*, or are true *derjugini*. Two birds collected in Sarikamis in the north-east, only a little inland of the range of *derjugini*, are distinctly paler than *derjugini*, near to *abietum/rufolateralis* or *michalowskii*; the bill is as in *abietum/rufolateralis*, but the wing is as long as in *michalowskii*: wing 68, 70.5 (n=2), bill 11.5, 12.4 (n=2), bill depth 3.8, 4.1 (n=2) (CSR). Apparently, *abietum/rufolateralis* grades into *michalowskii* in inland eastern Turkey, outside the zone influenced by the Black Sea. For measurements in Beynam forest (Central Plateau) of this and some other species of *Parus*, see Kiziroglu (1983).

References Kiziroglu, I. (1983) Biometrische Untersuchungen an vier Meisen-Arten (*Parus* spp.) in der Umgebung von Ankara. *Bonner zool. Beitr.* 34, 453-458.

PARUS CAERULEUS
Blue Tit   Mavi Bastankara

Habitat Open mature deciduous and mixed forest, riverine woodland and scrub, plantations, and orchards; at 0-750m in Thrace and Western Anatolia, at 100-1500m in the Black Sea Coastlands, but at 500-1700m in the Southern Coastlands and to 2000m in the East and South-East.

Distribution See map 88. Widespread but only locally common, absent or surprisingly rare in many areas which appear to have suitable habitat. Largely absent from the Central Plateau, from the Çukurova area, and from the East and South-East.

Geographical Variation

Subspecies described or recorded in the region:

(W)   ***P. c. caeruleus*** Linnaeus, 1758, Sweden. [Upperparts rather dark grey-green or green-grey (colour depending on abrasion); underparts greenish-yellow; rather large. Wing of ♂ from northern and central Europe 67.5 (65-71) (n=129), of ♀ 65.0 (62-67) (n=90) (Roselaar in Cramp & Perrins 1993); wing of ♂ from the south of former Yugoslavia, Bulgaria, Romania, and northern mainland Greece 67.9 (65-71) (n=32), of ♀ 64.9 (64-67) (n=15) (Stresemann 1920, Makatsch 1950, CSR).]

(W)   ***P. c. calamensis*** Parrot, 1908, Calamata (Kalámai, Pelopónnisos, Greece). [Upperparts

Songbirds of Turkey

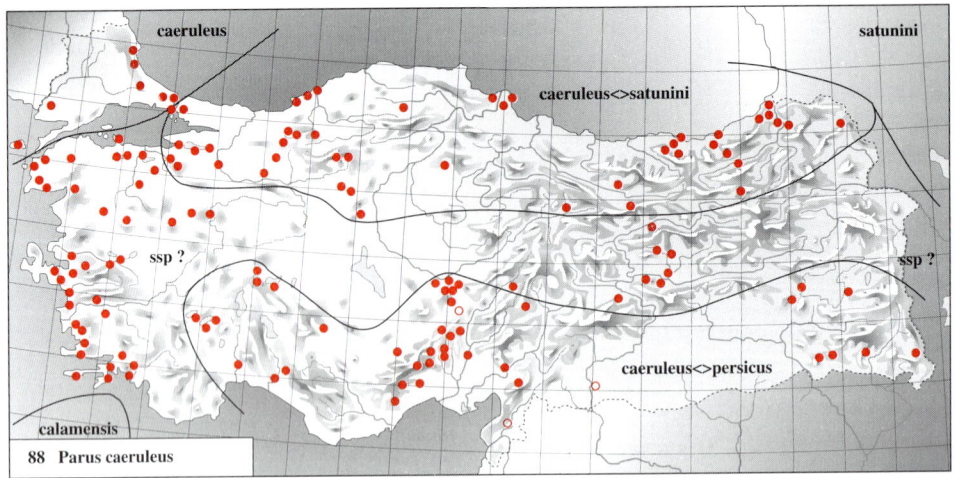

88 Parus caeruleus

slightly darker grey-green than in *caeruleus*, underparts slightly deeper yellow with faint grey-green tinge; smaller. Wing of ♂ from western and southern Greece (Kerkira, Peloponnisos, Crete, Rodhos) 63.2 (59-67) (n=38), of ♀ 61.3 (58-64.5) (n=9) (Stresemann 1920, Niethammer 1943, Vaurie 1957, CSR); wing of ♂ mainly 65 and less, of *caeruleus* mainly 66 and over.]

(E) **P. c. satunini** (Zarudny), 1908, Kumbashinsk (eastern Transcaucasia: south-east Azerbaijan). [Upperparts paler and greyer than in *caeruleus*, less green; underparts pure and rather bright pale yellow; rather large. Wing of ♂ from the Caucasus area and Iranian Azarbaijan 67.0 (64-70) (n=11), of ♀ 65.5 (63-68) (n=12) (Vaurie 1950, CSR).]

(E) **P. c. persicus** Blanford, 1873, Shiraz (Fars, south-west Iran). [Upperparts pallid bluish-grey, barely green, underparts pallid yellow with faint green tinge; rather small, bill slender. Wing of ♂ from Hamadan and Fars (south-west Iran) 65.8 (62-68) (n=37), of ♀ 63.6 (62-65) (n=12) (Paludan 1938, Vaurie 1950, CSR). Colour gradually darker from south-west to north-west Iran; in the north-west, *persicus* clinally merges into *satunini* (Vaurie 1950).]

Subspecies recognized in Turkey: difficult to establish due to the clinal character of the variation and the strong influence of bleaching and wear on the plumage; moreover, some of the birds examined were in juvenile plumage, for which racial characters as given above are not valid. Birds from Thrace are *caeruleus*; wing of birds of both sexes from here 65.2 (63.5-67) (n=3) (Stresemann 1920, Rokitansky & Schifter 1971). Birds examined from the Ulu Dag and the Black Sea Coastlands (from Zonguldak east to Rize) are best considered to be intermediate between *caeruleus* and *satunini*; wing of ♂ from this area 67.0 (64-69.5) (n=11), of ♀ 65.3 (63-67) (n=5) (Kummerlöwe & Niethammer 1934-35, CSR). Clearly paler birds with paler grey upperparts and paler yellow underparts occur from the Taurus mountains eastward to northern Iraq (and probably in the neighbouring Hakkari and Van areas of Turkey); though rather close to Caucasian *satunini*, they are slightly paler than in that subspecies and are probably best referred to as intermediates between *caeruleus* and *persicus*; wing of ♂ from this area (Burdur, Eregli, Belen pass, and Elazig) 67.7 (65-69) (n=5), of ♀ 63.5 (n=1) (CSR). Apparently, no birds have been collected in western Anatolia south of the Ulu Dag or in the easternmost Black Sea Coastlands, and the subspecies in these regions is not known. Either *caeruleus*, *calamensis*, or intermediates between these occur in Western Anatolia, either *satunini* or intermediates between *satunini* and *caeruleus* occur in the extreme north-east.

PARUS MAJOR

Great Tit   **Büyük Bastankara**

Habitat All kinds of forest, plantations, orchards, gardens, parkland, scrubland, cemeteries, villages with trees, and (locally) in parks within cities. Occurs from sea-level to the upper limit of trees, but mainly in valleys and on lower slopes, up to 1200m above Eber Gölü, to 1700m in the Ilgaz Daglari, to 1900m on the Ulu Dag and in the Taurus, and to 2200m in the east.

Distribution See map 89. Generally common to numerous, but scarce on the Central Plateau and in the mountain valleys of the East and absent from the treeless steppe area of the Central Plateau and the South-East.

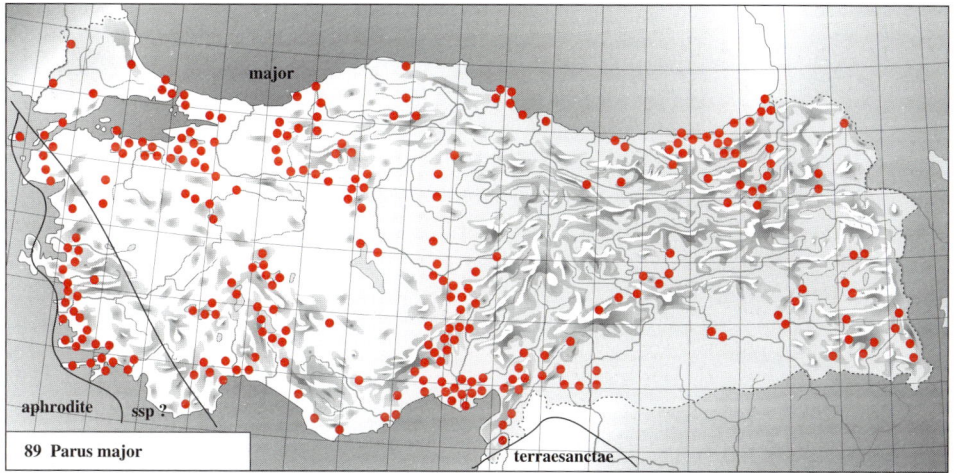

89 Parus major

Geographical Variation

Subspecies described or recorded in the region:

(W)   **P. m. major** Linnaeus, 1758, Sweden. [Mantle and scapulars green or olive-green; underparts bright lemon-yellow, flanks tinged grey-green; large. Wing of ♂ from central Europe 77.1 (73-81) (n=156), of ♀ 73.8 (70-78) (n=117) (Roselaar in Cramp & Perrins 1993); depth of bill at base 4.7 (4.5-5.1) (n=10) (CSR). Breeds in central Europe south to the north-east of former Yugoslavia and to Romania.]

(W)   **P. m. sulfureus** Kollibay, 1904, southern Dalmatia (Croatia). [Like *major*, but the underparts are purer light yellow; large. Wing of ♂ from Makedonija (in the south of former Yugoslavia) 76.2 (73-79) (n=42), of ♀ 73.4 (71-76) (n=25) (Stresemann 1920). Birds with these characteristics inhabit northern Italy, the west and south of former Yugoslavia, Albania, Bulgaria, and northern Greece.]

(W)   **P. m. aphrodite** Madarász, 1901, Larnaca (Cyprus). [Mantle and scapulars darker and more olive-grey than *major*, underparts somewhat variable, but mainly light yellow; small. Inhabits southern Italy, Sicily, Cyprus, and southern Greece, including most Aegean islands north to Lesbos and Samothraki, but not Crete where replaced by the cream-bellied *niethammeri*. Wing of ♂ from southern Greece and Cyprus 72.2 (70-78) (n=78), of ♀ 71.1 (68-77) (n=36) (Parrot 1905, Stresemann 1920, Delacour & Vaurie 1950, Jordans 1970, CSR).]

(S)   **P. m. terraesanctae** Hartert, 1910, Jerusalem (Israel). [Mantle and scapulars slightly paler

(E) **P. m. caucasicus** Domaniewski, 1933, Lagodekhi (eastern Georgia, Transcaucasia). [Mantle and scapulars duller grey-olive than *major*, less green; underparts paler yellow; large. Wing of ♂ from Iranian Azarbaijan 77.5 (74-81) (n=24), of ♀ 74.1 (72-76) (n=10) (Delacour & Vaurie 1950, Vaurie 1950).]

and more grey-green than *major*; underparts paler and purer yellow; small. Wing of ♂ from Syria and Israel 71.9 (69-76) (n=12), of ♀ 68.3 (67-70) (n=3) (Delacour & Vaurie 1950, CSR); bill rather slender, depth at base 4.5 (4.3-4.6) (n=6) (CSR).]

(E) **P. m. blanfordi** Prázak, 1894, Tehran (northern Iran). [Like *terraesanctae*, but even paler; size large, bill thicker than in other subspecies. Wing of ♂ from south-west and north-east Iran 76.3 (73-81) (n=85), of ♀ 72.8 (72-74) (n=3) (Paludan 1938, Vaurie 1950, CSR); bill depth at base 4.9 (4.7-5.2) (n=4) (CSR).]

Subspecies recognized in Turkey: hard to assess due to strong influence of bleaching and wear on plumage. In spring and summer, the upperparts of all subspecies become greyer, less green, and the underparts become paler; some of the subspecies mentioned above are apparently based on such worn birds and are thus invalid, as fresh birds are inseparable from *major*. Still, some variation in intensity of green of upperparts and yellow of underparts occurs, even in fresh plumage, and the small size of *aphrodite* and *terraesanctae* is a valid character. Most Turkish birds examined are inseparable from *major* of central Europe in colour and size; they do also not differ from the few Caucasus birds examined, and thus *caucasicus* seems to be a synonym of *major*. Wing of ♂ from Thrace 80 (n=1); of ♂ from the western Black Sea Coastlands (Bolu, Ankara, Devrek, Beycuma, Zonguldak) 76.6 (74.5-79) (n=8), of ♀ 72.8 (72-73.5) (n=3); of ♂ from the eastern Black Sea Coastlands (Sebinkarahisar, Rize) 78, 78 (n=2); of ♂ from the South-East (Tatvan) 79.5 (n=1); of ♂ from north-west Anatolia (Güngörmez, Manyas Gölü, Sogukpinar) 75, 79.5 (n=2), of ♀ 74.5, 75 (n=2); of ♂ from the western Taurus (Burdur) 76.6 (75.5-78) (n=7), of ♀ 71.5, 73.5 (n=2); of ♂ from the eastern Taurus and Amanus area (Pozanti, Osmaniye, Haruniye, Antakya) 78.1 (77-79) (n=4), of ♀ 73, 75.5 (n=2) (CSR, with some additional data of Kummerlöwe & Niethammer 1934-35; for the Ankara area, see also Kiziroglu 1983); bill depth at base for all adult birds from Turkey 4.6 (4.4-4.8) (n=10) (CSR). As *aphrodite* occurs on the Aegean islands off the coast of western Anatolia, this small subspecies can be suspected to occur at least in south-west Anatolia, from where no specimens were examined. The occurrence of this race in the south-west is supported by the small measurements of a single juvenile ♂ collected at Kasaba (near Kas, on the south coast) by Rokitansky & Schifter (1971), with wing 70mm. Neither *terraesanctae* nor *blanfordi* appear to occur in Turkey. However, 2 of the 5 birds examined from the eastern Taurus-Amanus area are almost as pale on the underparts as *terraesanctae*, though not as small, thus some influence of *terraesanctae* apparently exists.

References Parrot, C. (1905) Eine Reise nach Griechenland und ihre ornithologische Ergebnisse. *J. Orn.* 53, 515-556, 618-669. Delacour, J., & C. Vaurie (1950) Les Mésanges Charbonnières (Révision de l'espèce *Parus major*). *Oiseau* 20, 81-121. Jordans, A. von (1970) Die westpalaearktischen Rassen des Formenkreises *Parus major* (Aves, Paridae). *Zool. Abh. Staatl. Mus. Tierkde. Dresden* 31, 205-225.

SITTA KRUEPERI
Krüper's Nuthatch   Anadolu Sivacisi

Habitat Largely confined to stands of *Pinus brutia*, whether in dense forest or in parkland, but also frequent in *Cedrus*, *Juniperus excelsa*, *Pinus nigra*, *Abies cilicica*, and other conifers, sometimes mixed with deciduous trees such as oak *Quercus*. Mainly at 500-1000m, but locally down to 100m in the west and south and up to 1700m on the Ulu Dag and at 500-1800 (-2500) in the Taurus and the Black Sea Coastlands. See Frankis (1991) for association with *Pinus brutia*.

Distribution See map 90; also Kumerloeve (1958e). Largely confined to areas of *Pinus brutia* (see above), in the inland zone of the Black Sea Coastlands as well as in entire Western Anatolia and the Southern Coastlands.

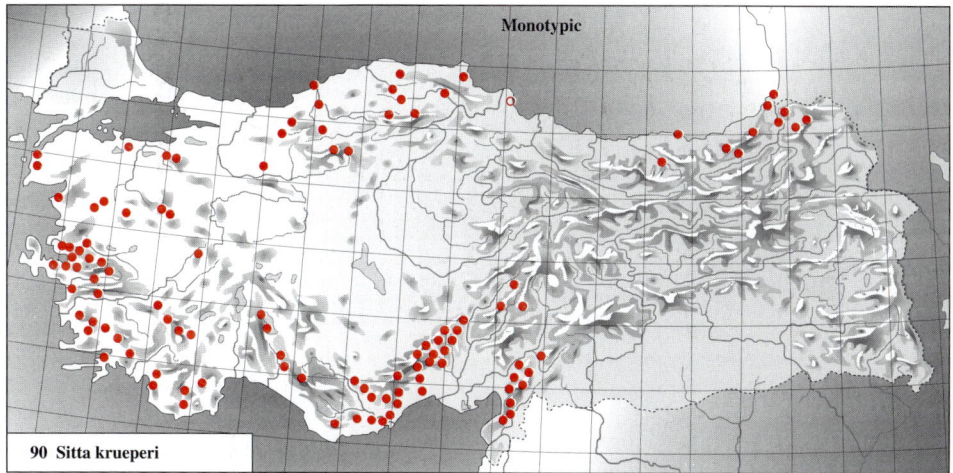

90 Sitta krueperi

Geographical Variation None. *S. krueperi* Pelzeln, 1863, was originally described from Izmir (Western Anatolia, Turkey). The species has a restricted distribution in Turkey, western Transcaucasia, and the western Caucasus; it also occurs on Lesbos (Greece). See Neufeldt & Wunderlich (1984) for details of the distribution. Wing of ♂ from Western Anatolia (mainly Izmir, including the type specimens) 74.5 (72-78) (n=21), of ♀ 72.0 (69-75.5) (n=8), bill of adult of both sexes 18.9 (17.9-20.7) (n=24); wing of ♂ from the Taurus 74.3 (72-76) (n=8), of ♀ 72.1 (69-73.5) (n=6), bill 19.0 (18.0-19.7) (n=12); wing of ♂ from the western Black Sea Coastlands 75, 77.5 (n=2), of ♀ 75.2 (74-76.5) (n=3), bill 18.9 (17.5-20.4) (n=4), wing of ♂ from the western Caucasus 78.5, bill 18.7 (CSR, with some additional wing data from Kummerlöwe & Niethammer 1934-35).

References Kumerloeve, H. (1958e) Sur la présence en Asie Mineure de la Sitelle naine de Krüper (*Sitta canadensis krüperi* Pelzeln). *Alauda* 26, 81-85. Neufeldt, I. A., & K. Wunderlich (1984) *Sitta krueperi*, in H. Dathe & I. A. Neufeldt (eds) Atlas der Verbreitung palaearktischer Vögel 12. Berlin. Frankis, M. P. (1991) Krüper's Nuthatch and Turkish pine *Pinus brutia*: an evolving association? *Sandgrouse* 13, 92-97.

SITTA EUROPAEA
Eurasian Nuthatch   Sivaci

Habitat Mainly deciduous and mixed forest at mid-levels, locally in pines higher up and in oaks and beech lower down, or in gardens and orchards; generally lower than *S. krueperi*. Mainly at 800-1800m in the Ilgaz Daglari, at 300-900m in the Taurus and the mountains of Western Anatolia, and below *c.* 1800m in the East and South-East

Distribution See map 91. Local but widespread in Thrace, the Black Sea Coastlands, the north of Western Anatolia, and the Taurus, scarcer in the remainder of Western Anatolia and in the Southern Coastlands, very local and restricted to a few lower valleys in the East and South-East.

Geographical Variation

Subspecies described or recorded in the region:

(W) **S. e. europaea** Linnaeus, 1758, Sweden. [Underparts largely cream or white. Large, wing of ♂ from Scandinavia 88.6 (86-93) (n=16), of ♀ 86.4 (84-89) (n=17), adult bill (both sexes) 21.4 (19.6-23.0) (n=32) (CSR).]

(W) **S. e. caesia** Wolf, 1810, Nürnberg (Bayern, Germany). [Underparts buffish-, ochre-, or orange-cinnamon. Fairly large, wing of ♂ from central Europe 87.2 (84-90) (n=56), of ♀ 84.6 (81-89) (n=47), bill 21.0 (19.2-22.7) (n=97) (Roselaar in Cramp & Perrins 1993); wing of ♂ from Makedonija (in the south of former Yugoslavia) 86.6 (85-89) (n=10), of ♀ 84.4 (82-87) (n=9) (Stresemann 1920). Intermediates between *europaea* and *caesia*, with variable underpart colour, occur in eastern Bulgaria (Jordans 1940, Rokitansky & Schifter 1971), as well as further north; paler variants among these have been separated as *homeyeri* Hartert, 1892, described from former East Prussia, darker birds as *sordida* Reichenow, 1907, described from eastern Germany. According to Voous and Van Marle (1953), the Bulgarian birds are intermediates between *caesia* and *levantina* and should be separated as *harrisoni*.]

(T) **S. e. levantina** Hartert, 1905, Taurus mountains (Turkey). [Underparts pale, pink-cinnamon when plumage fresh, pink-buff when worn, almost white on chin and cheeks; upperparts paler grey than in both previous subspecies, forehead and supercilium whitish, less grey; bill slightly shorter than in the previous subspecies, more slender, culmen almost straight. Wing of ♂ from the Taurus area (Elmali and Burdur eastward) 86.1 (84-89) (n=12), of ♀ 84.5 (83-86) (n=5), adult bill 19.2 (18.3-19.8) (n=10) (CSR); wing of ♂ from Western Anatolia (Aydin, Tire, and Bafa) 86.3 (85-88) (n=3), of ♀ 81 (n=1), bill 19.6 (18.9-21.3) (n=4) (CSR).]

(E) **S. e. caucasica** Reichenow, 1901, Nalchik (northern Caucasus, Russia). [Underparts warm tawny-cinnamon (brighter and deeper than in *caesia*); grey of upperparts dark, as *caesia*, but stripe along forehead often whitish; size rather small; bill short, but thick at base and blunt at tip, culmen curved. Wing of ♂ from the Caucasus area 85.5 (82-88)

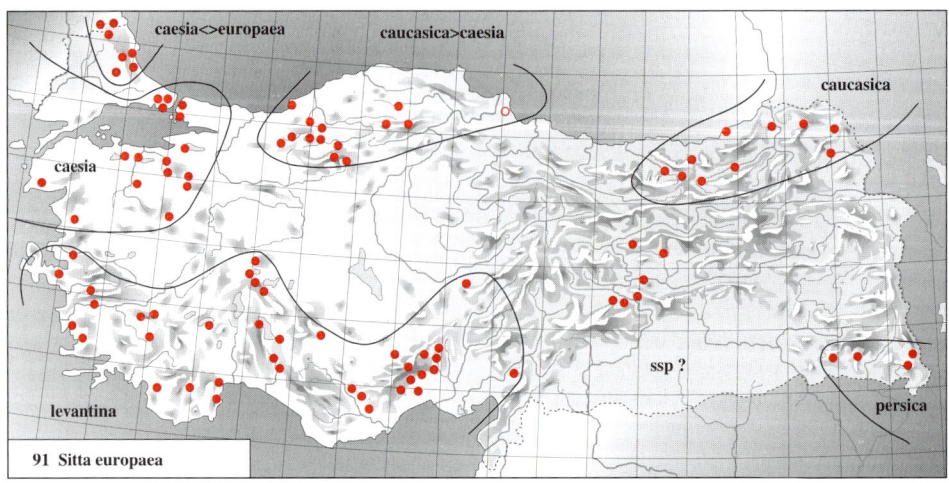

91 Sitta europaea

(n=22), of ♀ 84.2 (81.5-87.5) (n=7), bill 19.2 (18-21) (n=29) (Dunajewski 1934, CSR).]

(E) **S. e. persica** Witherby, 1903, 'Oakwoods of south-west Persia' (Zagros mountains, south-west Iran). [Underparts cream with yellow-ochre tinge, paler than *levantina*, chin and cheeks white; upperparts paler ashy blue-grey than in other subspecies, forehead and supercilium distinctly white; size small; bill slender, as *levantina*, but tip slightly more attenuated, culmen straight. Wing of ♂ from the Zagros mountains (south-west Iran) 84.5 (82-87) (n=21), of ♀ 82.1 (79-85) (n=6), bill 20.4 (19.5-21.5) (n=25) (Vaurie 1950, CSR). Intermediates between *levantina* and *persica* have been named *davidi* by Harrison (1955), described from the Amadiyah area in northern Iraq, close to the Turkish border; wing in northern Iraq 81, 83 (n=2), bill 18.5 (n=1) (Harrison 1955, Ctyroky 1987).]

Subspecies recognized in Turkey: 2 birds from the Istranca Daglari in Thrace are pale, similar to *europaea*, but they are on average smaller, wing of ♂ 84, 87 (n=2) (Rokitansky & Schifter 1971); they agree with *'homeyeri'* or *'harrisoni'*, but as these variable intermediates are preferably not recognized, they are best considered to be *europaea-caesia* or *caesia-levantina* intergrades. Birds from the Bosporus area and the Ulu Dag (Sogukpinar), though perhaps intergrades between *caucasica* and *levantina*, are inseparable from *caesia* (wing of ♂ 86; see also Grant 1975). Birds from the Black Sea Coastlands are *caucasica*, showing bright tawny-cinnamon underparts, though birds from the western half of this area verge somewhat towards *caesia* (or *levantina*) by showing a longer and more slender bill; wing of ♂ from the Black Sea Coastlands (Bolu, Devrek, Ilgaz Daglari, Kastamonu, Cayeli) 86.2 (84-88) (n=7), of ♀ 82, 83 (n=2), bill of adults of both sexes 20.0 (19.5-20.7) (n=5) (Kummerlöwe & Niethammer 1934-35, CSR). All birds from south-west Anatolia east to eastern Taurus are *levantina*. It is not known what race inhabits Western Anatolia between the Ulu Dag and Izmir, and it is not known what race breeds in the area near Malatya, Elazig, and the Munzur Daglari. The isolated birds occurring isolated in the extreme south-east probably resemble *'davidi'*, an apparent intermediate between *levantina* and *persica*. As *'davidi'* appears to be geographically isolated from *levantina*, but is close to the range of *persica* in Iran, it is probably best included in *persica*, following Vaurie (1959).

References Dunajewski, A. (1934) Die eurasiatischen Formen der Gattung *Sitta*. *Act. Orn. Mus. Zool. Polonici* 1, 181-251. Voous, K. H., & J. G. van Marle (1953) The distributional history of the Nuthatch (*Sitta europaea* L.). *Ardea* 41, extra nr., 1-68. Harrison, J. M. (1955) On the occurrence of the Nuthatch, *Sitta europaea* Linnaeus, in Iraq. *Sitta europaea davidi* ssp. nov. *Bull. Brit. Orn. Club* 75, 59-60. Grant, P. R. (1975) The classical case of character displacement. *Evol. Biol.* 8, 237-337.

SITTA TEPHRONOTA
Eastern Rock Nuthatch   Büyük Kaya Sivacisi

Habitat Recorded in the same habitat as *S. neumayer* (e.g., Beers 1982), but outside Turkey sometimes inhabits stony slopes with scattered trees, far from cliffs, crags, or boulders. Frequently feeds in shrubs and trees, and sometimes even breeds in trees, and thus not as strictly confined to stones and crags as *S. neumayer*. See Sarudny & Härms (1923).

Distribution See map 92. Restricted to the lower mountains along the northern fringe of the steppe country of the South-East, from west of Gaziantep east to the Zap and Nehil valleys and north to the Tunceli area; also, the Upper Aras region.

Geographical Variation
Subspecies described or recorded in the region:

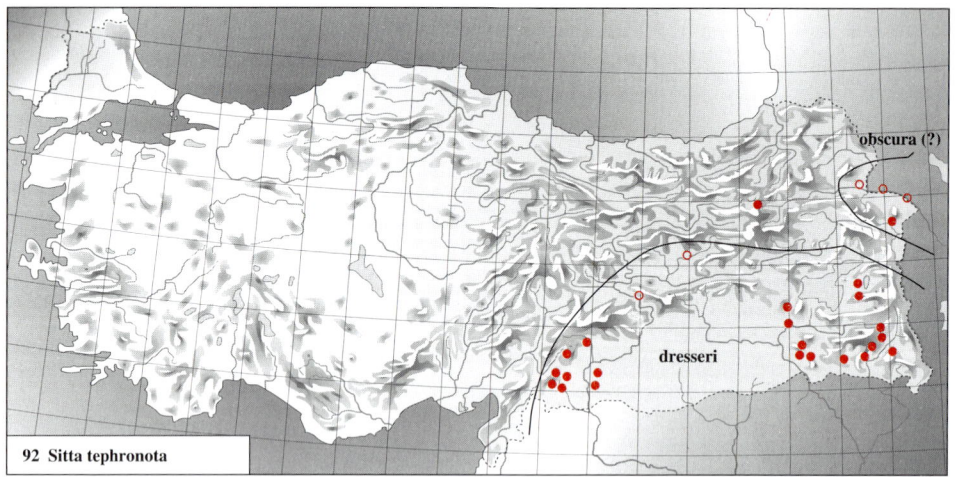

92 Sitta tephronota

- (S) **S. t. kurdestanica** Ticehurst, 1923, Tang-i-Dorq near Dihok (northern Iraq). [Larger than *dresseri* (see below), flanks and vent deeper rufous. Wing of birds of both sexes from Dihok (northern Iraq) and Mahabad (Saujbulagh, near Lake Rezaiyeh, north-west Iran) 94.2 (90-96.5) (n=10), bill 29.0 (27.2-30.7) (n=9) (Vaurie 1950, CSR).]
- (E) **S. t. dresseri** Zarudny & Buturlin, 1906, Shiraz (south-west Iran). [Upperparts light or pale grey; throat and breast pale cream, contrasting with the rufous of the remainder of the underparts. Large, but slightly less so than in *kurdestanica*. Wing in south-west Iran (Bisotun to Shiraz) 92.8 (90-98.5) (n=13), bill 27.9 (26.5-29.5) (n=12) (CSR).]
- (E) **S. t. obscura** Zarudny & Loudon, 1905, Elburz mountains (northern Iran). [Upperparts darker than in *dresseri*, dull medium bluish-grey; underparts either paler or darker rufous; smaller. Wing in birds from Nakhichevan (Transcaucasia) and the Elburz mountains (northern Iran) 88.8 (83-91.5) (n=16), bill 26.0 (23.6-28.4) (n=16) (Worobiev 1934, Vaurie 1950, Grant 1975, CSR).

Subspecies recognized in Turkey: only a single old skin examined, labelled 'Syria', collected at a time when 'Syria' (where not known to occur now) included the Gaziantep area of present-day South-East Turkey. This bird, with wing 96 and bill 28.2, is inseparable from *kurdestanica*, a subspecies described from northern Iraq, close to the Hakkari area of Turkey. However, as colours of underparts, though usually considered to have diagnostic value in identifying subspecies, appear to be largely dependant on abrasion and bleaching and are thus seasonal, and as size differences are not marked either, *kurdestanica* is best included in *dresseri*. Therefore, the birds inhabiting South-East Turkey, from the Gaziantep area east to the Van and Hakkari areas, are all *dresseri*. The darker and smaller subspecies *obscura* is known to inhabit the Aras valley in Nakhichevan and Armenia, and therefore this is probably the race recorded higher up in the Aras valley in Turkey.

References Sarudny, N., & M. Härms (1923) Bemerkungen über einige Vögel Persiens III, Gattung Sitta. *J. Orn.* 71, 398-418.

SITTA NEUMAYER
Western Rock Nuthatch   Kaya Sivacisi

Habitat Bare rocky surfaces of cliffs and steep slopes, strewn with boulders and stones, tolerating scattered growth of scrub or trees. Favours limestone. Mainly at 0-1500m in the west

Nuthatches

and centre of the country and to 2500m in the east, but locally to 2700 in the Taurus and to *c.* 3000m on the Nemrut and Süphan Dagi in the Van region.

Distribution See map 93. Widespread and often common over much of Turkey, but absent from Thrace and the coastal zone of the Black Sea Coastlands, and scarce and local in the South-East.

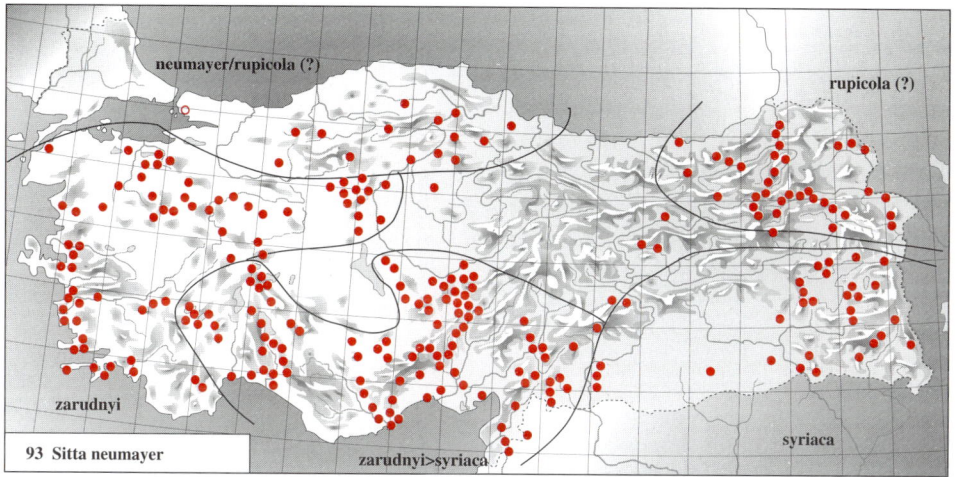

93 Sitta neumayer

Geographical Variation

Subspecies described or recorded in the region:

(W)   **S. n. neumayer** Michahellis, 1830, Dubrovnik (Croatia). [Upperparts light grey, flanks and belly pale rufous-cinnnamon or pink-cinnamon; rather large. Wing of birds (both sexes) from former Yugoslavia and mainland Greece 80.4 (75-85, mainly 78 and over) (n=34), adult bill 24.2 (23.0-26.0) (n=28) (CSR). Occurs from the Balkan countries to mainland Greece and the Ionian islands; the colour of birds from mainland Greece is paler than the colour of birds from further north; the colour of birds from southern Greece is close to that of *zarudnyi* (see below), but the size of the birds is larger.]

(T)   **S. n. zarudnyi** Buturlin, 1908, 'Aydin to the Taurus', here restricted to Aydin (Western Anatolia, Turkey). [A small pale race; upperparts light grey, as in *neumayer* or slightly paler; underparts paler, more extensively white, flanks and belly pink-buff. Wing of birds from Ankara, the Bolu area, and Western Anatolia 77.7 (75-80, exceptionally 82, generally 79 or less) (n=42), adult bill 23.1 (20.9-25.0) (n=19) (Weigold 1913-14, Kummerlöwe & Niethammer 1934-35, CSR).]

(S)   **S. n. syriaca** Temminck, 1835, Lebanon. [Underparts pale, as in *zarudnyi* or even paler, upperparts paler ash-grey; larger, wing usually over 79. Wing of birds from Lebanon and Syria 80.5 (79-85) (n=8), bill 23.8 (23.0-24.5) (n=8) (CSR).]

(E)   **S. n. rupicola** Blanford, 1873, the Elburz mountains north of Tehran (Iran). [Like *neumayer*, but upperparts slightly paler and less bluish; darker above and below than *zarudnyi* and *syriaca*; bill finer, tip laterally compressed. Wing in north-west Iran and the Elburz mountains 80.3 (78-82) (n=6), bill 22.8 (22.3-23.5) (n=6) (CSR).]

Subspecies recognized in Turkey: birds examined from Western Anatolia and the north-western fringe of the Central Plateau east to Elmali and Ankara are *zarudnyi*, those seen from the Van and Hakkari areas as well as from Dihok (just across the Turkish border in northern

Iraq) are *syriaca*. Wing of the latter birds is 80.5 (78-83.5) (n=3), bill 23.6 (22.8-24.4) (n=3) (CSR). Birds from the Taurus mountains are intermediate between *zarudnyi* and *syriaca* in size and colour, though closer to *zarudnyi*. Wing of Taurus birds (Burdur to Kilis near Gaziantep) 78.1 (74-81.5) (n=14), bill 23.8 (23.0-24.5) (n=9) (Kumerloeve 1961a, Vauk 1973, CSR). The subspecies of the birds inhabiting the interior of the Black Sea Coastlands and eastern Turkey north from the Aras valley is not known; as *rupicola* occurs in Armenia, one may expect this subspecies in north-east Turkey. Either *rupicola* or *neumayer* may occur further west in the Black Sea Coastlands: a single bird from the Bolu area measured by Rössner (1935) had a wing of 86, a size too large for *zarudnyi*. If Bolu birds are *neumayer*, this population of *neumayer* is apparently not in contact with *neumayer* in the Balkan area, as the species has not yet been recorded in Thrace.

TICHODROMA MURARIA
Wallcreeper   Duvar Tirmasigi

Habitat Rocky gorges and steep shaded precipices and cliffs, mainly at 2000-3200m (e.g., in the western and central Taurus and in the eastern part of the country), but at 2300-3900m on the Ala Dag.

Distribution See map 94. Restricted to extensive high mountain regions of 2000m and over, which occur in the Taurus, the eastern part of the Black Sea Coastlands, the East, and the South-East; absent from isolated high mountains in the west. Records near Antakya and Birecik in May and July perhaps refer to late migrants and early post-breeding dispersal, respectively.

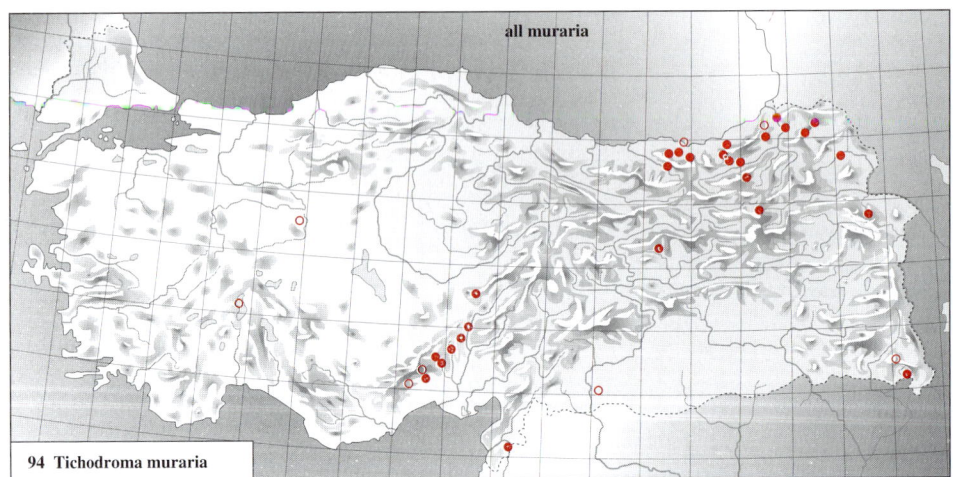

94 *Tichodroma muraria*

Geographical Variation
Subspecies described or recorded in the region:
- (W) **T. m. muraria** (Linnaeus), 1766, southern Europe. [Slightly larger. Wing of birds of both sexes from the Balkan countries 99.1 (94-102) (n=12), bill 33.0 (29.7-35.3) (n=11) (CSR).]
- (E) **T. m. longirostra** (S. G. Gmelin), 1774, mountains of northern Iran. [Slightly smaller. Wing of birds from the Caucasus area and north-west Iran 98.4 (94-102) (n=11), bill 31.8 (29-34) (n=11) (Vaurie 1950, CSR).]

Subspecies recognized in Turkey: the difference in size between *longirostra* and *muraria* is too small to warrant recognition of *longirostra* (see above). Furthermore there are no differences in colour either, thus only *muraria* is recognized for the western Palearctic, including Turkey. Wing of birds from north-east Turkey 100.4 (96-103) (n=4), bills not yet full-grown (Jordans & Steinbacher 1948, CSR).

CERTHIA FAMILIARIS
Eurasian Treecreeper   Orman Tirmasigi

Habitat Mainly in coniferous or mixed forest, locally in deciduous forest; at 1500-2000m in the inland western part of the Black Sea Coastlands, at 125-550m in the coastal zone of the eastern Black Sea Coastlands, and at 1800-2400m further inland in the north-east, but close to sea level in the Kizilirmak delta.

Distribution See map 95; also, Kumerloeve (1960, 1961a). Local in the Black Sea Coastlands and the neighbouring part of the Upper Aras region; also on the Ulu Dag, and recorded at Kiyiköy and Deretköy in Thrace in summer, hence probably breeding. Reported from the Karanfil Dag (Taurus) in October 1969 (Vittery 1972), but in general most records outside the known breeding range refer to misidentified *C. brachydactyla*.

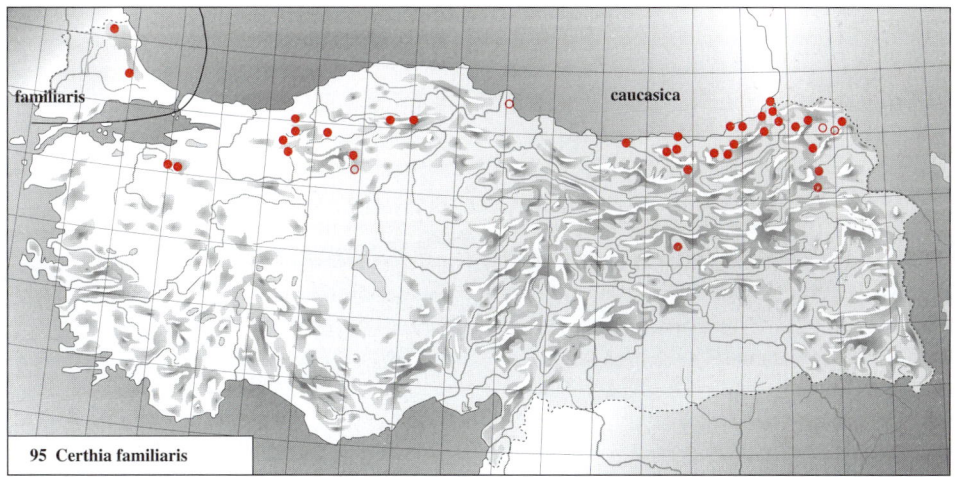

95 Certhia familiaris

Geographical Variation
Subspecies described or recorded in the region:

(W) **C. f. familiaris** Linnaeus, 1758, Sweden. [Cap dark brown, mantle and scapulars buffish olive-brown, marked with rather large elongated white spots; rump extensive light rufous-cinnamon; underparts silvery-white with faint and restricted cream-buff or pale rufous wash on lower flanks. Wing of both sexes from Romania 63.8 (60-69) (n=12), bill 16.4 (15.3-18.8) (n=11) (A. J. van Loon in Cramp & Perrins 1993).]

(E) **C. f. caucasica** Buturlin, 1908, Caucasus mountains. [Cap dark olive-brown, mantle and scapulars greyish olive-brown, marked with large white spots; rump with rather dull and restricted rufous-brown; underparts as in *familiaris*. Rather dark, like *macrodactyla* of central Europe, but cap of latter black-brown, mantle and scapulars dark (olive-) brown, both with smaller white spots. Wing in Caucasus 65.7 (62-68) (n=3), bill 17.3 (16-20) (n=14) (Vaurie 1957, CSR).]

(E) **C. f. persica** Zarudny & Loudon, 1905, Gilan, Mazandaran, and Astrabad (=Gorgan) (northern Iran). [Like *familiaris*, but bill longer. Wing in Iran 62.8 (60-66) (n=5), bill 18.2 (18.5-19.0, once 16.0) (n=5) (Stresemann 1928, Vaurie 1950, CSR).]

Subspecies recognized in Turkey: all birds from Asia Minor are apparently *caucasica*, showing rather dark upperparts with rather large and contrasting white spots on cap, mantle, and scapulars and with rather long bill. Wing of birds from northern Turkey 63.8 (60-68) (n=9), bill 16.2, 17.0 (n=2) (Kummerlöwe & Niethammer 1934-35, Rössner 1935, Jordans & Steinbacher 1948, CSR). The subspecies which breeds in Thrace is *familiaris*: a single ♂ collected Igneada on 15 May 1967, with wing 64, bill 16.8, and hind claw 9.3, is indistinguishable from *familiaris* from northern Europe, the Balkan area, and Russia. *C. f. persica* is restricted to the coastlands of the Caspian Sea in Iran and does not occur in Turkey.

References Kumerloeve, H. (1960) Sur la répartition des deux espèces de *Certhia* en Asie mineure. *Alauda* 28, 27-29.

CERTHIA BRACHYDACTYLA
Short-toed Treecreeper   Bahçe Tirmasigi

Habitat Deciduous, mixed, or coniferous forests or isolated stands of trees, up to the tree line; also, copses, plantations, gardens, and orchards. Occurs at 0-800m in Thrace and Western Anatolia, up to 1200m in the Black Sea Coastlands, and at 800-2000m in the Taurus.

Distribution See map 96; also, Kumerloeve (1960, 1961a). Widespread and locally common in the western half of the country, outside the treeless area of the Central Plateau, but apparently absent east from a line Samsun-Sivas-Erciyas Dagi-Mersin, though re-appearing in the south-west Caucasus.

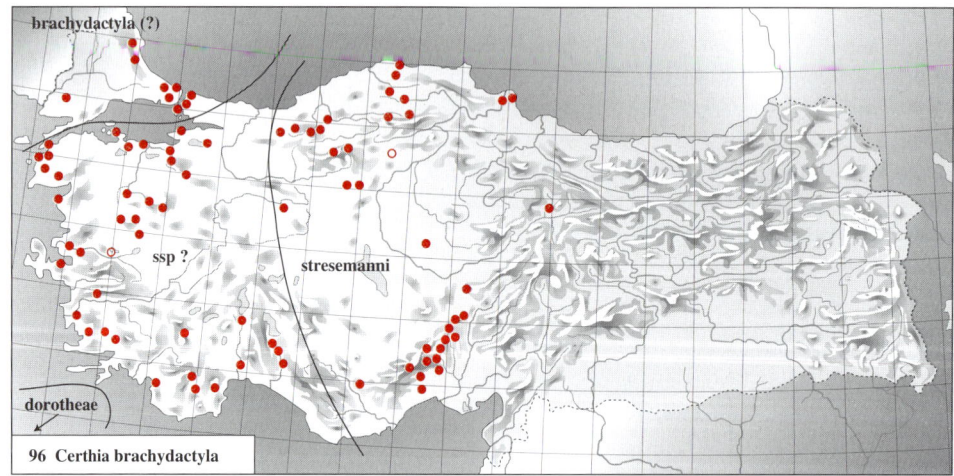

96 Certhia brachydactyla

Geographical Variation
Subspecies described or recorded in the region:

(W)   ***C. b. brachydactyla*** C. L. Brehm, 1820, Roda valley (Thüringen, Germany). [Ground-colour of upperparts greyish olive-brown, each feather with an elongated off-white spot, rump cinnamon-brown; underparts dirty white or light silvery-grey with restricted buff-grey on lower flanks. Wing of ♂ from Germany and Switzerland 64.0 (59-66) (n=19), of

♀ 61.9 (59-66) (n=19), bill of ♂ 18.8 (16.3-20.0) (n=6), of ♀ 17.1 (15.8-18.0) (n=4) (A. J. van Loon in Cramp & Perrins 1993).]

(W) *C. b. spatzi* Stresemann, 1926, Omalos Plateau, Crete (Greece). [Like *brachydactyla*, but off-white streaks on cap narrower and tail greyer, less brown. Wing of ♂ from Crete 63 (n=1), of ♀ 60, 60.5 (n=2), adult bill 20.1 (n=1); birds examined did not differ from *dorotheae* in colour and size, the differences in colour cited being due to differences in abrasion; thus, *spatzi* is a synonym of *dorotheae* (CSR).]

(T) *C. b. harterti* Hellmayr, 1901, Alemdag (above Üsküdar, in the western part of the Black Sea Coastlands, Turkey). [Rather similar to *megarhyncha*, a subspecies inhabiting western continental Europe, west of *brachydactyla*, which is characterized by more rufous upperparts; however, the upperparts of *harterti* are slightly duller rufous than in *megarhyncha*, especially the rump (Vaurie 1957, 1959). According to Kumerloeve (1961a), the type specimen is very similar to *megarhyncha* from the Rhine area of western Germany. Wing of the type 62, bill (to feathers?) 15.5 (Hartert 1903-10).]

(T) *C. b. stresemanni* Kummerlöwe & Niethammer, 1934, Kastamonu (western part of Black Sea Coastlands, Turkey). [Upperparts dull earth-brown, distinctly greyer than *harterti*, less rufous; flanks with a rather restricted amount of pale drab-olive; similar to *dorotheae* (below), but bill shorter (CSR). Similar to *brachydactyla*, but wing and bill longer and underparts slightly paler (Kummerlöwe & Niethammer 1934-35). Wing of ♂ from the western Black Sea Coastlands and the Taurus 64.7 (62-68) (n=3), of ♀ 61.2 (60-63.5) (n=3), bill of both sexes 19.1 (17.0-21.5) (n=6) (Kummerlöwe & Niethammer 1934-35, CSR).]

(S) *C. b. dorotheae* Hartert, 1904, Troodos mountains (Cyprus). [Like *brachydactyla*, but upperparts slightly duller greyish earth-brown and bill longer. Wing of ♂ from Cyprus 64.2 (60-68) (n=13), of ♀ 63.5 (n=1), adult bill of both sexes 19.8 (18-21.5) (n=14) (Vaurie 1957, CSR).]

Subspecies recognized in Turkey: Thrace is probably inhabited by *brachydactyla*, but the situation further east in Asia Minor is not clear. According to Sick (1939), the type of *harterti* is inseparable from a series of *stresemanni*, and all birds from Asia Minor should be named *harterti*, but according to Kumerloeve (1961a), 7 out of 8 birds from the Ilgaz Daglari, Ankara, and the Taurus are distinctly greyer (less rufous) than the type specimen of *harterti* from the Alemdag, and therefore *stresemanni* is valid, and *harterti* is possibly a synonym of *brachydactyla*. The latter suggestion is followed here, but further research is needed, and longer series of each population should be examined to check the validity of *harterti*, *stresemanni*, and *dorotheae*.

References Sick, H. (1939) *Certhia brachydactyla stresemanni* Kummerl. & Nieth. synonym von *C. b. harterti* Hellm. *Orn. Monatsber.* 47, 82-83.

REMIZ PENDULINUS
Penduline Tit   Çulhakusu

Habitat Restricted to marshy sites where branches of willows or poplars hang over water, as these are required for attaching nests. Mainly in the wetlands and river valleys in the lowlands of the west, but higher up on the Central Plateau and up to c. 2000m in the valleys of the East and South-East, where scrub alongside streams and at fringes of marshes is sufficient for breeding.

Distribution See map 97. Widespread and locally common in suitable habitat throughout the country.

Songbirds of Turkey

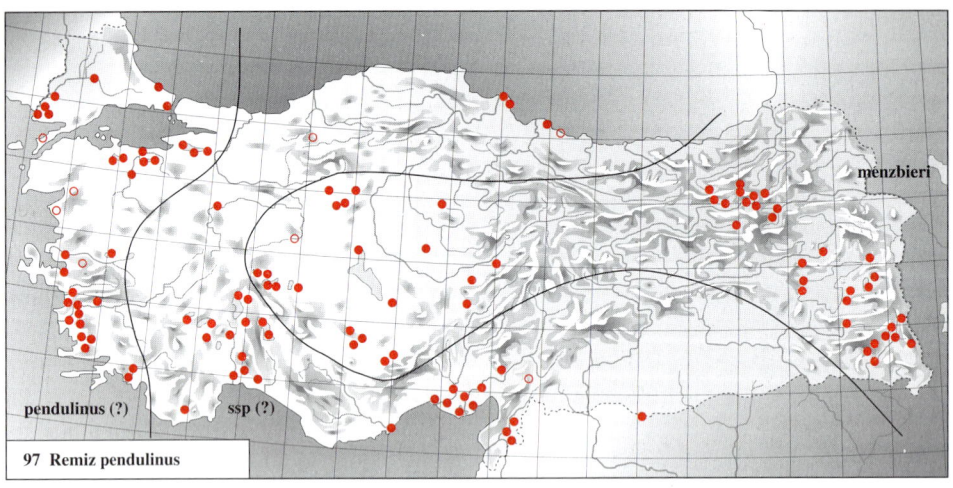

97 Remiz pendulinus

Geographical Variation

Subspecies described or recorded in the region:

- (W) **R. p. pendulinus** (Linnaeus), 1758, 'Poland, Lithuania, Hungary, Italy,...'. [Rather large; ♂ usually with a narrow chestnut bar on forehead above the black mask; mantle chestnut, rump tawny-buff; lesser upper wing-coverts tawny-buff. Wing of both sexes from central and southern Europe (Spain to the Balkan countries) 56.5 (54-59) (n=39), bill 11.3 (10.1-12.5) (n=39) (Vaurie 1950, Roselaar in Cramp & Perrins 1993).]
- (T) **R. p. persimilis** Hartert, 1918, Eregli (Central Plateau, Turkey). [Like *pendulinus*, but smaller, bill thinner; chestnut bar on forehead usually reduced or almost absent (as in *pendulinus*); rump and upper wing-coverts paler, more cream-buff, rump more contrasting with the chestnut of the mantle. As the width of the forehead bar is rather variable between individuals, in part depending on the age of the birds, this subspecies is best included in *menzbieri* (see below) (Vaurie 1950).]
- (E) **R. p. menzbieri** (Zarudny), 1913, Lower Karun River (south-west Iran). [Like *persimilis*, but the chestnut bar on the forehead is often slightly broader and more distinct than in *pendulinus* and *persimilis*. Wing of birds from Eregli and Kayseri (Turkey), Lenkoran (south-east Azerbaijan), Lake Rezaiyeh, Borujerd, and Dow Rud (Iran) 53.5 (52-57) (n=13), bill 10.9 (10-11.5) (n=8) (Hartert 1921-22, Vaurie 1950, 1957).]

Subspecies recognized in Turkey: both *pendulinus* and *menzbieri* occur in Turkey, but very few birds have been collected and the boundaries between the subspecies are hard to define. Birds from the Central Plateau (Eregli, Kayseri) are *menzbieri* (Vaurie 1950), as are probably those of eastern Turkey, near to Transcaucasia and western Iran where *menzbieri* is known to occur. A single juvenile bird examined from Solak (south of Elmali) was not full-grown and thus unidentifiable. Birds from Thrace and Western Anatolia are *pendulinus* according to Vaurie (1959), like those of Greece and Bulgaria further west, but it is not clear whether Vaurie examined specimens and, if he did so, from what localities. Birds from Amik Gölü were assigned to *pendulinus* after comparison with '*persimilis*' from the Levant by Meinertzhagen (1935), but no data on plumage or measurements were supplied and confirmation is needed.

ORIOLUS ORIOLUS

Golden Oriole   Sariasma

Habitat River valleys and plains with open forest, plantations, or copses of tall deciduous trees, especially poplar and oak, sometimes sycamore and beech; at 0-1200m in the west and centre of Turkey, up to 2200m in the mountain valleys of the east.

Distribution See map 98. Widespread and locally common all over the country, except for treeless steppe and mountainous areas, but the breeding range is hard to define due to the difficulty in distinguishing breeding birds from late spring migrants.

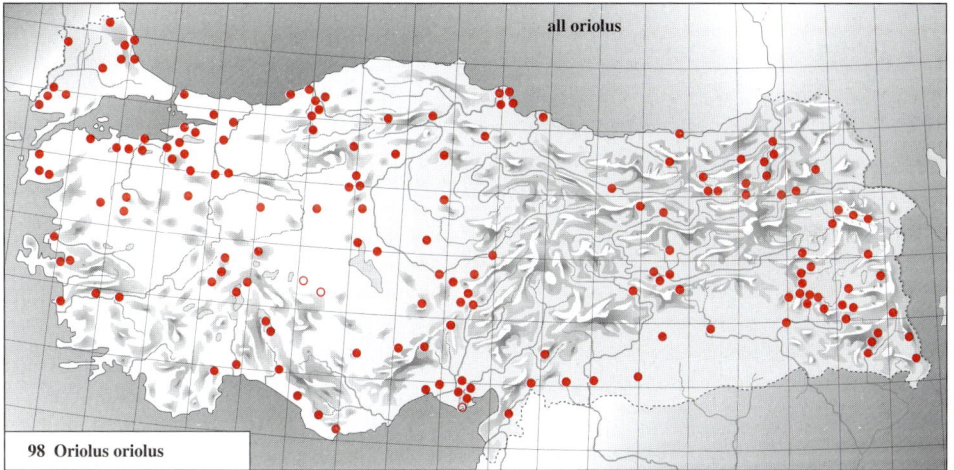

98 Oriolus oriolus

Geographical Variation

Subspecies described or recorded in the region:

(W)   ***O. o. oriolus*** (Linnaeus), 1758, Sweden. [A large subspecies; ♂ with the black on the side of the head restricted to the lores; yellow patches on tail-tip and primary coverts rather small. Wing of ♂ from central Europe (excluding juveniles) 153.5 (147-160) (n=65), of ♀ 151.0 (144-156) (n=42) (CSR), wing of ♂ from Caucasus 153.7 (150-158) (n=3) (Stresemann 1928).]

(E)   ***O. o. caucasicus*** Zarudny, 1918, Astrabad (Gorgan), Gilan, and Mazandaran (northern Iran). [Like *oriolus*, but smaller and ♂ deeper yellow. Wing of ♂ from northern Iran 148.3 (143-154) (n=9) (Hartert & Steinbacher 1932-38, Paludan 1940, Schüz 1959).]

Subspecies recognized in Turkey: the measurements and colour of birds from Turkey examined (not necessarily all local breeding birds) are inseparable from *oriolus* of central Europe; e.g., wing of ♂ from Turkey 152.6 (148.5-158) (n=7), of ♀ 149.1 (145-154) (n=5) (CSR, including some data from Weigold 1912-13, Kummerlöwe & Niethammer 1934-35, and Rokitansky & Schifter 1971). In fact, birds from Iran are not significantly different either, and *caucasicus* does not appear to be a valid subspecies. The range of *oriolus* thus extends from western Europe and North Africa east through Kazakhstan to Mongolia and through Turkey to Iran.

LANIUS COLLURIO

Red-backed Shrike   Çekirgekusu

Habitat Open forest with scattered shrubby undergrowth, forest edges, roadside scrub, hedgerows, or scrub and bushes in open country, always bordered by fields or open ground with low vegetation or bare soil; scrub preferably thorny with fairly dense foliage. Mainly at 0-1500m, but to 2300m in the eastern Taurus and at 1400-2000m in the mountain valleys of the East and North-East.

Distribution See map 99. Widespread and locally very common in Thrace, north-west Anatolia, the Black Sea Coastlands, the Taurus, and the valleys of the East; scarce or absent from the coastal zone of south-west Anatolia and the Southern Coastlands, from treeless steppe of the Central Plateau and the South-East, and apparently from the Kurdish Alps.

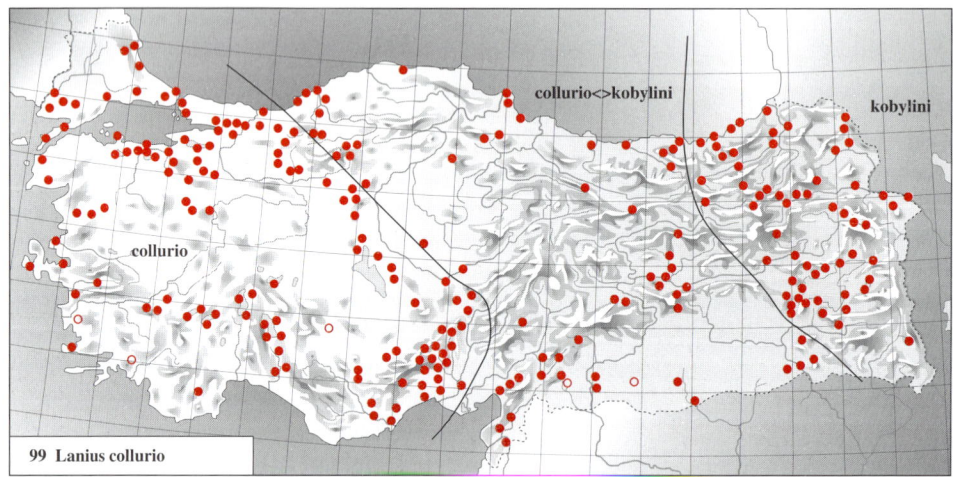

99 Lanius collurio

Geographical Variation

Subspecies described or recorded in the region:

(W) **L. c. collurio** Linnaeus, 1758, Sweden. [Lower mantle, scapulars, and back of ♂ rather extensively bright rufous-chestnut; rather large. Wing of both sexes from northern and central Europe 94.3 (89-99) (n=86) (Roselaar in Cramp & Perrins 1993).]

(E) **L. c. kobylini** (Buturlin), 1906, Kutaisi and Surami (Georgia, western Transcaucasia). [Rufous-chestnut of lower mantle, scapulars, and back of ♂ slightly duller, more chestnut-brown, and often more restricted, extending less far up on mantle and less far down on back; slightly smaller. Wing in the Caucasus 90.2 (87-94) (n=6) (CSR).]

Subspecies recognized in Turkey: the difference between *collurio* and *kobylini* is slight and the variation is strongly clinal, thus the boundary between both subspecies is hard to draw. Birds from the southern Balkan area and Thrace are smaller than *collurio*, but inseparable in colour; those from western Asia Minor are slightly larger but are also not different in colour. Birds from the Zonguldak-Beycuma area (western Black Sea Coastlands) and from the Anti-Taurus are variable, with some birds tending to *collurio*, others to *kobylini*, and these populations are considered to be transitional. The birds inhabiting the eastern Black Sea Coastlands (east from Rize) and probably those from elsewhere in the East and South-East (from where none were examined) are inseparable from *kobylini*. Wing of *collurio* from Greece and the south of former Yugoslavia 91.7 (87-96) (n=55) (Stresemann

1920, Makatsch 1950, CSR), from Thrace 89.1 (88-94) (n=4) (Rokitansky & Schifter 1971), from western Asia Minor (east to the Sapanca Gölü, Ankara, and the Taurus) 92.8 (87-99) (n=8) (Kummerlöwe & Niethammer 1934-35, Kumerloeve 1961a, Rokitansky & Schifter 1971, CSR); wing of intermediates from the Zonguldak-Beycuma area and the Anti-Taurus 92.3 (89-96.5) (n=14) (Kumerloeve 1963a, CSR); wing of *kobylini* from Rize and Varsambeg (Verçenik) 90.2 (89-93) (n=6) (CSR); wing of spring migrants Beysehir (subspecies unknown) 91.8 (87-97) (n=42) (Vauk 1973), wing of migrants of both races Urfa 92.0 (88-100) (n=12) (Weigold 1912-13).

LANIUS MINOR
Lesser Grey Shrike   **Karaalin Çekirgekusu**

Habitat Drier and more open areas than the habitat of *L. collurio*, inhabiting copses, bushes, and isolated trees in fields and pastures; sometimes in open forest with scanty undergrowth. Mainly at 0-1500m in the west and centre, up to *c.* 2500m in the valleys of the east.

Distribution See map 100; also, Kumerloeve (1961a). Apparently widespread; many birds are conspicuously present along roads up to early June and from late July onwards, and these are presumed to breed by travelling ornithologists and thus are recorded as such in the literature. As the species arrives late, departs early, and breeds rather secretively, at least some of the records shown on the map may refer to migrants. However, confirmed breeding has been recorded for all regions.

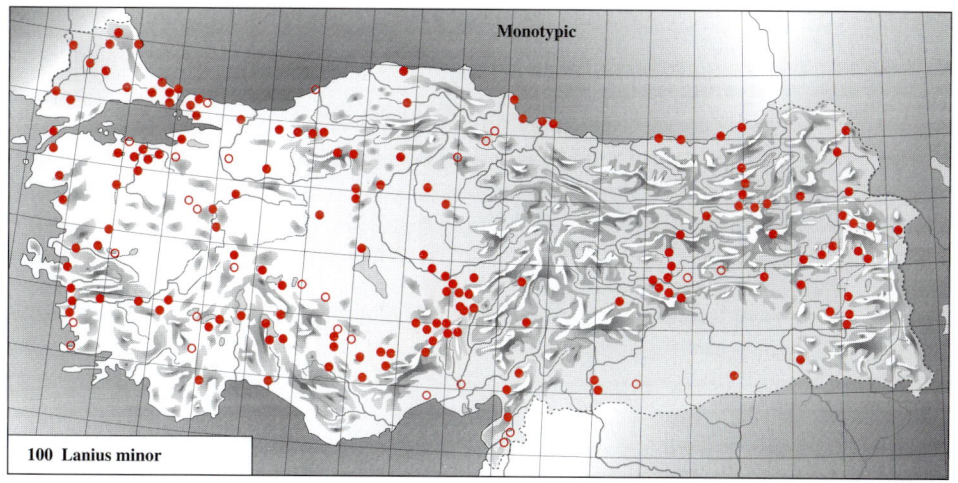

100 Lanius minor

Geographical Variation None recognized here. *L. minor* Gmelin, 1788, was originally described from Italy. Birds inhabiting central Asia are sometimes separated as *turanicus* Fediushin, 1927, described from Fergana (Uzbekistan). This latter subspecies is stated to be larger and the juvenile is said to show a more sandy tinge on the upperparts, but the differences in size and colour are slight and all populations show much individual variation. Wing of both sexes from central and eastern Europe 116.0 (110-122) (n=52), from central Asia 119.5 (115-123) (n=15) (Keve & Rokitansky 1966); wing in Greece and Romania 116.7 (113-120.5) (n=11) (CSR), in Turkey 116.8 (113-123) (n=15) (Kummerlöwe & Niethammer 1934-35, Rössner 1935, Rokitansky & Schifter 1971, CSR), in Iran 117.7 (112-123) (n=19) (Stresemann 1928, Paludan 1940, Schüz 1959, Diesselhorst 1962, Érard &

Etchécopar 1970, CSR); some of these samples undoubtedly include migrants.

References Keve, A., & G. Rokitansky (1966) Die Vögel der Almásy-Ausbeute, 1901 und 1906. *Ann. naturhist. Mus. Wien* 69, 225-283.

LANIUS SENATOR
Woodchat Shrike   Kizilbasli Çekirgekusu

Habitat Hedgerows and bushes along fields and roads, mixed with tall deciduous trees, as well as open mature deciduous or mixed forest with ample open ground, maquis with isolated trees and open patches, old orchards, olive groves, parkland, and forest edges, mainly in plains and valley bottoms. At 0-800m in Thrace and Western Anatolia, at (100-)500-1200m in the Southern Coastlands, and at 500-1500 (-2000)m in the East and South-East.

Distribution See map 101; also, Kumerloeve (1961a). Restricted to the western and southern third of the country, with a few additional breeding sites in the Kars area. Common especially in the western part of Western Anatolia and between the Ceyhan and upper Firat rivers, scarce on the lower slopes of the Taurus and in the East and South-East.

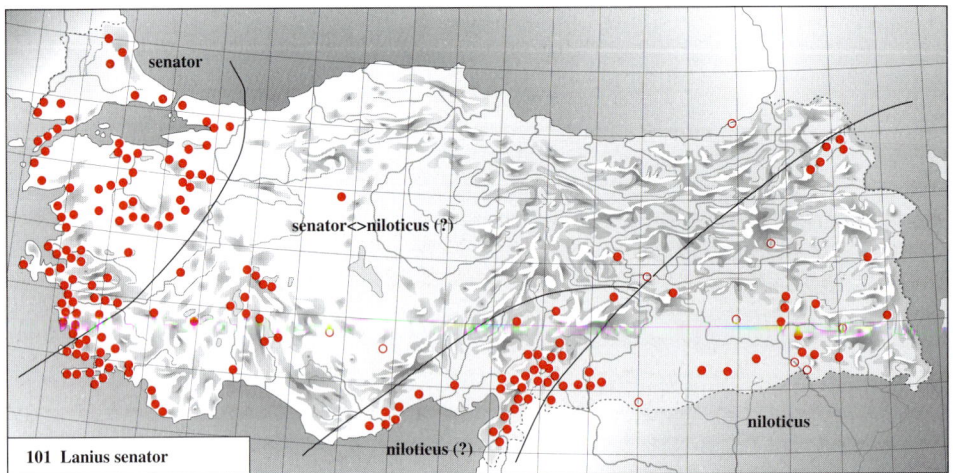

101 *Lanius senator*

Geographical Variation
Subspecies described or occurring in the region:

(W)   ***L. s. senator*** Linnaeus, 1758, Rhine valley. [Central pair of tail feathers of adult fully black or with a tiny amount of white hidden at the extreme base; breast and flanks extensively washed pale pink-cinnamon or cream-buff; maximum length of the white patch at the base of the primaries of the male 13 (10-16) mm (n=64). Wing of adults of both sexes from central Europe 100.2 (97-103) (n=35) (CSR), from the south of former Yugoslavia 99.8 (96-103) (n=18) (Stresemann 1920).]

(E)   ***L. s. niloticus*** (Bonaparte), 1853, White Nile (Sudan, on migration). [Basal part of central tail feathers white for 25-35 mm; underparts extensively cream-white, flanks with a restricted amount of buff or pink; white patch at the base of the primaries of the male 19 (17-21) mm long (n=13). Wing in Syria, Transcaucasia, and Iran 100.2 (95-104) (n=24) (Paludan 1938, Vaurie 1955, Schüz 1959, Érard & Etchécopar 1970, CSR).]

Subspecies recognized in Turkey: birds from Thrace and Western Anatolia south to at least Izmir are *senator*. Some birds from here show a little white at the base of the central tail feathers, but generally much less than in *niloticus*, and in other plumage characters they are close to *senator*. Birds occurring east from Gaziantep and Birecik are *niloticus*. Very few data are available from the Southern Coastlands. The species is rather scarce here, except in the Anti-Taurus were it is common. A single bird from Mugla in the south-west is *niloticus*, another from Aksehir is *senator* (Kumerloeve 1970b), but the latter was perhaps still on migration. Either all birds from the Southern Coastlands are *niloticus*, or they are a mixture of both subspecies; those east from the Göksu delta are especially likely to be *niloticus*. Wing of *senator* from western Turkey 98.2 (94-103) (n=5) (Kumerloeve 1970b, Rokitansky & Schifter 1971, CSR), wing of *niloticus* from southern Turkey 98.8 (96-102) (n=7) (Weigold 1912-13, Kumerloeve 1964a, 1970b, CSR).

LANIUS NUBICUS
Masked Shrike   Maskeli Çekirgekusu

Habitat Mature open woodland, light coniferous forest, plantations (e.g., of pines or *Eucalyptus*), isolated trees at the fringe of the maquis, or (occasionally) dry olive groves, always with more large trees and with less scrub than the habitat favoured by *L. collurio*, and usually on more arid sites than the similar habitat of *L. senator*. Mainly in lowlands or on the lower slopes of valleys, but at 800-1100 (-2000)m in the Taurus.

Distribution See map 102; also, Harrison & Pateff (1937) and Kumerloeve (1961a, 1961b). Restricted to Thrace, Western Anatolia, and the Southern Coastlands, where widespread but only locally common; scarce and local in the mountain valleys of the South-East, and a few records at the northern fringes of Central Anatolia.

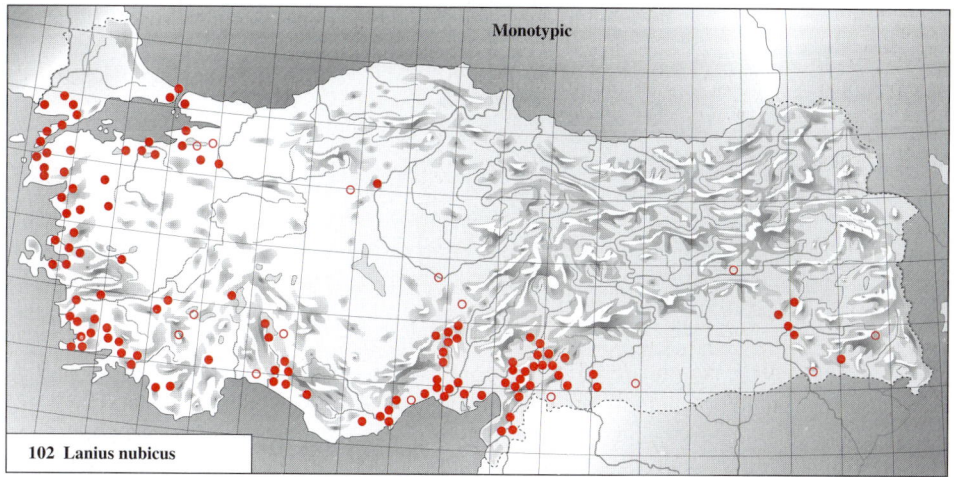

102 Lanius nubicus

Geographical Variation None. *L. nubicus* Lichtenstein, 1823, was originally described from birds migrating through or wintering in Nubia (northern Sudan). Wing of birds of both sexes from former Yugoslavia, Bulgaria, and Greece 91.4 (91-92) (n=4) (Stresemann 1920, CSR), from Izmir, Aydin, and Burdur (western Turkey) 90.7 (86-96) (n=40) (CSR), from Urfa (South-East Turkey) 90 (Weigold 1912-13), from Iran 91.8 (89-94) (n=4) (Paludan 1938, Érard & Etchécopar 1970), from migrants and wintering birds from Cyprus, the Levant, and north-east Africa 91.5 (84-96) (n=41) (CSR).

References Kumerloeve, H. (1961b) Sur la distribution en Turquie de la Pie-grièche masquée, *Lanius nubicus* Lichtenstein. *Alauda* 29, 134-137.

Songbirds of Turkey

GARRULUS GLANDARIUS

Eurasian Jay   **Kestane Kargasi**

Habitat All kinds of trees and scrub, whether closed stands or isolated copses, favouring oak, beech, pine, and hazel; also, olive groves, orchards, and (occasionally) parks. In the west, at 0-1600m, but to 1900m in the Ilgaz area, at 400-2100m in the Taurus, and to the tree line at *c.* 2500m in the east.

Distribution See map 103. Common and widespread, absent only from treeless steppe and semi-desert.

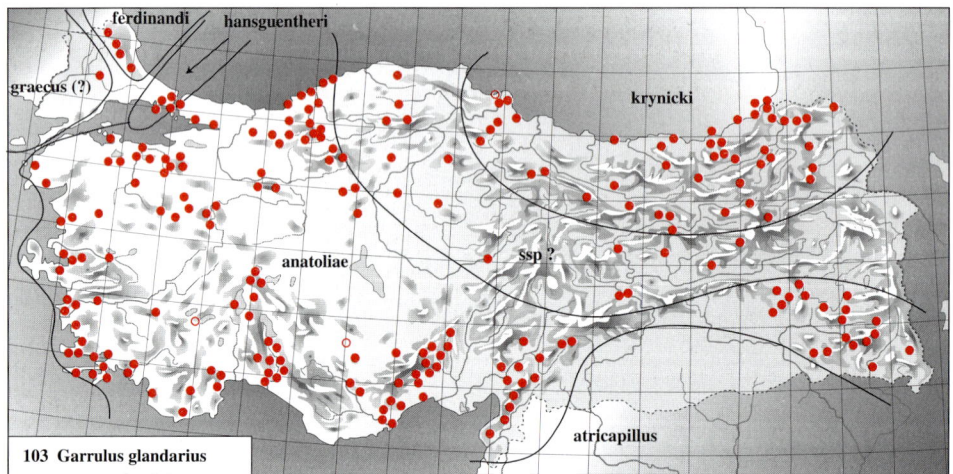

103 Garrulus glandarius

Geographical Variation

Subspecies described or recorded in the region:

(W)   ***G. g. glandarius*** (Linnaeus), 1758, Sweden. [Cap streaked black-and-white, hindneck vinous, mantle and scapulars vinous with slight grey tinge, sides of head white with vinous tinge, underparts white with strong vinous cast on breast and flanks. Wing of both sexes from northern and central Europe 184.4 (175-196) (n=55) (CSR). Occurs in northern and central Europe south to the north and east of former Yugoslavia and western Bulgaria]

(W)   ***G. g. graecus*** Kleiner (=Keve), 1939, Taïyetos mountains, Pelopónnisos (Greece). [Like *glandarius*, but mantle and scapulars more vinous-grey, contrasting somewhat with the purer vinous of the hindneck; vinous of breast tinged grey, contrasting with the white of the belly. Wing in Makedonija and Greece 180.0 (169-192) (n=36) (Stresemann 1920, Kleiner 1939a, 1939b, Niethammer 1943, Makatsch 1950, CSR). Restricted to mainland Greece, the south of former Yugoslavia, and southern Bulgaria. More data on this and other subspecies can be found in the revision of the species by Keve (1973).]

(W)   ***G. g. ferdinandi*** Keve, 1943, Skef, near Burgas (Bulgaria). [Like *glandarius*, but upperparts paler vinous-pink; underparts more extensively white, vinous paler and more restricted. Wing in south-east Bulgaria 178.5 (169-183) (n=9) (Keve-Kleiner 1943). Occurs in eastern Bulgaria.]

(W)   ***G. g. zervasi*** Kleiner [=Keve], 1939, Mytilíni, Lésvos (Lesbos) island (Greece). [Like *anatoliae* (see below), but paler and smaller, wing 179.8 (172-185) (n=10) (Kleiner 1939b, Keve 1973).]

(W) ***G. g. chiou*** Kleiner [=Keve], 1939, Khíos island (Greece).[Like *anatoliae*, but slightly paler and smaller, wing in 2 birds 168 and 174 (Kleiner 1939b), 185 in another (Keve 1973). None seen; perhaps not separable from *zervasi*, *anatoliae*, or *atricapillus*.]

(W) ***G. g. samios*** Kleiner [=Keve], 1939, Vathí, Samos island (Sporadhes, Greece). [Dark, like *krynicki* or even darker, but mantle, scapulars, breast, and flanks with rufous-vinous admixed. Wing on Samos and Ikaría 183.6 (179-185) (n=5) (Kleiner 1939b, Keve 1973).]

(W) ***G. g. rhodius*** Salvadori & Festa, 1913, Attaviros mountain, Rodhos (Rhodes) island (Greece). [Like *atricapillus*, but distinctly greyer above and below. Wing of 2 birds 185 and 189 (Keve 1973). Probably this race on Kos.]

(T) ***G. g. hansguentheri*** Keve, 1967, Tasköprü-Büyükçekmece area (eastern Thrace, Turkey). [Like *anatoliae* or even paler, near to *atricapillus* (see below), but black of cap with varying amount of white streaking (usually narrow). Wing 183.5 (176-190) (n=11) (Keve 1973). Restricted to both sides of the Bosporus, from Belgrat Ormani in the west to Üsküdar in the east, but outside the breeding season recorded in the Istranca Daglari further north-west in Thrace (Keve 1973).]

(T) ***G. g. anatoliae*** Seebohm, 1883, 'Asia Minor'; a neotype was selected from Gozna, north of Mersin in the Taurus, by Keve (1973). [Forehead white with black spots, cap black, mantle and scapulars vinous with a grey tinge; sides of head white with a vinous tinge or streaks; breast and flanks vinous, sometimes slightly tinged grey on the breast, white on belly rather restricted. Wing in western Asia Minor (western Black Sea Coastlands and Western Anatolia) 183.8 (178-200) (n=34), wing in the Taurus mountains (Eregli and Silifke to Maras) and in north-west Syria 187.3 (177-198) (n=11) (CSR). *G. g. lendlii* Madarász, 1907, described from the Taurus mountains, and *G. g. susianae* Keve, 1973, described from the Zagros mountains in south-west Iran, are synonyms.]

(S) ***G. g. atricapillus*** Geoffroy St. Hilaire, 1832, Lebanon mountain (Lebanon). [Forehead broadly white, cap black, mantle and scapulars vinous-pink, hardly grey; sides of head largely white; underparts extensively white, vinous on breast and flanks pale and restricted. Wing in Lebanon, southern Syria, Israel, and Jordan 184.0 (173-194) (n=23) (CSR).]

(E) ***G. g. krynicki*** Kaleniczenko, 1839, Georgiyevsk (northern Caucasus, Russia). [Forehead white with black spots or almost fully black, cap black; vinous of mantle, scapulars, and breast strongly tinged grey; sides of head strongly tinged vinous. Wing in the Caucasus area 188.6 (183-197) (n=16), wing in north-east Turkey (Sebinkarahisar eastward) 193.3 (187-195, once 213) (n=8) (CSR). *G. g. nigrifrons* Buturlin, 1906, described from Batumi (south-west Georgia) and Artvin (north-east Turkey), is a synonym.]

Subspecies recognized in Turkey: Two clearly defined groups of subspecies occur, the *glandarius*-group with a streaked cap and the *atricapillus*-group with a uniform black cap. The latter group occurs from Iran and the Caucasus west to Asia Minor and to the islands in the eastern Aegean Sea, the former group in Europe south to (among other islands) Crete and central Thrace. Within each group, many subspecies are described (see Keve 1973 for a survey); the differences between the subspecies are often small, but perceptible when a series of skins can be compared. In Turkey, the *glandarius*-group is confined to north-west Thrace; here, *ferdinandi* occurs in the Istranca Daglari, while *graecus* may occur in the south-west (e.g., in the Kesan area). Both *ferdinandi* and *graecus* are sometimes included in *glandarius*, *ferdinandi* being slightly paler than *glandarius*, and *graecus* slightly darker; though the difference of each from *glandarius* is slight, they are easily recognizable when compared to each other. *G. g. hansguentheri* from the Bosporus area is close to *anatoliae* or *atricapillus* of the black-capped group, differing in often showing some white streaking on the cap, apparently due to introgression of characters of the *glandarius*-group; probably *ferdinandi* and *hansguentheri* grade into each other in eastern Thrace. Much of Asia Minor is inhabited by *anatoliae*, a subspecies intermediate between *krynicki* and *atricapillus*, but easily separable from both when a series of skins is compared. Pure *krynicki,* indistinguishable from Caucasian birds, occurs in north-east Turkey, south to Agri, Erzurum, and Sebinkarahisar, west to Samsun but not further west, *contra* Keve (1973). Influence of any of the subspecies of the Aegean islands on birds from Western Anatolia has not yet been proved: only birds from the Izmir area have been examined, which are large and rather pale, like *anatoliae*, not small and pale like *zervasi* and *chiou* from Lésbos and Khíos, nor large and dark like *samios* and *rhodius* from Samos, Ikaría, Kos, and Rodhos. The validity of these subspecies requires confirmation; most of the type specimens no longer exist.

References Kleiner, E. [=Keve, A.] (1939a) Systematische Studien über die Corviden des Karpathen-Beckens, nebst einer Revision ihrer Rassenkreise II. *Garrulus glandarius* L. *Aquila* 42-45, 141-226. Kleiner, E. (1939b) Ergänzung zur systematischen Revision des Eichelhähers. *Aquila* 42-45, 542-549. Keve-Kleiner, A (1943) Ein neuer Eichelhäher aus Südost-Bulgarien - *Garrulus glandarius ferdinandi* ssp. n. *Aquila* 50, 369-370. Keve, A, (1973) Über einige taxonomische Fragen der Eichelhäher des Nahen Ostens. *Zool. Abh. Staatl. Mus. Tierkde. Dresden* 32, 175-198.

PICA PICA
Common Magpie   Saksagan

Habitat Open plains and valleys, with at least some bushes or trees for breeding, though (locally) structures such as telegraph poles or pylons will suffice; also, steep scrub-clad slopes near cultivation or steppe, as well as orchards, plantations, gardens, and (locally) city parks. Occurs mainly at 0-1300m, but up to 2200m in the east and to 2500m near Eleskirt and on the Süphan Dagi near Van Gölü.

Distribution See map 104. Virtually everywhere and often numerous, but largely absent from the coastal fringe of the Black Sea Coastlands and from Hatay, and rather scarce along the coastal fringe of the Southern Coastlands. Even occurs in open steppe and semi-desert if fissures in rocks or man-made structures are available for nest-sites.

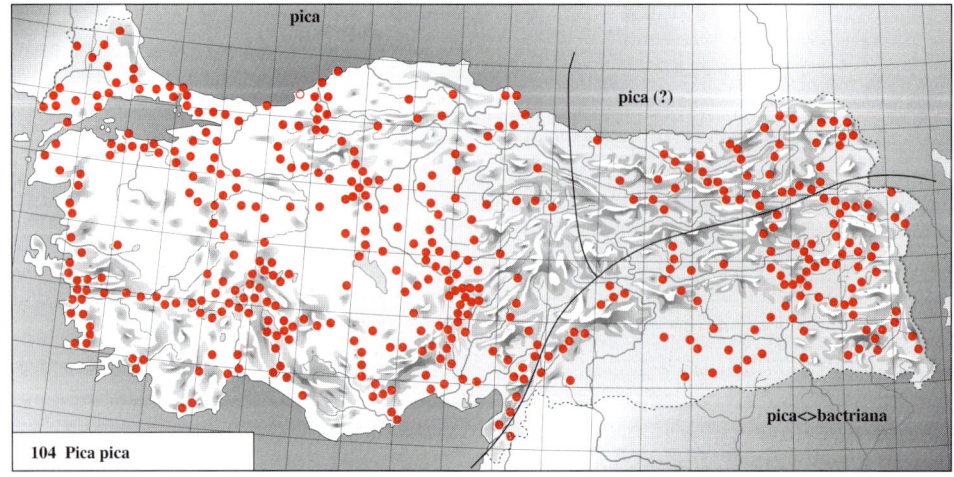

104 *Pica pica*

Geographical Variation

Subspecies described or recorded in the region:

(W) ***P. p. pica*** (Linnaeus), 1758, Sweden. [Rump light grey, but occasionally white and rarely dark grey; white on centres of each primary of adult fully hidden when wing closed, even on 3rd-5th innermost primary; tips and inner webs of tertials glossy blue, of secondaries deep blue; rather small. Wing of adult ♂ from southern Scandinavia 204.6 (195-213) (n=12), of adult ♀ 192.4 (183-203) (n=8), tail of adult ♂ 260.3 (240-277) (n=12), of adult ♀ 244.9 (233-259) (n=8) (CSR). Occurs throughout Europe, south to Greece, Cyprus, the Caucasus, and Transcaucasia, except for northern Fenno-Scandinavia and northern Russia, where replaced by the larger white-rumped subspecies *fennorum*, as well as in the plains of south-east European Russia, where replaced by *bactriana*. In this and other samples, wing and tail of juveniles and first-year birds are ignored, as these are considerably shorter than those of adults.]

(E) ***P. p. bactriana*** Bonaparte, 1800, Kandahar (Afghanistan). [Rump always white; white on centre of each primary of adult extensive, black fringe of each primary narrow, white visible on tips of 3rd-4th (-5th) innermost primaries when wing closed; tertials and secondaries mainly glossed green; tail often lighter green or brass-coloured than in *pica*; larger. Wing of adult ♂ from west-central Asia 219.9 (204-220) (n=8), of adult ♀ 204.5 (195-216) (n=6), tail of adult ♂ 287.6 (266-302) (n=9), of adult ♀ 262.7 (226-279) (n=6) (CSR). Occurs from south-east European Russia and Iran eastwards.]

Subspecies recognized in Turkey: birds from western Turkey (east to Tokat in the north and to Osmaniye in the south, or to about 37°E) are inseparable in plumage and size from *pica*. Wing of adult ♂ from this area 198.6 (186-207) (n=15), of adult ♀ 193.0 (185-200) (n=12), tail of adult ♂ 264.8 (256-279) (n=12), of adult ♀ 253.0 (244-270) (n=5) (Kummerlöwe & Niethammer 1934-35, Kleiner 1939c, Jordans & Steinbacher 1948, Rokitansky & Schifter 1971, CSR). East from Gaziantep and Malatya and south from Agri, adult birds often have some white visible on the primaries at rest, and they average slightly larger in size. Thus, wing of adult ♂ from Agri and from the Van and Hakkari areas is 203, 210 (n=2), of adult ♀ 196.7 (192-201) (n=3), tail of adult ♂ 258, 263 (n=2), of adult ♀ 264.8 (247-282) (n=3) (CSR). Wing and tail lengths are not as long as in *bactriana*, however, and the gloss on the wing is more similar to *pica*. Therefore birds of south-east Turkey are considered to be intermediate between *pica* and *bactriana*, grading into true *bactriana* in Iran. No birds from north-east Turkey have been examined; these are probably *pica*, like those of Transcaucasia further east and like those of the western Black Sea Coastlands further west.

References Kleiner, E [=A. Keve] (1939c) Systematische Studien über die Corviden des Karpathen-Beckens, nebst einer Revision ihrer Rassenkreise I. *Pica pica* L. Aquila 42-45, 79-140.

PYRRHOCORAX GRACULUS
Alpine Chough   Sarigaga Dagkarkasi

Habitat Breeds on cliffs and crags of high mountains, foraging on fields, grassy slopes, and alpine meadows lower down; occurs from (800-) 1300m to at least 3500m.

Distribution See map 105. Only on the higher mountains of the Southern Coastlands and in the eastern third of the country, as well as on the Ulu Dag. Birds observed or collected in winter near Izmir and Söke were probably involved in winter dispersal, but the nearest known breeding site in the Taurus is rather far away and perhaps the species breeds in small numbers nearer to the Anatolian west coast, e.g. at the Boz or Aydin Sira Daglari or above Denizli. Observed near Yesilce (north-west of Gaziantep) on 29 July (Frost & Hornbuckle 1992) and may breed nearby (e.g., above Maras?).

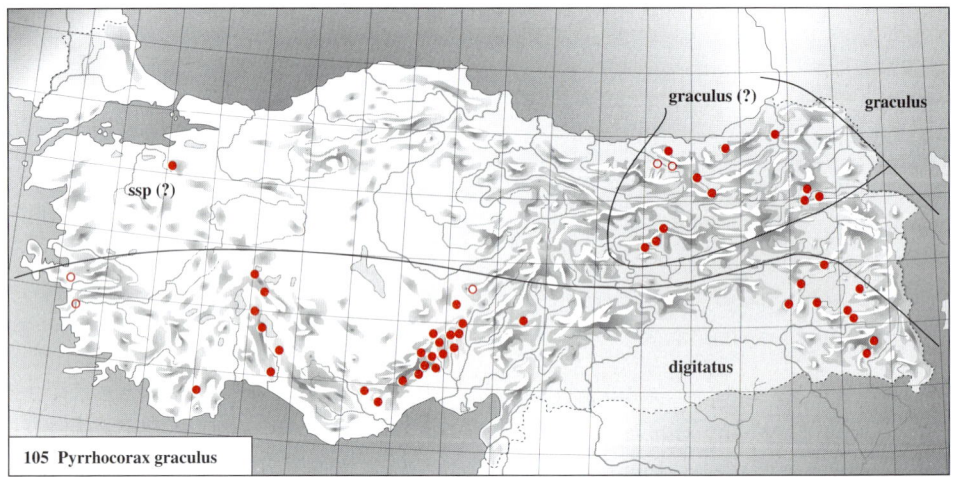

105 Pyrrhocorax graculus

Geographical Variation

Subspecies described or recorded in the region:

(W/E) ***P. g. graculus*** (Linnaeus), 1766, Alps of Switzerland. [Wing, bill, and tarsus relatively short, bill and feet slender. Wing of adult ♂ from the Pyrenees, Alps, former Yugoslavia, and Greece 277.1 (264-284) (n=16), of ♀ 259.8 (251-267) (n=13), bill (both sexes) 34.4 (32.2-36.8) (n=29), tarsus 46.0 (43.4-48.5) (n=30), bill depth at middle of nostril 9.5 (8.5-10.6) (n=36); wing of adults (both sexes) from the Caucasus and northern Iran 271.0 (263-280) (n=8), bill 35.1 (33.6-37.5) (n=7), tarsus 44.5 (39.8-46.5) (n=7) (CSR). Sexual difference in measurements is marked, especially in wing length and bill depth, but data of both sexes are combined in some populations when many of the available specimens were unsexed. Also, wing (and tail) of juveniles and 1st-year birds are markedly shorter than in adults in this species (and in most other corvids), and therefore data of these birds are not included in the samples given. Measurements of bill, leg, and foot of *graculus* are similar to those of *forsythi* from central Asia, but the wing of the latter is longer: wing of adult ♂ from central Asia is generally over 285, of adult ♀ over 265.]

(S) ***P. g. digitatus*** Hemprich & Ehrenberg, 1833, Lebanon. [Wing, bill, and tarsus longer than in *graculus*, foot and bill heavier; wing as in *forsythi* from central Asia, but bill and leg of latter about as small as in *graculus*. This subspecies is restricted to southern Turkey, the Lebanon, and south-west Iran. Wing of both sexes of adult *digitatus* from this area 284.6 (263-280) (n=8), bill 37.5 (35.4-38.8) (n=8), tarsus 50.3 (48.6-54.8) (n=8), bill depth 10.9 (10.2-12.0) (n=8) (CSR).]

Subspecies recognized in Turkey: bill and leg of birds from the Taurus and of a single unsexed bird from Izmir are heavy and the wing is long, similar to two birds from the Lebanon, one of which is the type specimen of *digitatus*. Birds from the Zagros mountains in south-west Iran are also *digitatus*. In contrast to this, birds from the Caucasus and northern Iran have a shorter wing and more slender bill and feet, and these birds are inseparable from *graculus* from the Alps. As no birds from the Ulu Dag and eastern Turkey have been examined, it is impossible to say whether these are *digitatus* or *graculus*. Perhaps those of the Ulu Dag and the north-east (north of the Aras valley) are *graculus*, thus more or less connecting the range of *graculus* in the Balkans with that in the Caucasus, and perhaps those of the Van and Hakkari areas are *digitatus*, but confirmation is required.

PYRRHOCORAX PYRRHOCORAX
Red-billed Chough   Kizilgaga Dagkargasi

Habitat Cliffs and crags near alpine meadows or near steppe country, mainly at (1000-) 1800-3500m, but up to 4400m in the mountains around Van Gölü.

Distribution See map 106. Distribution very similar to that of *P. graculus*, but generally more common and more widespread, breeding in virtually all mountains which reach over 2000m, as well as locally lower down in the steppe country in the South-East. Bred on cliffs near Bozüyük in the last century (Danford 1877-78) and perhaps also in the Samsun Dag near Priene, where recorded by Weigold (1913-1914) in early spring.

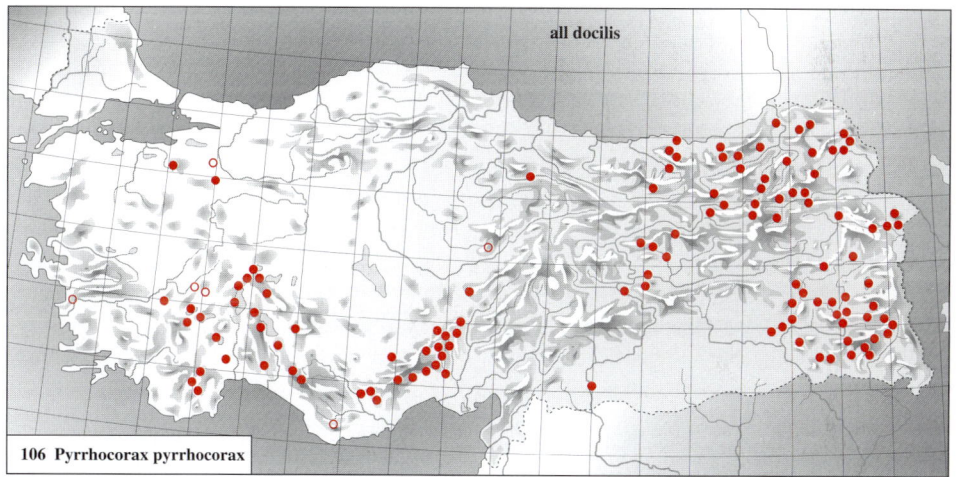

106 Pyrrhocorax pyrrhocorax

Geographical Variation
Subspecies described or recorded in the region:

(E)   **P. p. docilis** (S. G. Gmelin), 1774, 'Tschurdost', Gilan (northern Iran). [Gloss of body and upper wing-coverts green; size large. Size as in *erythrorhamphus* from southern Europe (east to the north-west of former Yugoslavia), but the latter is glossed bluish; *pyrrhocorax* from Britain, Ireland, and north-west France is smaller and shows purplish-blue gloss. Wing of adult ♂ *docilis* from the Caucasus and south-west Iran (including the type specimen of *pontifex* Stresemann, 1928, from Pish Kuh, Gilan, a synonym of *docilis*) 315.5 (293-327) (n=9), of adult ♀ 304.7 (292-317) (n=8), bill of ♂ 54.3 (49.9-59.2) (n=5), of ♀ 50.6 (46.9-54.1) (n=4), tarsus of ♂ 56.6 (53.7-58.5) (n=5), of ♀ 52.8 (50.6-55.2) (n=5); wing of adult ♂ from Crete 306.6 (299-311) (n=4), of adult ♀ 295.7 (289-315) (n=6); bill of ♂ 55.6 (54.3-57.4) (n=3), of ♀ 50.6 (48.3-52.5) (n=4), tarsus of ♂ 56.7 (55.7-57.2) (n=4), of ♀ 52.1 (50.7-53.2) (n=5) (CSR; wing of birds from Iran includes data of Diesselhorst 1962 and Érard & Etchécopar 1970).]

Subspecies recognized in Turkey: all birds examined from Turkey are *docilis*. Wing of adult ♂ from Elmali, Berendi, Güsle Dag, 'Taurus mountains', and Varsambeg (Verçenik) 318.8

(299-331) (n=9), of adult ♀ 301.0 (294-309) (n=4), bill of ♂ 55.4 (53.5-58.9) (n=9), of ♀ 50.4 (49.0-52.3) (n=3), tarsus of ♂ 55.6 (54.6-56.8) (n=8), of ♀ 53.3, 54.1 (n=2) (CSR). No birds were examined from the steppe country of the South-East, but a single ♂ from the semi-desert of nearby Syria (where once common but now apparently extinct) does not differ in size or colour from Turkish birds: e.g., wing 310, bill 55.7, tarsus 57.0 (CSR).

CORVUS MONEDULA
Eurasian Jackdaw   Cüce Karga

Habitat Arable land, open or with scattered bushes, trees, or hedgerows, with nest sites available nearby in tree holes, in fissures of rock walls, or between boulders and stones; in more open areas, often depends on structures such as buildings or ruins for nesting, and thus absent where these are not available. Occurs at 0-1500m in the west and centre of the country, to *c.* 2500m in the east.

Distribution See map 107. Common to abundant over much of the country, but scarcer in the coastal zone of the Black Sea Coastlands and in the Southern Coastlands; in the steppe of Central Anatolia and the South-East, it is largely restricted to villages.

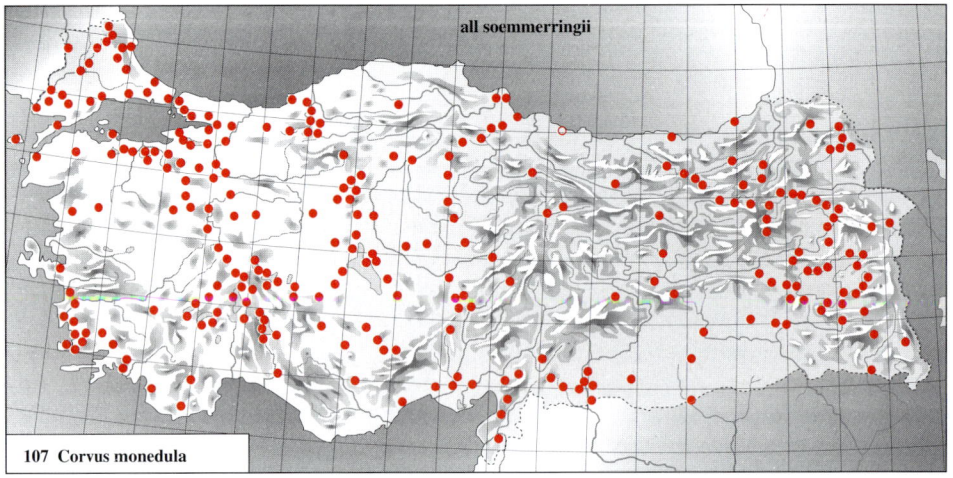

107 *Corvus monedula*

Geographical Variation
Subspecies described or recorded in the region:

(W/E) ***C. m. soemmerringii*** Fischer, 1811, Moskva (Moscow, central European Russia). [Grey of sides of head and neck pale, that of underparts rather pale; adult with a distinct white crescent or half-collar on the side of the neck. Some influence of bleaching and wear, sides of head and neck as well as the feather fringes on the underparts of adults becoming silvery-white when plumage is worn in spring, but the head, neck, and underparts of first-year birds often become more uniform dull grey, without the white crescent on the side of the neck. Occurs from eastern European Russia, the Balkan countries, and Greece east to central Asia. *C. m. collaris* Drummond, 1846, described from Macedonia, and *C. m. pontocaspicus* Kleiner, 1939, described from Cyprus, are synonyms; the differences between these as cited in, e.g., Kleiner (1939d) are largely due to differences in abrasion and bleaching. Wing of adult ♂ *soemmerringii* from the south of former Yugoslavia 236.9 (223-244) (n=25), of adult ♀ 229.9 (221-243) (n=17) (Stresemann 1920).]

Subspecies recognized in Turkey: all birds examined are *soemmerringii*, with (in adults) a distinct white crescent along the side of the neck of *c.* 4-5 cm long and 0.5 cm wide; in worn plumage, the grey of the head, neck, and underparts becomes silvery-white, apparently more so than in *soemmerringii* from the Balkans and eastern European Russia. Wing of ♂ *soemmerringii* from Turkey (ages combined) 235.8 (233-243, once 220) (n=10), of ♀ 225.4 (220-234, once 211) (n=11) (Kummerlöwe & Niethammer 1934-35, Jordans & Steinbacher 1948, Kumerloeve 1961a, 1964a, 1969a, 1970b, CSR). The measurements of all races of Jackdaw are markedly constant throughout the species' range. Due to inclusion of short-winged 1-year-olds in some samples and exclusion of these in others, the averages seem to show some variation, but this is not apparent when only birds of one age-group are measured.

References Kleiner, A [=A. Keve] (1939d) The Jackdaws of the Palearctic region, with description of 3 new races. *Bull. Brit. Orn. Club* 60, 11-14.

CORVUS FRUGILEGUS
Rook   Ekinkargasi

Habitat Cultivated open steppe, depending on tall scrub or trees for nesting, such as lines of trees along rivers, copses amidst arable fields, and mature trees on graveyards, in parks, or in gardens near towns and villages. Occurs from sea level to *c.* 2500m.

Distribution See map 108; also, Roer (1962), Kumerloeve (1967c), and Kasparek (1989, 1992). Virtually restricted to steppe country and to open cultivated valleys; especially common in the east.

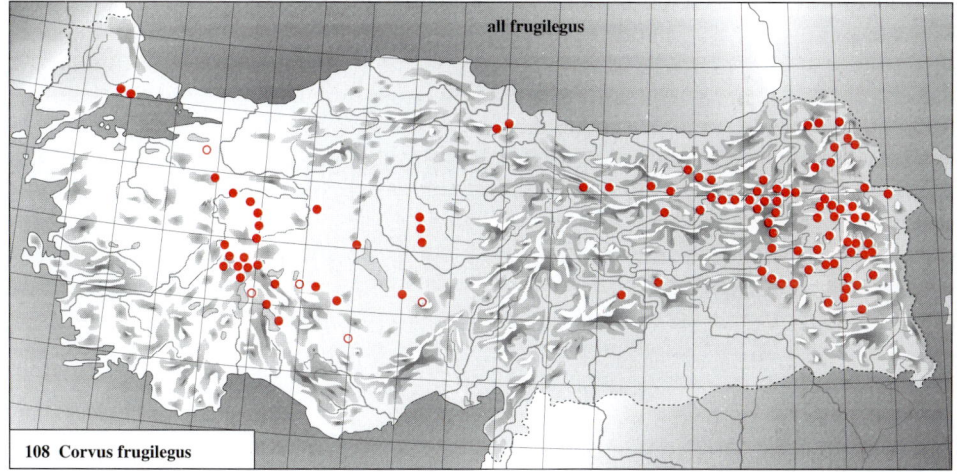

108 Corvus frugilegus

Geographical Variation
Subspecies described or recorded in the region:

(W/E) *C. f. frugilegus* Linnaeus, 1758, Sweden. [Lores and chin bare; bill rather heavy at base. Wing of adult ♂ from western and central Europe 322.9 (311-335) (n=47), of adult ♀ 306.6 (297-320) (n=54), bill depth (at middle of nostril) of ♂ 18.1 (17.2-19.2) (n=45), of ♀ 16.8 (15.5-17.7) (n=35) (CSR). Breeds east to central Asia.]

Subspecies recognized in Turkey: all birds from Turkey are *frugilegus*, as would be expected, as the only other valid subspecies of Rook, *pastinator*, which shows a feathered chin and slender bill, is restricted to east Asia. Wing of adult ♂ from eastern Romania, Turkey, and the Caucasus area 324.3 (322-329) (n=9), of adult ♀ 306.2 (298-322) (n=16), bill depth of ♂ 18.1 (17.4-18.5) (n=9), of ♀ 16.8 (15.8-17.7) (n=16) (CSR).

References Roer, H. (1962) Saatkrähe (*Corvus frugilegus*) Brutvogel in der europäischen Türkei. *J. Orn.* 103, 494. Kumerloeve H. (1967c) Zum Brutvorkommen der Saatkrähe in der Türkei. *Ardea* 55, 138-140. Kasparek, M. (1989) Breeding distribution of the Rook *Corvus frugilegus* in Turkey. *Sandgrouse* 11, 89-95.

CORVUS CORONE
Carrion Crow   Leskargasi

Habitat All types of open country, whether cultivated or not, from open steppe to glades in dense forest, providing tall scrub or trees are available for nesting; less readily adapted to nesting near humans than *C. monedula* (or less tolerated), and thus largely absent from treeless areas. Mainly in lowlands and level valleys and plateaux, but to *c.* 2000m in the Taurus and to over 2200m in the east.

Distribution See map 109. Common and widespread in the northern half and the western third of the country as well as in the Taurus, but scarce and local in the coastal zone of the Southern Coastlands, in the steppe country of the Central Plateau and in the South-East; greatly outnumbered by *C. frugilegus* in the East.

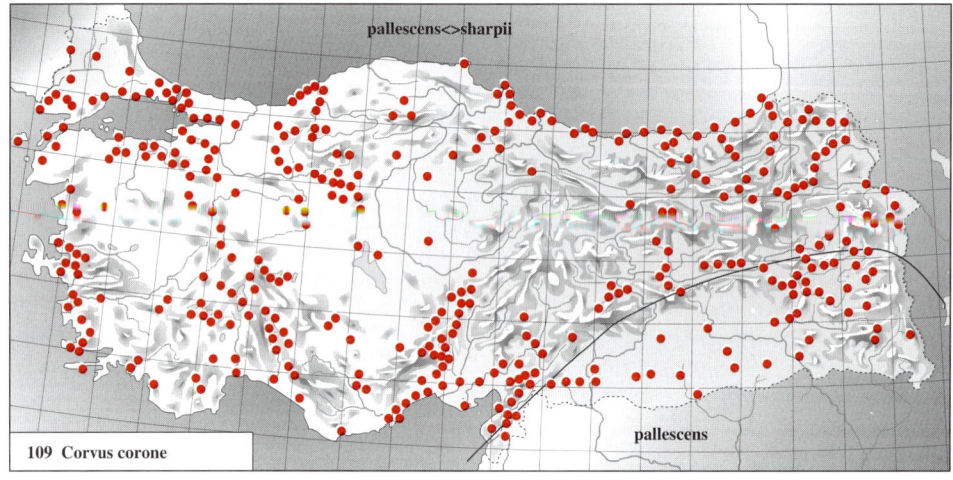

109 Corvus corone

Geographical Variation
Subspecies described or recorded in the region:

(W)   ***C. c. cornix*** Linnaeus, 1758, Sweden. [Grey of body in fresh plumage rather dark; size large. Wing of adult ♂ from northern and central Europe 328.3 (305-349) (n=38), average bill depth of ♂ (at middle of nostril) 19.3 in northern Europe, 18.3 in central Europe (CSR). In this and other races of *C. corone*, wing and bill of adult ♀ are on average shorter than in adult ♂ and the bill is more slender, while wing of juvenile and 1st-year birds is shorter than in adults; data on these latter birds are excluded from the samples presented here, but they show trends comparable to those of adult ♂.]

(W)   ***C. c. sardonius*** Kleinschmidt, 1903, Sardinia. [Grey of body intermediate between that of *cornix* and *sharpii*; size rather large. Wing of adult ♂ from Sardinia 326.9 (309-338) (n=11), average bill depth of ♂ 18.5 (CSR).]

(E) ***C. c. sharpii*** Oates, 1889, Mardan (Punjab, India; in winter). [Grey of body light; size large. Wing length of adult ♂ from south Siberia and Kazakhstan 329.2 (309-345) (n=9), average bill depth of ♂ 18.5 (CSR).]

(S) ***C. c. pallescens*** Madarász, 1904, Cyprus. [Grey of body pale, even paler than in *sharpii*, but not as pale milky-grey as in the very large *capellanus* from central and southern Iraq; size small. Wing of adult ♂ from Cyprus 310.3 (308-314) (n=3), average bill depth of ♂ 17.5 (CSR).]

Subspecies recognized in Turkey: difficult to establish due to the complicated variation within the species and the strong influence of bleaching and wear on the grey of the body. Only data from fresh or moderately worn birds are used here. The difference in colour of grey between the extremes is worthy of recognition (e.g., the difference between *cornix* and *sharpii/pallescens*), as is the difference in size between extremes (e.g., the difference between *cornix/sharpii* and *pallescens*), but many populations are intermediate in size, colour, or both and are difficult to place; although subspecies names are available for a number of them, the characters they show are rarely distinctive. Here, the name *cornix* is restricted to the large dark birds of northern and central Europe. *C. c. sardonius* is included in *sharpii*; this subspecies as recognized here includes large pale birds which breed Sardinia, Corsica, mainland Italy, the north of former Yugoslavia, Romania, as well as from the Ukraine, Crimea, and the Caucasus eastwards. The small pale *pallescens* is restricted to Cyprus, Crete, and the Levant; the fairly small and pale birds breeding in Greece and Egypt are best considered to be intermediate between *pallescens* and *sharpii*. In Turkey, the birds from the northern and western half of the country (Rize, Kastamonu, Beycuma, Zonguldak, Bolu, Nallihan, Thrace, Uluborlu, Tekir Dag, Tarsus, Kadirli, Haruniye, Iskenderun, Amanus mountains, Malatya) are also best referred to as intermediate between *pallescens* and *sharpii*, even though the grey is slightly darker than in both these subspecies; wing of adult ♂ from this area 320.2 (311-330) (n=20), bill depth of ♂ on average 18.3 (CSR). Birds from the south-east (Urfa, Van, Hakkari) are best referred to as *pallescens:* wing of adult ♂ 314.6 (308-318) (n=4), average bill depth of ♂ 17.6. They are very similar to birds from the Levant, which have wing of adult ♂ 313.6 (304-324) (n=3) and bill depth 17.3, and are slightly paler grey than birds from Egypt, which have wing of adult ♂ 317.9 (297-330) (n=9) and bill depth 17.3 (CSR). The all-black eastern subspecies *orientalis* is occasionally reported to occur in Turkey (e.g., Beaman 1975); as with the all-black West European subspecies *C. c. corone*, it is easily confused with juveniles of *C. frugilegus*. *C. c. orientalis* breeds from eastern Iran and Afghanistan eastwards and is largely non-migratory.

CORVUS RUFICOLLIS
Brown-necked Raven
Habitat Replaces *C. corax* in deserts and semi-deserts, breeding on cliffs, man-made structures, and in isolated acacias or other trees and scrub.

Distribution No map. Once recorded in the breeding season, at Cizre, where a flock of 7 birds was observed on a rubbish dump on 9 July 1985 (Jakobsen 1986). No evidence of breeding. Flocks of non-breeding 1-year-old corvids are known to wander, and hence occurrence is not entirely unlikely, as the species is known to occur in the Syrian desert nearby. Note that 1-year old Common Ravens *C. corax* of the Turkish subspecies *laurencei* show a very brown nape and neck in summer, and thus are easily misidentified as *C. ruficollis*. Also, a flock of 7 Common Ravens *C. corax* was seen in July at the rubbish dump at Cizre (Dufourry 1990).

Geographical Variation None. *C. ruficollis* Lesson, 1831, was originally described from the Cape Verde Islands. No measurements are available for Turkey. Wing of adult ♂ of the population known to occur nearest to Turkey (viz., Sinai and Levant) 378, 390 (n=2), of ♀ 386.4 (374-397) (n=6), bill of ♂ 65.9, 67.5 (n=2), of ♀ 63.7 (61.4-66.4) (n=6). Bill depth at nostril of ♂ (all populations combined) 22.6 (21.0-24.0) (n=16), of ♀ 21.1 (19.5-23.0) (n=19), distinctly more slender than the bill of *C. corax corax* and *C. c. laurencei*, which have an average depth of 27.5 in ♂ and of 26.4 in ♀ (CSR).

References Jakobsen, O. (1986) Occurrence of Brown-necked Raven, *Corvus ruficollis*, at Cizre in eastern Turkey. *Zool. Middle East* 1, 32-33.

CORVUS CORAX
Common Raven   Karakarga

Habitat As in *C. corone*, occurs in all types of open country, if cliffs or large trees are available for nesting, from sea coasts up to mountain tops; occurs at 0-1700m in the north-west, to 2900m in the Taurus, and probably even higher in the east.

Distribution See map 110. Widespread, though in small numbers, generally avoiding human habitation when nesting; very scarce or absent from south-west Anatolia, the south coast, the heart of the Central Plateau, and the steppe of the South-East.

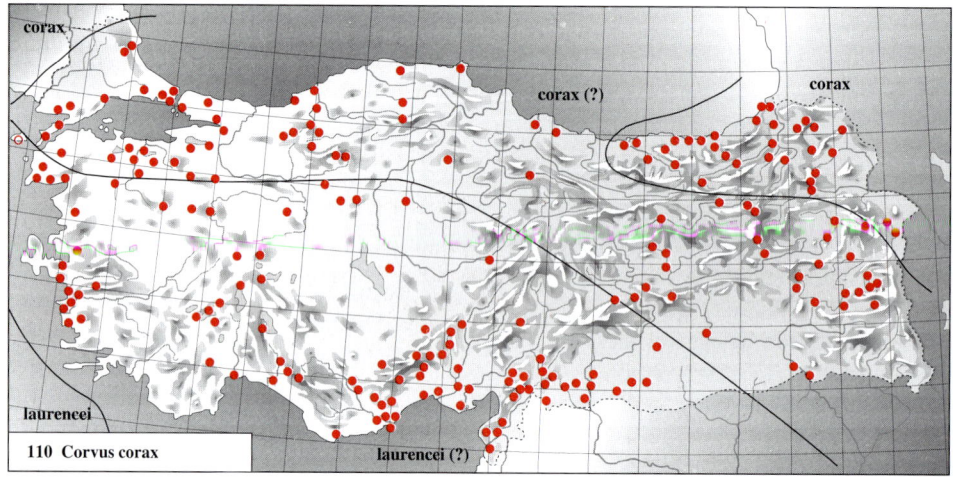

110 Corvus corax

Geographical Variation
Subspecies described or recorded in the region:

(W/E) **C. c. corax** Linnaeus, 1758, Sweden. [Plumage of adult strongly glossed purple-blue, feathers of throat lanceolate; large, but tarsus relatively short. Wing of adult ♂ from northern and central Europe 430.4 (407-452) (n=15), of adult ♀ 412.4 (400-439) (n=10), bill of ♂ 78.1 (73.0-83.4) (n=24), of ♀ 74.5 (70.9-80.3) (n=19), tarsus of ♂ 70.7 (66.2-75.0) (n=26), of ♀ 68.6 (65.0-72.3) (n=20); wing of adult ♂ from former Yugoslavia and western mainland Greece 416.2 (400-438) (n=6), ♀ 421.2 (416-427) (n=3), bill 74.7 (73.6-76.2) (n=5), of ♀ 71.1 (n=1), tarsus of ♂ 70.5 (69.8-71.3) (n=5), of ♀ 67.5 (CSR; wing measurements from Yugoslavia include data of Stresemann 1920). Occurs Europe and west Siberia south to Bulgaria, western mainland Greece, the Caucasus, Transcaucasia, and northern Iran.]

(S) ***C. c. laurencei*** Hume, 1873, Punjab (India). [Gloss of plumage of adult slightly duller and more oily blue; tips of throat feathers often bifurcated; large, tarsus relatively long. Wing of adult ♂ from Crete and the Levant 436.2 (423-450) (n=6), of adult ♀ 424, 432 (n=2), bill of ♂ 80.4 (76.0-85.0) (n=6), of ♀ 76.2, 78.2 (n=2), tarsus of ♂ 73.4 (72.0-75.7) (n=6), of ♀ 74.0, 74.8 (n=2) (CSR). Occurs from south-eastern mainland Greece, Aegean islands, Crete, Cyprus, and the Levant east through Iraq and Iran (except north) to central Asia. *C. c. subcorax* Severtzov, 1872, is an invalid name for the bird now known as *laurencei*.]

Subspecies recognized in Turkey: only 3 birds have been examined, all from the north-east (Rize, Verçenik), and these are inseparable from *corax*: bill of ♂ 78.9, 79.4 (n=2), of ♀ 74.4 (n=1), tarsus of ♂ 67.4, 72.8 (n=2), of 71.0 (n=1). As Turkey is bordered in the west and south by the range of *laurencei*, one may expect this subspecies in western and southern Turkey; *corax* of extreme north-east Turkey and Transcaucasia may extend further west in the Black Sea Coastlands and in the Van area, but there is no proof, and further data are needed.

STURNUS VULGARIS
Common Starling   Sigircik

Habitat Open country, in particular soft pastures, with trees or human structures (bridges, sheds, houses) nearby, the latter providing holes for nesting. Mainly in flat or gently sloping plains and valley bottoms, up to c. 1500m, but to c. 2700m in the valleys of the east.

Distribution See map 111. Generally common to abundant, but apparently scarce or absent in parts of Western Anatolia, scarce and local in the Black Sea Coastlands and in the Taurus, and virtually absent from the south coast and Hatay.

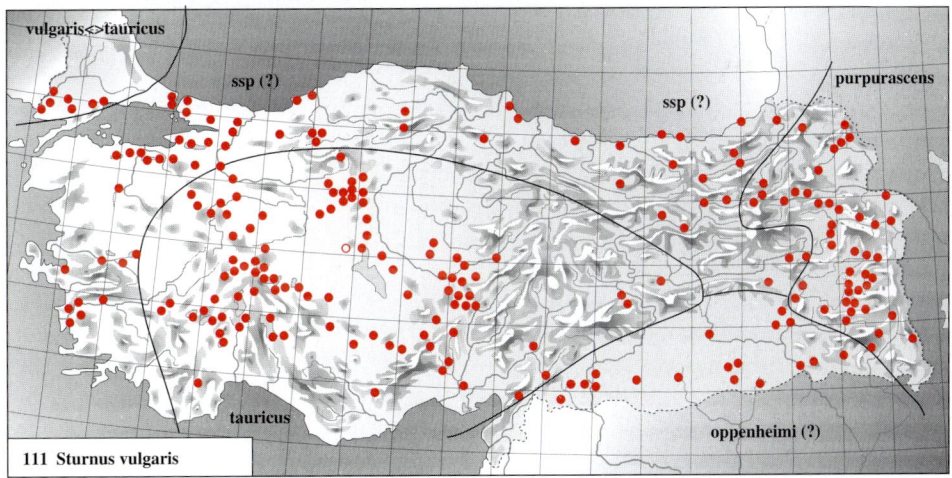

111 Sturnus vulgaris

Geographical Variation
Subspecies described or recorded in the region:

(W) ***S. v. vulgaris*** Linnaeus, 1758, Sweden. [Cap, chin, and upper throat of adult glossed green in worn plumage; ear-coverts green; nape, side of neck, and lower throat (bronzy) purple; hindneck to upper tail-coverts green (sometimes partly bluish or purplish); underparts (bluish-) green; upper wing-coverts and fringes of tertials and secondaries blue-green to purple-blue. Underwing not glossy, dark grey with broad cream-buff feather fringes of 1-2 mm wide. Wing of adult ♂ from north, west, and central Europe 133.0 (127-141) (n=187), of adult ♀ 129.8 (123-137) (n=147) (CSR). Populations intermediate between *vulgaris* and *tauricus* (sometimes separated as *'balcanicus'* Buturlin & Härms, 1909) inhabit eastern Greece (east of the Nestos river), eastern Bulgaria, and eastern Romania east to the lower Dnepr river in Ukraine; these are too variable individually to deserve recognition as a separate subspecies. Typically, they have cap, chin, and upper throat glossed purple; ear-coverts purple; nape, side of neck, and lower throat green; hindneck to upper tail-coverts green; underparts (bluish-) purple; upper wing-coverts (bluish-) purple; underwing as *vulgaris*. See Pateff (1947) and Munteanu (1967) for distribution and individual variation of these intermediate populations.]

(W?)   *S. v. tauricus* Buturlin, 1904, Crimea. [Cap, chin, and upper throat of adult glossed green in worn plumage; ear-coverts green; nape, side of neck, and lower throat reddish-purple; hindneck to upper tail-coverts mainly reddish- or bronzy-purple; underparts bronzy-purple; wing-coverts bronze-green (sometimes purple). Underwing not glossy, blackish with narrow off-white fringes 0.5-1 mm wide. At present, inhabits 2 disjunct areas, one in the steppe country along the northern shore of the Black Sea (from the lower Dnepr east to the Sea of Azov), the other in the steppe of Central Anatolia in Turkey (Stegmann & Stresemann 1935). Originally, both populations may have been connected with each other along the western side of the Black Sea, but intrusion of *vulgaris*, apparently better adapted to habitat changes, resulted in mixed populatons along this western shore (*'balcanicus'*, see above). A subspecies morphologically very similar to *tauricus*, named *porphyronotus*, breeds in the steppe and semi-desert of central Asia. Wing of adult ♂ of *tauricus* from central Turkey (Ankara, Eskisehir, Afyon, Eber Gölü, Aksehir, Beysehir, Burdur, Elmali, Sivas, Elazig, and Hazer Gölü) 135.9 (132-142) (n=16), of adult ♀ 132.7 (126-139) (n=12) (CSR).]

(T)   *S. v. purpurascens* Gould, 1868, Erzurum (Turkey). [Cap, chin, and upper throat of adult glossed purple in worn plumage; ear-coverts green; nape, side of neck, and lower throat green; hindneck to upper tail-coverts (bluish- or purplish-) green; underparts purple; upper wing-coverts (purple-) bronze; underwing as in *tauricus*. Restricted to eastern Turkey and neighbouring parts of Transcaucasia (Armenia and southern Georgia east to Tbilisi) and north-east Iraq. Wing of adult ♂ from eastern Turkey (Erzurum, Sarikamis, Taslicay, Ercis, Tatvan, Erçek, Van, Baskale, Yüksekova) 133.4 (128-137) (n=15), of ♀ 132.1 (130-136.5) (n=8) (CSR, including some data of Érard & Etchécopar 1970).]

(E)   *S. v. caucasicus* Lorenz, 1887, northern Caucasus. [Cap, chin, and upper throat of adult glossed green, ear-coverts green; nape, side of neck, and lower throat purple; hindneck to upper tail-coverts green; underparts bluish- or reddish-purple; upper wing-coverts bluish-purple; underwing as in *tauricus*. Occurs from the Volga delta south to the Caucasus, eastern Transcaucasia, and northern and south-west Iran. Wing of adults of both sexes from the Caucasus and eastern Transcaucasia 130.9 (128-134) (n=6) (CSR).]

Subspecies recognized in Turkey: birds from Thrace are intermediates between *vulgaris* and *tauricus* (*'balcanicus'*), those of the Central Plateau and its fringes are *tauricus* (see above), those of eastern Turkey *purpurascens* (see above). Some birds from Elazig are *tauricus*, some from Eleskirt are *purpurascens*, but others from both these localities are intermediates between *tauricus* and *purpurascens*, and these subspecies probably grade into each other in the entire upper Murat valley; wing of intermediate ♂ from Eleskirt 133 (130-136) (n=12) (Érard & Etchécopar 1970). The species is common in the Bosporus area and along the shores of the Sea of Marmara, but it is surprisingly rare elsewhere in

Western Anatolia and in the Black Sea Coastlands. No birds were examined from these areas, but the habitat here may suit *vulgaris* or *vulgaris/tauricus* intermediates better than pure *tauricus*. Perhaps, *vulgaris*-like birds are still spreading into these areas. The birds from the steppe in the South-East (Nizip and Birecik to Cizre) are usually considered to be intermediate between *tauricus* and *purpurascens* or between *tauricus* and *nobilior* (the latter a subspecies restricted to eastern Iran, southern Turkmenistan, and Afghanistan). However, unlike the intermediates from elsewhere (like *'balcanicus'*, above), the characters of the population of the South-East appear to be uniform, all birds having the cap, chin, and upper throat green, ear-coverts purple, nape, side of neck, and lower throat green, hindneck to upper tail-coverts purple with much green on mantle, and underparts purple with a bronze tinge on the flanks. Perhaps they form a separable subspecies, for which the name *oppenheimi* Neumann, 1915, described from Tall Halaf (near Ras-el-Ain in present-day Syria, very close to the Turkish border at Ceylanpinar) may be available (Neumann 1915, Sushkin 1933; but see Meinertzhagen 1924). Wing of ♂ from the steppe area of the South-East 132.0 (131-134) (n=7), of ♀ 126, 132.5 (n=2) (CSR). Though *caucasicus* occurs in north-west Iran, no birds with the characters of *caucasicus* were found in Turkey.

References Neumann, O. (1915) Über eine kleine Vogelsammlung von Nord-Mesopotamien. *J. Orn.* 63, 118-124. Meinertzhagen, R. (1924) Notes on a small collection of birds made in Iraq in the winter of 1922-23. *Ibis* (11)6, 601-625. Sushkin, P. (1933) Notes on some eastern forms of *Sturnus vulgaris*. *Ibis* (13)3, 55-58. Stegmann, B., & E. Stresemann (1935) *Sturnus vulgaris purpurascens* und *St. v. tauricus*. *Orn. Monatsber.* 43, 29-31. Pateff, P. (1947) On the systematic position of the Starlings inhabiting Bulgaria and the neighbouring countries. *Ibis* 89, 494-507. Munteanu, D. ('1965'=1967) Révision systematique des Étournaux, *Sturnus vulgaris* L., des environs de la Mer Noire. *Larus* 19, 179-203.

STURNUS ROSEUS
Rose-coloured Starling   Pembesigircik

Habitat Extensive open fields covered with rough grasses and scattered trees (e.g., natural steppe, pastures, or wasteland amidst cultivation) with an abundant food supply of grasshoppers, near to piles of stones, boulder scree (e.g., at foot of slopes, in steep gullies, or in quar-

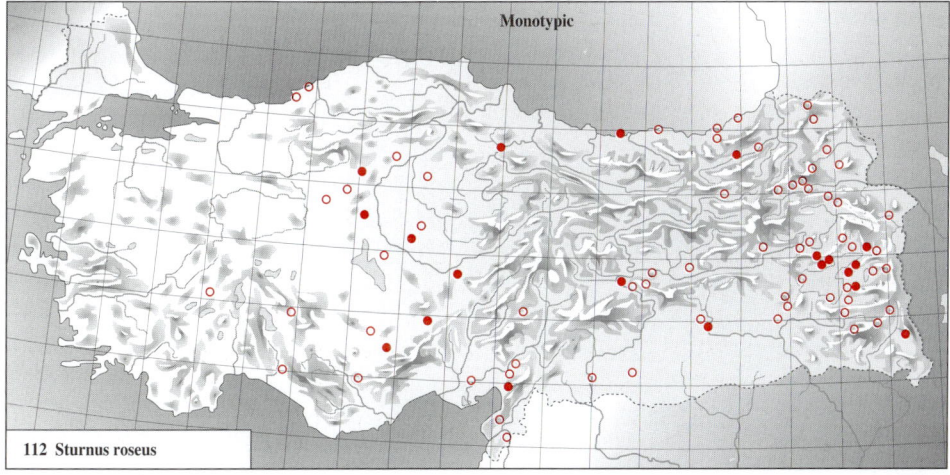

112 *Sturnus roseus*

ries), or holes in rocky faces which provide numerous breeding sites for this colonial species. Sometimes nests in fissures of walls in or near human habitation, especially in areas where it is held in high esteem due to its destruction of grasshoppers. Presence of fruit-bearing trees (e.g., mulberries, grapes, figs) are an extra incentive for settling in an area.

Distribution See map 112. Highly erratic in its nesting, depending on mass occurrence of grasshoppers. Wandering flocks occur widely in May and from mid-July onwards, and these are not mapped. Known breeding sites are shown in black (these are not necessarily occupied annually); other records in June and early in July are shown in white (these may point to local breeding, but may also refer to wandering non-breeding flocks).

Geographical Variation None. *S. roseus* (Linnaeus), 1758, was originally described from wanderers to Lapland and Switzerland. Wing of adult ♂ from Turkey 131.0 (128-132) (n=7), of ♀ 125.0 (118-128.5) (n=5), of 1st year ♂ 126, 128 (n=2) (Kummerlöwe & Niethammer 1934-35, Jordans & Steinbacher 1948, CSR); adult ♂ from elsewhere in the entire geographical range of the species 132.3 (126-139) (n=44), of ♀ 128.6 (120-135) (n=20) (CSR).

PASSER DOMESTICUS
House Sparrow Ev Serçesi

Habitat Near human habitation (such as towns, villages, isolated houses, or ruins), surrounded by cultivated fields; however, also often nests away from humans (but near cultivation) in tangled scrub or in creepers in woodland. Mainly in lowlands and valleys, but in mountains up to at least 2300m wherever arable land exists.

Distribution See map 113. Common to abundant everywhere, except dense woodland and mountain tops away from cultivation.

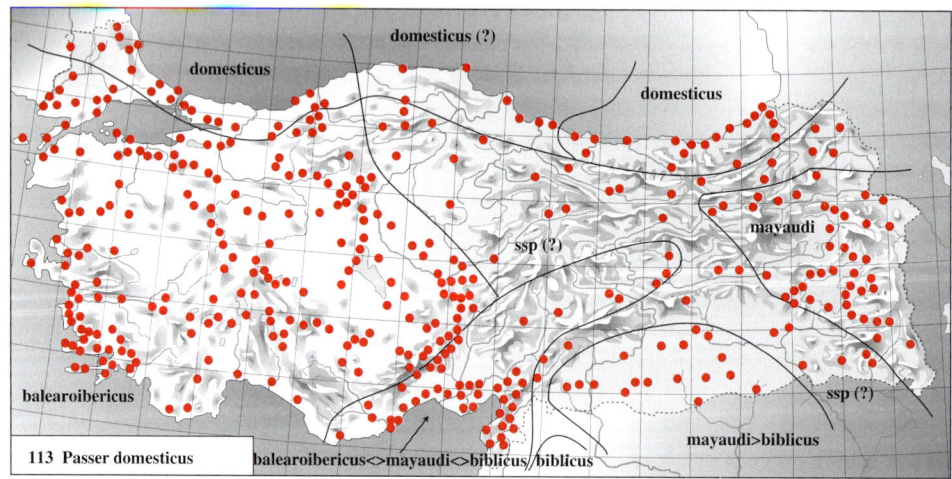

113 Passer domesticus

Geographical Variation
Subspecies described or recorded in the region:

(W) **P. d. domesticus** (Linnaeus), 1758, Sweden. [Cap and rump of ♂ dull grey, fringes of feathers of mantle and scapulars rufous-chestnut, ear-coverts, upper cheeks, and lower side of neck dull ash-grey (gradually becoming whiter when plumage becomes worn), sides of breast and flanks extensively dull ash-grey; lesser upper wing-coverts deep chestnut. Wing of ♂ from Sweden 79.4 (75-85) (n=101), of ♀ 76.5 (73-80) (n=46), bill of both sexes 15.6 (14.9-16.2) (n=10); wing of ♂ from Romania 79.5 (78-82) (n=16), of ♀ 75, 76 (n=2), bill 15.4 (14.2-16.3) (n=18) (CSR). Breeds in central and northern Eurasia, south to, among others, the northern Caucasus and eastern Transcaucasia.]

(W) **P. d. balearoibericus** Von Jordans, 1923, Valldemosa, Mallorca (Spain). [Cap and rump of ♂ paler than in *domesticus*, medium ash-grey, fine dark streaks on the cap more contrasting with the grey ground-colour; rufous of mantle and scapulars paler, more pink-cinnamon when worn; grey of ear-coverts, sides of breast, and flanks paler, light ash-grey, less washed brown or olive, more restricted on the side of the breast and on flanks, showing a broader white border between the black of the bib and the grey side of the breast, as well as a more extensively white belly; lower cheek and lower side of neck almost white. Size as in *domesticus*, e.g., wing of ♂ from southern Greece 78.7 (78-81) (n=12) (Niethammer 1943, CSR). Breeds Spain (except the north-west), the Balearic islands, Mediterranean France, the southern Balkan countries, and Greece.]

(T) **P. d. colchicus** Portenko, 1962, Artvin (eastern Black Sea Coastlands, Turkey). [Similar to *domesticus*, but rufous of upperparts slightly darker. Wing of ♂ from north-east Turkey and western Georgia 77.0 (74-79) (n=10) (Portenko 1962). Breeds also in the south-west Caucasus area.]

(T) **P. d. mayaudi** Kumerloeve, 1969, Van (eastern Turkey). [Cap purer grey than in *biblicus*, rufous of upperparts saturated rufous-brown, less pale as *biblicus*, less dark as *domesticus*; ear-coverts and cheeks white, not grey, in contrast to *domesticus* and *biblicus* (Kumerloeve 1969a). Close to *balearoibericus*, but upperparts slightly darker, near *domesticus*, and ear-coverts paler. Larger: wing of ♂ from Erzurum, Agri, Igdir, Ercis, and Van, including the type specimen, 83.0 (81-85) (n=21), of ♀ 79.5, 79.5 (n=2), bill of both sexes 15.6 (14.6-16.0) (n=13), bill depth at base 8.5 (7.9-9.1) (n=13) (CSR).]

(S) **P. d. biblicus** Hartert, 1904, Suweima (Jordan). [Paler than *balearoibericus*, cap and rump of ♂ light ash-grey, extensively washed sandy-buff in fresh plumage; rufous of mantle and scapulars paler, rufous-cinnamon; ear-coverts and lower cheeks pale ash-grey; sides of breast and flanks with a paler and more restricted pale ash-grey wash than *balearoibericus*, underparts extensively white; lesser upper wing-coverts rufous-chestnut. Wing of ♂ from Cyprus and the Levant 80.4 (78-84.5) (n=10), of ♀ 76.2 (73.5-79) (n=5), bill 15.6 (15.2-16.5) (n=9), bill depth 8.7 (8.3-9.1) (n=9) (Diesselhorst 1968, CSR); wing of ♂ from the Levant, Syria, Iraq, and Qasr-i-Shirin and Kermanshah in western Iran 81.2 (78.5-85) (n=26), of ♂ from Azarbaijan and Hamadan (north-west Iran) 82.2 (79-85) (n=18) (Vaurie 1949).]

Subspecies recognized in Turkey: difficult to summarize. When influence of bleaching and wear on plumage is accounted for, most birds from Turkey are closely similar in general colour, except for the colour of the ear-coverts and cheeks (which tend to vary both individually and geographically, and which are also strongly influenced by wear, becoming gradually whiter during spring and summer). Only birds from the north-west (east to Zonguldak) and the north-east (Rize and Borçka) are clearly darker in general colour of plumage than other Turkish birds, and are inseparable from *domesticus*; 'colchicus' from the north-east does not appear to be a valid subspecies. Further inland (south from Sogukpinar, Bolu, Devrek, Ankara, Erzurum, and Agri), the general colour of all birds is paler, comparable with *balearoibericus*. These paler birds are not uniform in size or in cheek colour: size is smaller in Western Anatolia and on the Central Plateau, intermediate in the Taurus, the Anti-Taurus, the Amik Gölü area, and in Elazig, and large in the steppe of the South-East and in the mountains of the east; cheek colour, though strongly variable individually, is on average greyer in Western Anatolia, the Central Plateau, and the Taurus and whiter in the mountains of the east. The smaller, greyer-cheeked birds from the west and centre of the country are included in *balearoibericus*; the larger, whiter-cheeked birds from the mountains of the east (Agri, Erzurum, Igdir, Ercis, and Van) are *mayaudi*. No Turkish birds are similar to the greyer-cheeked intermediate-sized *biblicus* from the Levant, but birds of intermediate size and variable cheek colour occurring in the Çukurova area (Mersin, Tarsus), the Anti-Taurus (Osmaniye, Haruniye), the Amik Gölü area (Bedirge, Gölbasi, Amik Gölü), and Elazig are more or less intermediate between *balearoibericus, biblicus* and *mayaudi,* while birds from the steppe of the South-East (Birecik, Urfa, Ceylanpinar) are intermediate between *biblicus* and *mayaudi,* though nearer to *mayaudi*. Wing of ♂ of *domesticus* from the north-west 75.5, 76 (n=2), of ♀ 75 (n=1) (Rokitansky & Schifter 1971, CSR), of ♂ from the north-east 78.0 (75-81) (n=3) (CSR). Wing of *balearoibericus* from Western Anatolia and the Central Plateau north to Bolu and Devrek 78.9 (76-81.5) (n=12), of ♀ 76.2 (74-81) (n=4) (Kummerlöwe & Niethammer 1934-35, CSR), of ♂ from the western and central Taurus (Burdur, Aksehir, Beysehir, Solak, and Eregli) 80.3 (78-82) (n=19), of ♀ 77, 77.5 (n=2) (CSR). Wing of intermediate ♂ from the Çukurova area, the Anti-Taurus, the Amik Gölü area, and Elazig 80.0 (77-84) (n=15), of ♀ 75.7 (75-76.5) (n=3), of intermediate ♂ from the steppe of the South-East 82.0 (80-84) (n=8), of ♀ 75 (n=1) (CSR).

PASSER HISPANIOLENSIS
Spanish Sparrow   Bataklik Serçesi
Habitat Mainly in lowlands, locally to over 1000m, nesting in tangled scrub (e.g., poplars, willows, tamarisks) in fields, along water, along roads, or in bottoms of nests of storks and eagles, close to arable land.

Distribution See map 114. Common to abundant in suitable habitat in Thrace, north-west Anatolia, and the Çukurova area, locally common to infrequent in the remainder of Western Anatolia, the Southern Coastlands, the Central Plateau, and the plateaux of the South-East, virtually absent from the Black Sea Coastlands (but common in the Kizilirmak delta) and from the mountain valleys of the East and South-East.

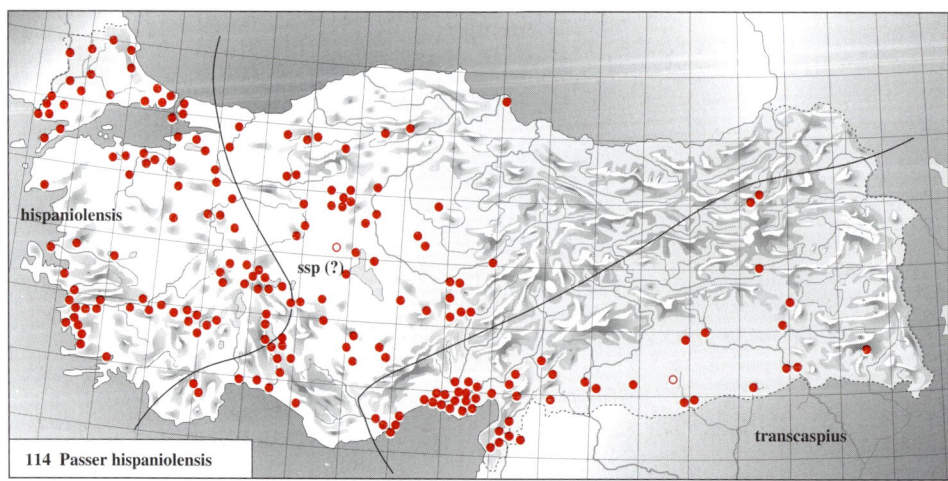

114 Passer hispaniolensis

Geographical Variation
Subspecies described or recorded in the region:
- (W) **P. h. hispaniolensis** (Temminck), 1820, Algeciras (southern Spain). [In fresh plumage, borders of flight feathers, tertials, greater upper wing-coverts, scapulars, and feathers of mantle of male deeper rufous-pink, tips of feathers of mantle and scapulars duller sandy olive-grey. Wing of ♂ from Spain, Sardinia, and North Africa 78.5 (77-81) (n=27), of ♀ 75.9 (73-78) (n=13), bill of both sexes 15.5 (14.5-16.3) (n=32) (CSR).]
- (E) **P. h. transcaspicus** Tschusi, 1902, Iolotan (eastern Turkmenistan). [In fresh plumage, rufous of upperparts and wing of male paler rufous-pink and tips of feathers of mantle and scapulars paler grey-buff; in worn plumage, chestnut of head slightly less rich maroon; size slightly larger. In all plumages (but especially when worn), colour differences are subtle, even when series of skins can be compared. Wing of ♂ from Iran and Transcaspia eastwards 81.3 (78.5-84) (n=14), of ♀ 78.1 (75-82) (n=4), bill 15.6 (14.8-16.8) (n=11) (CSR).]

Subspecies recognized in Turkey: birds from Turkey are more or less intermediate in colour between *hispaniolensis* and *transcaspicus*. When size is also accounted for, the smaller birds from Thrace and Western Anatolia (east to Beyşehir and Akşehir) are best included in *hispaniolensis*, while the larger ones from the Çukurova delta eastwards are *transcaspicus*. No birds were examined from most of the Central Plateau nor any of the western part of the Southern Coastlands, and the subspecies inhabiting these regions is unknown. Wing of adult males from Thrace 77.2 (75-81) (n=4) (Rokitansky & Schifter 1971), from Burdur 79.2 (77-81.5) (n=22), from Akşehir and Beyşehir Gölü 78.8 (76-82) (n=8), from Adana, Amik Gölü, and Jerablus (Syria, just south of Birecik) 80.1 (78-82) (n=10), and from Ceylanpınar 80.5 (78-83.5) (n=21); wing of adult ♀ from Burdur, Akşehir, and Beyşehir 75.3 (73-77) (n=10), from Ceylanpınar 75 (n=1) (CSR, with some additional data for Beyşehir from Vauk 1973).

PASSER MOABITICUS
Dead Sea Sparrow   Ölüdeniz Serçesi
Habitat Dense stands of willows or poplars near cultivation, usually those bordering rivers or lakes, but sometimes in willows in marshy reedbeds or in scrub up to 10 km from a river.

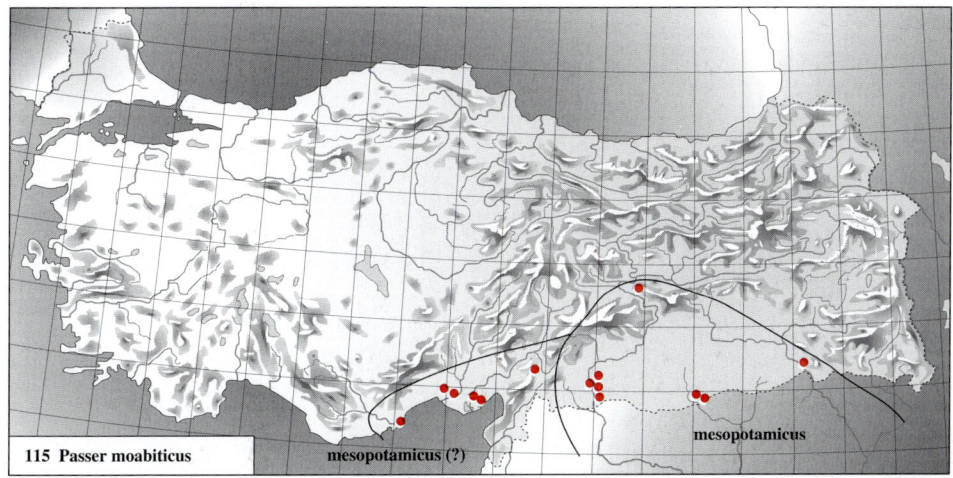

115 Passer moabiticus

Distribution See map 115. Restricted to a few localities in the eastern part of the Southern Coastlands and in the South-East, where its occurrence is relatively recent. For colonization of and expansion in Turkey, see Kumerloeve (1965, 1969g, 1978), Érard & Etchécopar (1968), and Cramp (1971). The small colonies are sometimes hard to find amongst the dense ones of *P. hispaniolensis,* which often share the same sites (Berk & Letschert 1988).

Geographical Variation

Subspecies described or recorded in the region:

(S)   ***P. m. moabiticus*** Tristram, 1864, Palestine. [Smaller, wing of ♂ from the Dead Sea Valley 62.3 (60-64) (n=32), of ♀ 59.5 (58-62) (n=19) (CSR).]

(E)   ***P. m. mesopotamicus*** Zarudny, 1904, Mochammera (=Khorramshahr, south-west Iran). [Larger, wing of ♂ from Iraq and south-west Iran 65.2 (63-67) (n=11), of ♀ 62.8 (62-64) (n=6) (CSR).]

Subspecies recognized in Turkey: birds examined from Ceylanpinar and Birecik clearly belong to *mesopotamicus,* as wing of ♂ from here is 65.6 (64-67) (n=10), of ♀ 62.5 (n=1) (CSR). The other Turkish populations likely belong to *mesopotamicus* also. Usually, no subspecies other than *moabiticus* and the very yellow *yatii* from south-east Iran and south-west Afghanistan are recognized in this species, but *mesopotamicus* is a valid race, combining the size of *yatii* with the colour of *moabiticus.* Less than 10% of the non-yellow birds cannot be assigned to either *moabiticus* or *mesopotamicus* on size. The species clearly has colonized Turkey from Iraq through the Firat valley. Thereafter, the Karasu and Ceyhan valleys were reached (which flow into the Mediterranean), after crossing the watershed north-west of Gaziantep (Érard & Etchécopar 1968). Until the subspecies of the birds of the Çukurova and Göksu deltas (and Cyprus) is known, the crossing of the watershed near Gaziantep cannot be proved.

References Kumerloeve, H. (1965) Le Moineau moabite, *Passer moabiticus* Tristram, près Birecik sur l'Euphrate. *Alauda* 33, 257-264. Kumerloeve, H. (1969g) The Dead Sea Sparrow: a second breeding place on Turkish and the first-known breeding place on Syrian territory. *Ibis* 111, 617-618. Cramp, S. (1971) The Dead Sea Sparrow: further breeding places in Iran and Turkey. *Ibis* 113, 244-245. Kumerloeve, H (1978) Situation des moineaux moabites nicheurs en Turquie. *Alauda* 46, 181-182.

PASSER MONTANUS

Tree Sparrow   **Dag Serçesi**

Habitat Cultivated country, breeding in dense stands of willow- or poplar-scrub or copses of deciduous forest, either near human habitation or away from humans; sometimes in parks, gardens, and plantations. Occurs mainly in lowlands, on plateaux, and in bottoms of valleys, at 0-1400m.

Distribution See map 116; also, Kumerloeve (1961a). Widespread but very local, mainly north of 40°N, on the fringes of the Central Plateau, and at the foot of the western Taurus, with a few scattered breeding season records elsewhere.

Geographical Variation

Subspecies described or recorded in the region:

(W)   ***P. m. montanus*** (Linnaeus), 1758, Bagnacavallo (Ravenna, northern Italy). [General colour dark; size small to intermediate, bill intermediate to strong. Wing of adult ♂ from northern and central Europe 71.3 (68-74) (n=81), of adult ♀ 68.6 (66-72) (n=51), bill (both sexes) 13.2 (12.3-14.3) (n=86) (CSR). Occurs Eurasia, in Europe south to mainland Greece and to the centres of Bulgaria, Romania, Ukraine, and European Russia. Replaced

Sparrows and Snowfinches

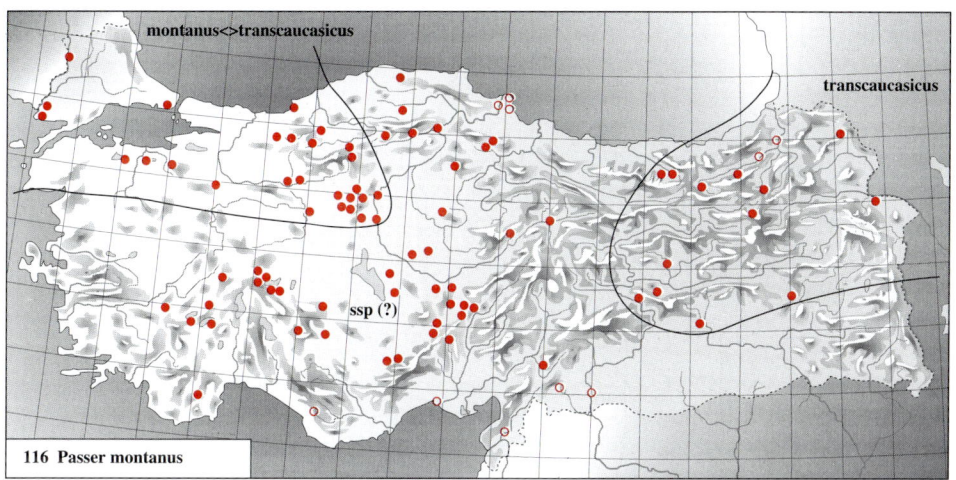

116 Passer montanus

by birds intermediate between this and the next subspecies in eastern Bulgaria and Romania, in the southern Ukraine and Crimea, and along the northern fringe of the Volga-Ural steppe. Wing of ♂ from eastern Bulgaria and eastern Romania 70.9 (68-72.5) (n=4), of ♀ 69.5, 72.5 (n=2), bill 12.8 (12.2-13.5) (n=6) (CSR).]

(E) **P. m. transcaucasicus** Buturlin, 1906, Akhaltsikhe (Georgia, south-west Transcaucasia). [General colour slightly paler and brighter, cap paler, underparts whiter; general size as in *montanus*, but bill shorter, to skull *c*. 1.3 mm less (Vaurie 1949), to nostril *c*. 0.5 mm less (Keve & Kohl 1978). Wing of both sexes from the Caucasus and northern Iran 68.5 (64-73) (n=31) (Vaurie 1949, Schüz 1959, Diesselhorst 1962, CSR), bill in birds from the Caucasus 12.9 (12.2-13.5) (n=4) (CSR). Occurs from the lower Volga south to the Caucasus, Transcaucasia, and northern Iran.]

Subspecies recognized in Turkey: although birds from Istanbul are said to be referable to *transcaucasicus* by Keve (1978) in his review of the geographical variation of the species, birds examined from Ankara further east are closely similar to *montanus* from central Europe, with wing of ♂ 71 and 73.5, of ♀ 70, and bill 13.7 and 13.9 (CSR). The colour of these birds is only very slightly paler than in *montanus*, with the borders of the mantle and scapulars slightly paler cinnamon and the rump more sandy-buff, less buffish-olive; they are closely similar to birds from eastern Bulgaria and eastern Romania, which are also rather like *montanus* in size but somewhat paler in plumage. All birds from eastern Bulgaria, eastern Romania, and north-west Turkey are here considered to be intermediate between *montanus* and *transcaucasicus*. Birds examined from eastern Turkey (Elazig, Erzurum) are pure *transcaucasicus* (wing of ♂ 71.5 and 72.5, of ♀ 68; bill 12.0, 13.2, and 13.6: CSR).

References Keve, A. ('1977'=1978) Revision der Unterarten des Feldsperlings [*Passer montanus* (Linné, 1758)]. *Zool. Abh. Staatl. Mus. Tierkde. Dresden* 34, 245-273. Keve, A., & S. Kohl (1978) Variations-statistische Untersuchungen an den unterarten des Feldsperlings (*Passer montanus* Linné, 1758). *Nymphaea* 6, 583-606.

GYMNORIS XANTHOCOLLIS
Yellow-throated Sparrow   Sarigirtlak Serçe
Habitat Arid cultivated steppe and semi-desert, dotted with stands of acacias, palms, tamarisks, or hedgerows, often close to human habitation.

Distribution See map 117. Restricted to some scattered localities in the South-East, where occurrence is apparently recent.

Geographical Variation

Subspecies described or recorded in the region:

(S) ***G. x. transfuga*** Hartert, 1904, Bahu Kalat (south-east Baluchestan, Iran). [A pale race; upperparts brown-grey, less dark drab-grey than *xanthocollis* from India; rump buffish-drab instead of brown-grey; lesser upper wing-coverts rufous-cinnamon instead of chestnut. Wing of ♂ of *transfuga* from Iraq and Iran 84.4 (81.5-88) (n=10), of ♀ 78, 81.5 (n=2) (CSR), or, in another sample from Iran, wing of ♂ 81.5 (79-85) (n=14), of ♀ 78, 78 (n=2) (Diesselhorst 1962, Érard & Etchécopar 1970).]

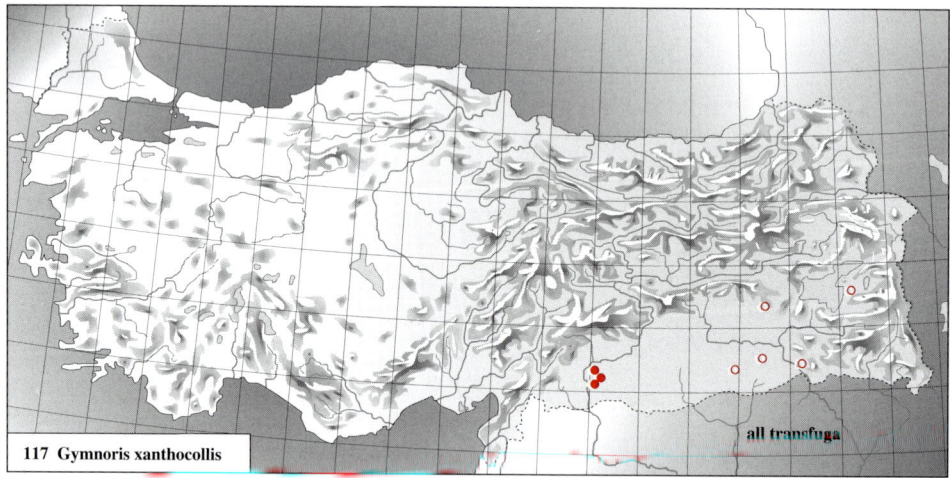

117 *Gymnoris xanthocollis*

Subspecies recognized in Turkey: apparently never collected in Turkey, but unlikely to be anything other than *transfuga*, which is the subspecies breeding nearby in Iraq and Iran.

Note: Yellow-throated Sparrow and its 3 Afrotropical relatives differ markedly from Rock Sparrow *Petronia petronia* in plumage, structure, habitat and behaviour, warranting recognition of a separate genus *Gymnoris* (see Cramp & Perrins 1994a).

PETRONIA PETRONIA

Rock Sparrow   **Kayalik Serçesi**

Habitat Dry stony slopes and plains with scattered trees, near walls of porous rock or loam, boulder scree, or neglected buildings which provide holes which are needed for nesting; mainly at 150-1200m in the west, in the Black Sea Coastlands, and on the Central Plateau, but at 200-2500 (-3000)m in the Taurus mountains and in the East and South-East.

Distribution See map 118. Widespread but rather local in all areas where rocks occur near cultivation; apparently absent from Thrace and from areas close to coasts, common only in some mountain valleys of the Taurus and the East.

Geographical Variation

Subspecies described or recorded in the region:

(W) ***P. p. petronia*** (Linnaeus), 1766, northern Italy. [Ground-colour of upperparts greyish-brown; stripes on cap and on mantle and scapulars (brown-) black; yellow of spot on central throat deep sulphur; wing and bill length rather short, bill rather thick at base. Wing of ♂ from eastern Spain, Italy, central Europe, former Yugoslavia, and Greece 98.7 (95-102) (n=28), of ♀ 95.8 (92-100) (n=19), bill of both sexes 17.2 (16.2-18.5) (n=40) (CSR; wing measurements include some data of Stresemann 1920).]

(E) **P. p. exigua** (Hellmayr), 1902, Rostov-on-the-Don (Rostov-na-Donu, southern European Russia). [Ground-colour of upperparts duller and greyer than in *petronia*, rump paler and greyer; stripes on cap dark sepia-brown; streaks on mantle and scapulars black and sharply defined, but shorter and less extensive than in *petronia*; throat spot deep sulphur-yellow; wing and bill length slightly larger, bill rather thick at base. Wing of ♂ from Rostov-na-Donu and the Caucasus 100.2 (97.5-102.5) (n=7), of ♀ 97.5, 98.5 (n=2), bill 17.2 (16.0-18.5) (n=9) (CSR; data include those of the type specimen). Occurs also in Transcaucasia and northern Iran.]

(S) **P. p. puteicola** Festa, 1894, Madaba (western Jordan). [Ground-colour of upperparts sandy-brown; stripes on cap greyish olive-brown; streaks on mantle and scapulars restricted in width and extent, sepia-brown or pale grey-brown; throat spot pale yellow; wing and bill long, bill very thick at base. Wing of ♂ from Israel, Lebanon, Jordan, and Syria 101.6 (98-106) (n=20), of ♀ 99.1 (95-103) (n=17), bill 18.7 (17.1-20.0) (n=26) (CSR).]

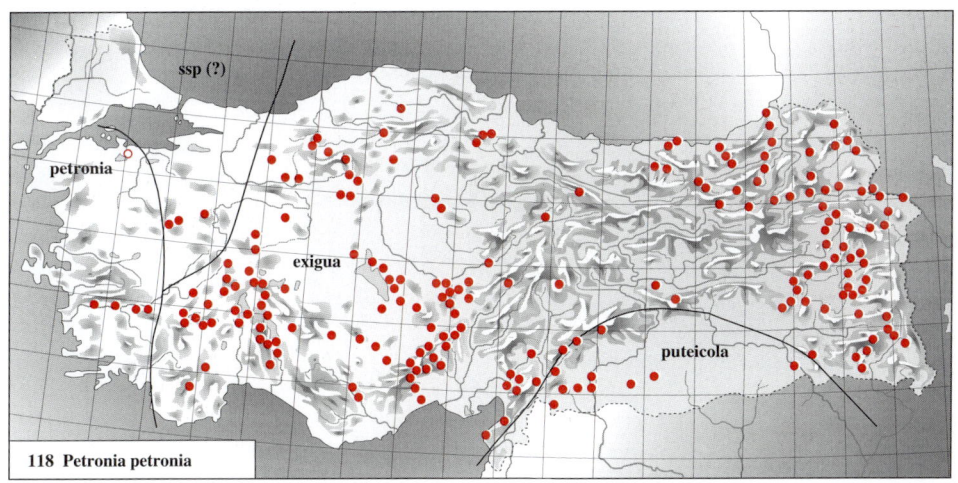

118 *Petronia petronia*

Subspecies recognized in Turkey: birds from Western Anatolia are *petronia*, as far as can be ascertained from a single ♂ examined from Aydin with wing 96.5 and bill 17.5 (CSR). The population inhabiting the area round Gaziantep, Birecik, and Urfa in the South-East belong to *puteicola*; wing of ♂ from this area 101.3 (100-103) (n=5), of ♀ 95, 95 (n=2) (Bird 1937, CSR). The remainder of Turkey, east from at least Burdur, Mudurnu, the Ankara area, and Kastamonu, is inhabited by *exigua*. Wing of ♂ of *exigua* from central Turkey (Burdur, Taurus mountains, Mudurnu, Çankiri, and Kastamonu) 100.9 (98-103) (n=8), of ♀ 96.3 (95-98) (n=5), bill 18.3 (17.0-19.5) (n=13); wing of ♂ of *exigua* from eastern Turkey (Kars and Van areas) 101.8 (98.5-106) (n=7), bill of ♂ 18.0 (17.1-18.5) (n=7) (CSR).

CARPOSPIZA BRACHYDACTYLA

Pale Rock Sparrow   **Tas Serçesi**

Habitat Arid rocky slopes or flat stony plains in steppe or semi-desert, patchily covered with thorny scrub, some isolated trees, or other scanty vegetation. Mainly at 500-1200m in the hills of the South-East, but to *c.* 2000m in the Van and Dogubayazit areas. Makes a neatly woven cup-nest in dense scrub, in contrast to *Passer*, *Gymnoris*, and *Petronia*, and this, together with the colour of the eggs and some peculirities in voice and behaviour, points to cardueline relationships rather than to a position in the sparrow family.

Distribution See map 119; also, Kumerloeve (1975c). Locally common but rather erratic in the South-East, and not all sites are occupied annually. Mainly in hills on the Mesopotamian plateau as well as along its border, with scattered breeding season records (June-July) in the surroundings of Van, Hakkari, and Dogubayazit, close to breeding sites in Armenia. Occurrence above Ispir (where a single bird observed on 21 July 1987 - Vermeulen 1987) requires confirmation.

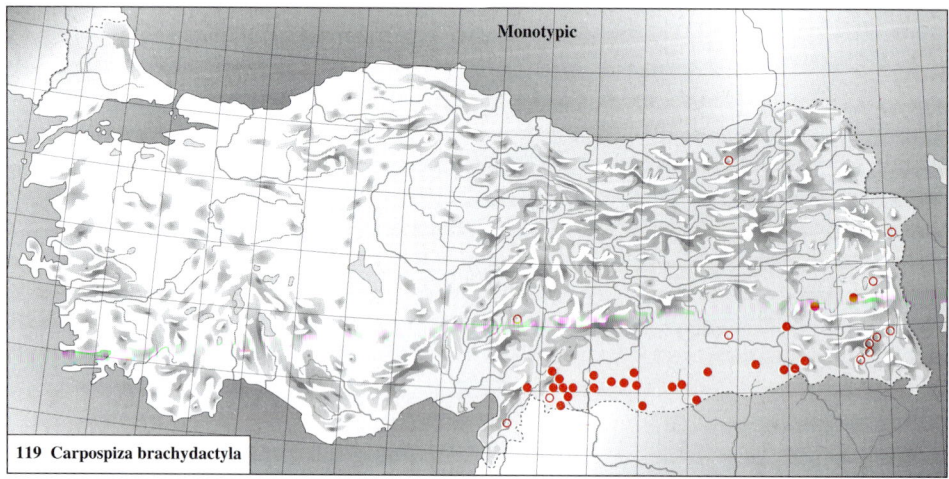

119 *Carpospiza brachydactyla*

Geographical Variation None. *C. brachydactyla* (Bonaparte), 1850, was originally described from a migrant collected at Al Qunfidhah (south-west Saudi Arabia). *C. b. psammochroa* Reichenow, 1916, described from Kusha (Sarhad area, eastern Iranian Baluchistan) and said to be paler and larger, is based on a bleached juvenile (type specimen examined), which does not differ in characters from bleached juveniles from elsewhere in the species' range. No geographical variation in colour or size is apparent. Wing of ♂ from Turkey (Gaziantep and Ceylanpinar) 97.2 (96-98) (n=3), of ♀ 89.5, 94.5 (n=2), bill of both sexes 14.9 (14.4-15.4) (n=4); wing of ♂ from elsewhere 97.3 (93-102) (n=26), of ♀ 93.2 (89-96) (n=25), bill 15.2 (14.4-16.9) (n=29) (CSR); wing of ♂ from Dzhulfa (Nakhichevan, Transcaucasia) 94.4 (91.5-96) (n=7) (Worobiev 1934).

References Kumerloeve, H. (1975c) Expansion du Moineau soulcie pale *Petronia brachydactyla* et du Roselin cramoisi *Carpodacus erythrinus* en Turquie. *Alauda* 43, 324-325.

MONTIFRINGILLA NIVALIS

Snowfinch   Kar Ispinozu

Habitat Alpine slopes with scree and stones, covered with sparse cushions of low vegetation, close to patches of snow. Mainly at 1800-2900m, in the mountains along the fringes of the Central Plateau sometimes down to *c.* 1000m. In late summer, occurs up to 3500m, but apparently does not breed there.

Distribution See map 120. On all mountains which reach over 2000m, but in the north-west and west only recorded from the Sultan Daglari and from the mountains near Elmali, and perhaps less common here than elsewhere.

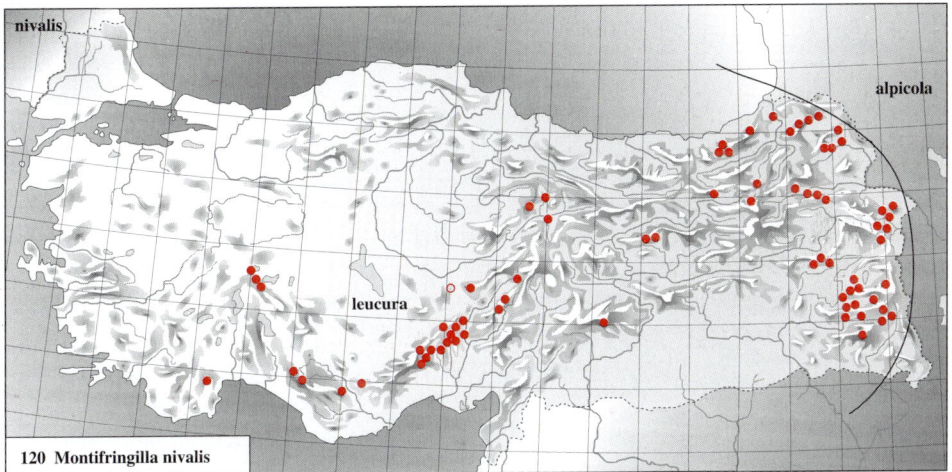

120 *Montifringilla nivalis*

Geographical Variation

Subspecies described or recorded in the region:

(W)   ***M. n. nivalis*** (Linnaeus), 1766, Switzerland. [Top of head and neck of ♂ blue-grey, contrasting with the deep sepia-brown of mantle and scapulars; breast and flanks washed grey; upper wing-coverts and secondaries white, except for a little black hidden at the extreme base; large, but bill relatively short. Wing of ♂ from the Alps 121.6 (117-127) (n=32), of ♀ 116.4 (113-122) (n=19), bill of both sexes 16.8 (15.4-18.3) (n=38) (CSR). Wing of ♂ from Greece 119 (116-120) (n=4), of ♀ 116 (n=1), bill (apparently to feathers instead of to skull) 13.6 (13.5-14) (n=5) (Watson 1961).]

(T)   ***M. n. leucura*** Bonaparte, 1855, Erzurum (eastern Turkey). [Top of head and neck of ♂ brown-grey, rather similar to the grey-brown of mantle and scapulars, (the latter are paler than in *nivalis*); underparts more extensively white, grey of breast and flanks more restricted; upper wing-coverts and secondaries largely white, as in *nivalis*, but lesser coverts sometimes with black basal half and outermost secondary often with some black at base on outer web; wing slightly shorter but bill longer than in *nivalis*. Wing of ♂ from eastern Turkey (Kars and Baskale areas) 115.9 (112-120) (n=7), of ♀ 113.0 (111-114) (n=3), bill 18.2 (18.0-18.7, once 16.4) (n=10) (CSR). *M. n. fahrettini* Watson, 1961, described from Ak Dag (above Kas in the west of the Southern Coastlands, Turkey), was originally compared with *'gaddi'* from the Zagros mountains in south-west Iran (a subspecies very similar to or identical with *alpicola*); it was stated to differ from *'gaddi'* in paler and greyer mantle and scapulars, less reddish-tan than those of *'gaddi'*. Thus, it is apparently very similar to *leucura*, and G. E. Watson (in Kumerloeve 1968) was indeed not able to separate birds from eastern Turkey from his series of *fahrettini*. Therefore, *'fahrettini'* becomes a synonym of *leucura*. Wing of ♂ of *'fahrettini'* from the area round Kas and Antalya 114.0 (111-117) (n=5), of ♀ 110.8 (106-116) (n=4), bill (apparently measured to the feathers) 14.1 (13.5-14.5) (n=9) (Watson 1961).]

(E) **M. n. alpicola** (Pallas), 1831, 'alpine peaks of the Caucasus Mountains'. [Similar to *leucura*, but mantle and scapulars of ♂ darker drab-grey with sepia tinge. Colour of mantle and scapulars approaches that of *nivalis*, but the grey-brown of the head is less contrasting with mantle and scapulars than in *nivalis*, underparts are whiter, and some more black is present on the lesser coverts and the outermost secondary; wing and bill long. Wing of ♂ from the Caucasus and northern Iran 120.7 (117-125) (n=13), of ♀ 112.8 (110-117) (n=9), bill 18.5 (16.4-19.5) (n=22) (CSR); wing of ♂ from south-west Iran 119.3 (116-126) (n=19), of ♀ 114.3 (112-116) (n=4) (Vaurie 1949).]

Subspecies recognized in Turkey: all birds from Turkey are *leucura:* see the discussion above. The European subspecies *nivalis* does not reach further east than Greece and Bulgaria; *alpicola*, occurring from the Caucasus south to northern and south-western Iran, apparently does not reach Turkey.

FRINGILLA COELEBS
Chaffinch   Ispinoz

Habitat Open deciduous, mixed, or pine forests, olive groves, orchards, plantations, and (locally) city parks. At 0-2100m in the west and north, but from *c.* 500m upwards elsewhere.

Distribution See map 121. Common and widespread in the wooded regions of the Black Sea Coastlands, Thrace, Western Anatolia, and the Taurus, more local in forest relicts at the fringes of the Central Plateau and in the valleys of the mountainous parts of the East and South-East.

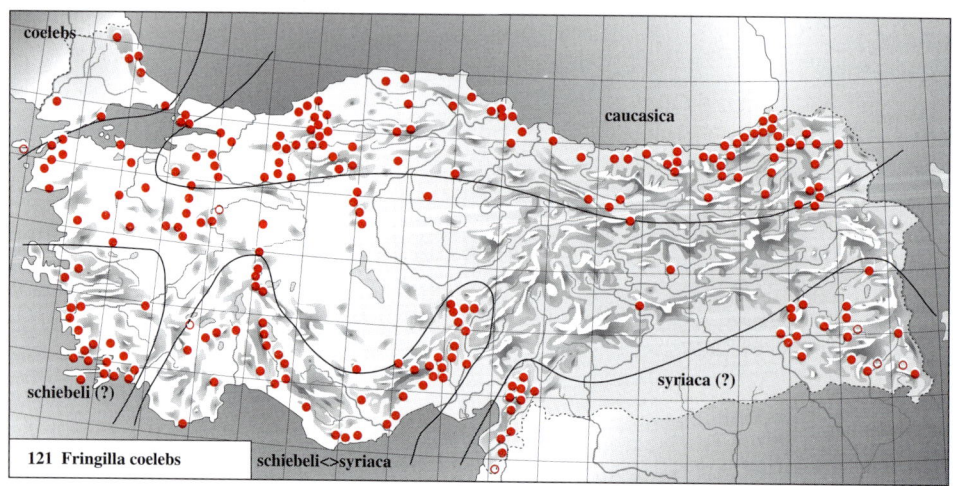

121 Fringilla coelebs   schiebeli<>syriaca

Geographical Variation
Subspecies described or recorded in the region:

(W)   **F. c. coelebs** Linnaeus, 1758, Sweden. [Mantle and scapulars of ♂ umber- or tawny-brown, rump bright green, sides of head deep rufous-cinnamon with vinous cast, chin to breast vinous-cinnamnon, belly pink-vinous; underparts of ♀ extensively olive-drab. Wing of ♂ from Scandinavia and central Europe 89.5 (86-95) (n=72), of ♀ 83.7 (81-86) (n=28), bill of both sexes 14.9 (14.0-16.0) (n=90), bill depth at base of ♂ 7.4 (6.8-8.0) (n=46); wing of ♂ from Bulgaria 91.2 (90-94) (n=6), of ♀ 84.3 (83-87) (n=3), bill 14.5 (13.5-15.3) (n=6), bill depth of ♂ 7.6 (7.2-7.8) (n=4) (CSR).]

(W) **F. c. schiebeli** Stresemann, 1925, near Canea (Crete, Greece). [Like *coelebs*, but paler; mantle and scapulars of ♂ mixed coppery-orange and green, side of head more orange-cinnamon, breast light vinous-pink. Wing of ♂ from Crete 87.5 (82-91) (n=12), of ♀ 79.5, 83 (n=2), bill 15.3 (14.0-16.0) (n=15), bill depth of ♂ 7.5 (7.2-8.3) (n=13) (CSR).]

(S) **F. c. syriaca** Harrison, 1945, Bcharre (Lebanon). [Mantle and scapulars of ♂ paler tawny- or orange-brown than in *coelebs*, rump yellowish-green, sides of head deep vinous-cinnamnon, underparts almost uniform pale pinkish-mauve or pale vinous-pink; upperparts of ♀ paler drab-brown than in ♀ of *coelebs*, dark stripes on cap paler brown, underparts paler with more restricted light ash-grey tinge on sides of breast and flanks. Wing of ♂ from Cyprus 90.3 (87-93) (n=9), of ♀ 85.0 (84-86) (n=3), bill 15.4 (14.8-16.3) (n=11), bill depth of ♂ 7.5 (7.3-7.8) (n=8) (CSR). Occurs also in Levant and perhaps this subspecies in the Zagros mountains of western Iran.]

(E) **F. c. caucasica** Serebrovski, 1925, Zakataly (eastern Caucasus, Azerbaijan). [Like *coelebs*, but mantle and scapulars of ♂ mixed dull umber-brown and green, rump rather dull olive-green, and breast slightly paler vinous-pink; bill heavy at base, but not as heavy as in *solomkoi* from the Crimea and the south-west Caucasus (bill depth of a single ♂ examined from Crimea 10.2). Breeds in the eastern Caucasus and in Transcaucasia.]

Subspecies recognized in Turkey: difficult to establish, as the differences mentioned above are slight and strongly influenced by bleaching and wear. In colour, birds from northern Turkey are like the darker subspecies *coelebs* and *caucasica* cited above, those from the Taurus and a single bird examined from Çatak (near Van) are paler, closer to the intermediate race *schiebeli* and the pale race *syriaca*. The bill of birds from the Ulu Dag and the western part of the Black Sea Coastlands is longer and heavier than that of *coelebs*, more similar to the bill of *caucasica*, and therefore all northern birds (except those of Thrace) are included in *caucasica* : wing of ♂ 89.6 (87-91) (n=10), of ♀ 84.5 (81-88) (n=11), bill (both sexes, excluding juveniles) 15.6 (14.9-16.3) (n=12), bill depth of ♂ 7.7 (7.4-8.4) (n=8) (CSR). Birds of Thrace are *coelebs*; wing of ♂ from Thrace 88.2 (84.5-93) (n=3), bill 15.2 (14.9-15.7) (n=3), bill depth 7.8 (7.5-7.9) (n=3) (CSR). Birds from the Taurus and Van areas are intermediate between *schiebeli* and *syriaca* in colour and size; perhaps those of south-west Anatolia are nearer to *schiebeli* and those of the eastern Taurus and Van are nearer to *syriaca*, but the skins available are too few to established on this. Wing of ♂ from the Taurus and Van 88.8 (86-92) (n=3), of ♀ 82.3 (81-83) (n=3), bill 15.0 (14.4-15.8) (n=6), bill depth of ♂ 7.5 (7.3-7.7) (n=3) (CSR).

SERINUS PUSILLUS
Red-fronted Serin   Kizilalin Iskete
Habitat *Berberis* and *Juniperus* scrub on otherwise quite bare and rough rocky slopes and ravines at the upper tree line; at 2000-2400m on the Ulu Dag, at 1600-2800 (-3900)m in the Taurus, at 1500-2000m in the Munzur Daglari, and at 2300-3300m in the valleys near Van Gölü.

Distribution See map 122. Occurs on most mountains which reach over 2000m, though not

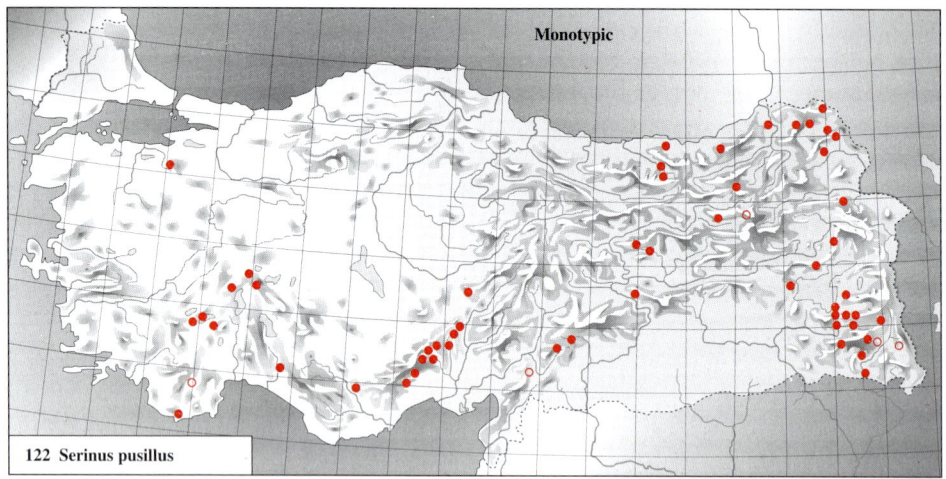

122 Serinus pusillus

recorded on many of those in the north-west and west. For the Ulu Dag, see e.g. Géroudet (1963) and Kumerloeve (1966c).

Geographical Variation None. *S. pusillus* (Pallas), 1811, was originally described from 'near the Caucasus and the Caspian Sea'. Wing of ♂ from Turkey (mainly from the Taurus, a few from the east) 74.3 (72-77) (n=12), of ♀ 72.4 (68-75.5) (n=7) (A. J. van Loon, CSR); wing of ♂ from the Caucasus 76.5 (74.5-79) (n=10), of ♀ 72.8 (70-75) (n=12) (CSR, with some additional data from Vaurie 1949), wing of ♂ from northern Iraq (Dihok) and Iran 75.3 (73-78.5) (n=10), of ♀ 72.8 (70.5-75.5) (n=5) (A. J. van Loon, CSR).

References Géroudet, P. (1963) Le Serin nain *Serinus pusillus* (Pallas) et le Roselin cramoisi *Carpodacus erythrinus* (Pallas) à l'Ulu Dag (Turquie). *Alauda* 31, 241-245. Kumerloeve, H. (1966c) Tendances expansives chez des espèces de *Carpodacus*, *Rhodopechys* et *Serinus* en Asie mineure. *Nos Oiseaux* 28, 284-287.

SERINUS SERINUS

European Serin   Kanarya

Habitat Light deciduous, mixed, or coniferous forest, rows of trees, plantations, orchards, scrub, and parkland. Mainly at 600-2000m, locally to sea level in the north-west and west. Frequently up to the tree line, but rarely above this and thus largely below the favoured habitat of *S. pusillus*, though both overlap at 1600-1800m in the central Taurus (Kumerloeve 1964a).

Distribution See map 123. Generally common in the western part of the Black Sea Coastlands, Thrace, Western Anatolia, and the southern foothills of the Taurus, scarce and local in the eastern part of the Black Sea Coastlands and on the fringe of the Central Plateau, apparently entirely absent east from the Amanus mountains in the South-East and East.

Geographical Variation None. *S. serinus* (Linnaeus), 1766, was originally described from Switzerland. Wing of ♂ from Turkey 72.7 (70-75) (n=10), of ♀ 71.1 (69-74) (n=5), bill of both sexes 10.4 (10.0-11.2) (n=7), bill depth at base 6.0 (5.5-6.3) (n=7) (CSR, with additional wing data from Kummerlöwe & Niethammer 1934-35 and Jordans & Steinbacher 1948). Wing of ♂ from Crete, Cyprus, and northern Iraq 74.3 (72-75, once 78) (n=8), of ♀ 70 (n=1), bill 10.4 (9.8-11.0) (n=9), bill depth 5.7 (5.4-6.3) (n=5); wing of ♂ from the remainder of the species' range 71.9 (68-76) (n=37), of ♀ 68.2 (66-72) (n=18), bill 10.3 (9.7-11.3) (n=62), average bill depth 5.8 (n=18) (A. J. van Loon, CSR).

Finches

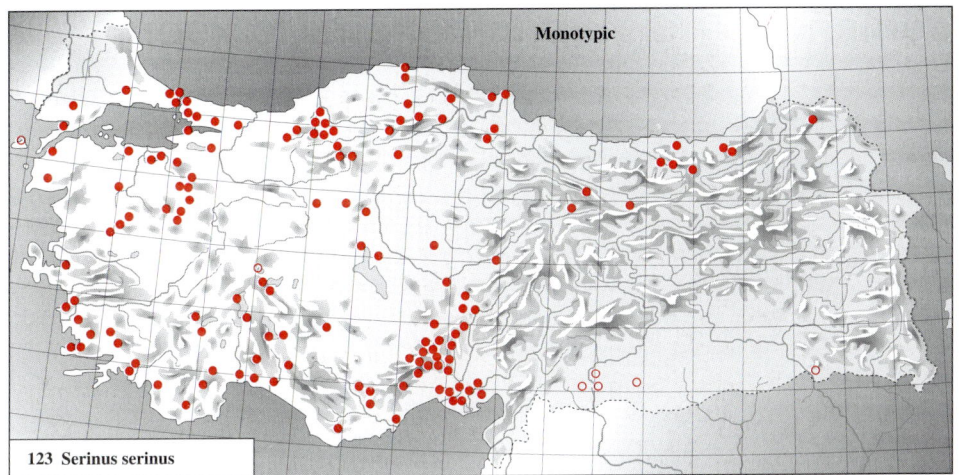

123 Serinus serinus

CARDUELIS CHLORIS
European Greenfinch   Florya

Habitat Open deciduous, mixed, or coniferous forest, as well as orchards, olive groves, parks, and gardens; mainly from 0-1000m in the north and west, but up to the tree line at 2000m in some mountains, and in the Taurus sometimes up to 2600m.

Distribution See map 124. Widespread but often rather scarce, and sometimes surprisingly absent from suitable habitat. Distribution closely similar to that of *Serinus serinus*, but more widespread in the extreme north-east and also occurs in the Amanus mountains. May breed at a few localities in the valleys of the South-East.

Geographical Variation
Subspecies described or recorded in the region:

(W)   **C. c. muehlei** (Parrot), 1905, Greece. [Size intermediate; worn upperparts of ♂ olive-green (less brown than *chloris* from central and northern Europe), feathers fringed grey when plumage fresh (less brown); forehead of worn ♂ rather contrastingly green-yellow, rump yellow-green; breast yellow-green, throat and belly rather bright yellow, but the latter nearer to *chloris* than to *aurantiiventris*. Wing of ♂ from Serbia, Romania, and Bulgaria 88.2 (83-91) (n=34), of ♀ 86.5 (83-92) (n=15) (Beretzk *et al.* 1969). Breeds in southern and eastern Romania, Bulgaria, Moldova, and mainland Greece (except the west and south).]

(W)   **C. c. aurantiiventris** (Cabanis), 1851, southern France. [Size small; worn upperparts of ♂ bright yellow-green, hardly contrasting with the yellow of forehead and rump; feather fringes of fresh upperparts light grey; underparts of ♂ as in *muehlei*, but yellow of throat and belly deeper and more extensive. Wing of ♂ from Crete 83.5 (82-86) (n=9), of ♀ 81.7 (79-84) (n=5), bill of both sexes 16.1 (15.2-17.0) (n=12), average bill depth at base 10.4 (n=12) (CSR); wing of ♂ from southern mainland Greece 83.5 (82-86) (n=6) (Niethammer 1943, CSR); wing of ♂ from Karpathos and Rodhos 83.8 (82-85.5) (n=4), of ♀ 78.5 (n=1), bill 15.7 (14.3-16.9) (n=5), average bill depth 10.4 (n=5) (CSR). Breeds in southern and eastern Spain, southern France, southern Italy, the coastal region of former Yugoslavia, Albania, western and southern Greece (including Crete and Karpathos), and Cyprus.]

(S)   **C. c. chlorotica** (Bonaparte), 1850, Syria. [Size small; upperparts as in *aurantiiventris* or paler yellow-green; underparts of worn ♂ deep golden-yellow with yellow-green breast and flanks (the latter less grey-green as *aurantiiventris*); pale fringes of flight feathers pale

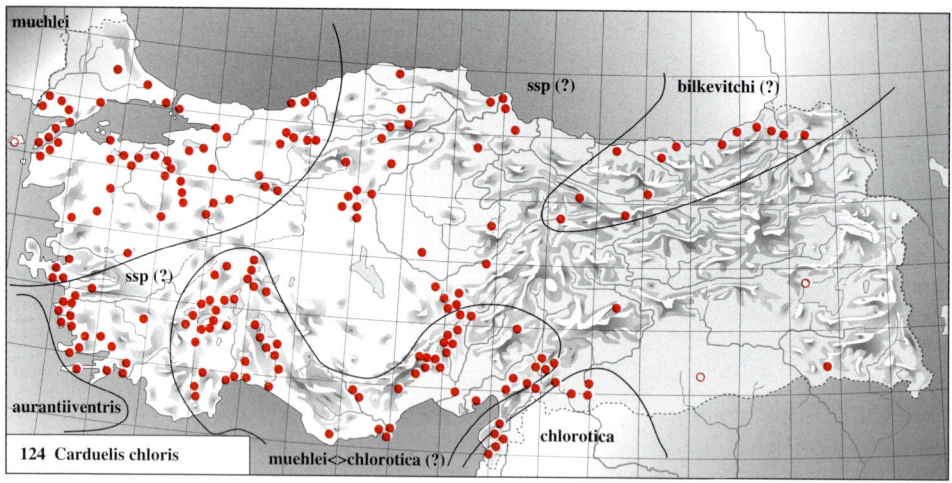

124 Carduelis chloris

yellow, less bright yellow. Wing in Levant and north-east Egypt 83.2 (80-87) (n=27), of ♀ 81.1 (78-84) (n=9), bill 15.7 (14.7-16.8) (n=29), average bill depth 10.2 (n=22) (A. J. van Loon, CSR). Breeds from northern Egypt through the Levant to Syria.]

(E) ***C. c. bilkevitchi*** (Zarudny), 1911, Ashkhabad (southern Turkmenistan). [Size intermediate; ♂ like *chlorotica*, but greyer, less yellow, underparts paler yellow. Wing of ♂ from the Caucasus and northern Iraq 86.6 (84-90) (n=7), of ♀ 83.7 (80.5-86) (n=5), bill 15.5 (14.3-16.6) (n=12), average bill depth 10.5 (n=4) (A. J. van Loon, CSR). Breeds in the Crimea, the Caucasus, Transcaucasia, and from northern Iraq through Iran to western Turkmenistan. This race is often combined with *turkestanicus* from central Asia (e.g., in Vaurie 1959), but birds from central Asia are much larger: wing of ♂ 93.1 (91-95) (n=4), of ♀ 91, 91 (n=2), bill 18.1 (17.1-18.8) (n=6), average bill depth 11.5 (n=6) (CSR). The heavy bill of *turkestanicus* is equalled only by *voousi* from the Atlas mountains of Morocco and Algeria (see Roselaar1993).]

Subspecies recognized in Turkey: the birds from Turkey examined (from Kesan, Sapanca Gölü, Afyon, Izmir, Burdur, Sogukpinar, Ulu Dag, Bolu, Zonguldak, and Sumela above Trabzon) are intermediate in size, like *muehlei* and *bilkevitchi:* wing of these birds is 88.2 (86-91) (n=10) in ♂, 84.3 (82-87) (n=6) in ♀, with bill of both sexes 17.0 (16.0-17.6) (n=9) and average bill depth 11.1 (n=10) (CSR). In colour, birds from the north-west are indistinguishable from *muehlei*, but the bird from Sumela is perhaps *bilkevitchi* (unfortunately it could not be compared directly with *bilkevitchi*). The birds from the Burdur area are brighter yellow, like *aurantiiventris* or (presumably) intermediates between *muehlei* and *chlorotica*, but Burdur birds are as large as *muehlei*. No birds were examined from south-west Anatolia (south of Izmir), which may show *aurantiiventris* characters, and none were seen from the eastern Taurus and the Amik area, where birds may verge towards *chlorotica*; a bird examined by Kumerloeve (1961a) from Haruniye (eastern Taurus; male, wing 87.5) was intermediate between *muehlei* and *chlorotica*, but it may still have been on migration.

References Beretzk, P., A. Keve, & M. Marián (1969) Taxonomische Bemerkungen zum Problem der Grünlings-Population des Karpathenbeckens. *Bonner zool. Beitr.* 20, 50-59. Roselaar, C. S. (1993) New subspecies of Fan-tailed Raven and Greenfinch. *Dutch Birding* 15, 258-262.

CARDUELIS CARDUELIS

European Goldfinch   Saka

Habitat Rocky slopes with sparse trees, wasteland, glades in forest, orchards, plantations, trees along roads, or farmland with hedges or rows of trees; generally, in areas where trees alternate with a mixture of herbaceous plants and open ground. Mainly at 0-1500m, but to 2000 (-2300)m in the Taurus and east.

Distribution See map 125. Widespread and often common, but in smaller numbers in the well-wooded parts of the Black Sea Coastlands and absent from the treeless country of the Central Plateau and the South-East.

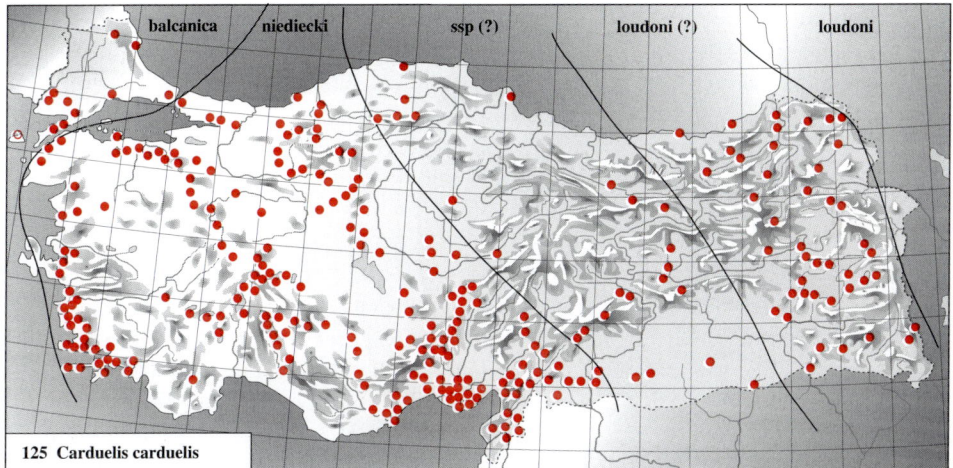

125 Carduelis carduelis

Geographical Variation

Subspecies described or recorded in the region:

(W)   ***C. c. balcanica*** Sachtleben, 1919, Kaluckovo (Makedonija). [Upperparts dull cinnamon-brown, slightly tinged grey; sides of breast with a fairly large cinnamon-brown patch, sometimes extending into a narrow breast band. Wing of ♂ from Albania, Bulgaria, and former Yugoslavia 78.9 (76-83) (n=16), of ♀ 76.4 (73-81) (n=11), bill of ♂ 15.7 (14.3-16.5) (n=16), of ♀ 14.9 (13.5-16.0) (n=11) (A. J. van Loon, CSR); wing of ♂ from southern Yugoslavia and Greece 78.7 (75-83) (n=46), of ♀ 75.6 (73-79) (n=19) (Stresemann 1920, Niethammer 1943).]

(T)   ***C. c. niediecki*** Reichenow, 1907, Eregli (Central Plateau, Turkey). [Upperparts and patch at sides of breast paler and more drab-brown, less cinnamon, clearly paler and greyer than in *balcanica* when plumage worn; patch at sides of breast on average smaller, often partly bordered yellow. Wing of ♂ from the Taurus and the Amanus area (Elmali, Burdur, Beysehir, Eregli, 'Taurus', Haruniye, and Gaziantep) 79.1 (76-81.5) (n=17), of ♀ 76.6 (74-80) (n=12), bill of both sexes 15.8 (14.3-17.5) (n=25) (A. J. van Loon, CSR); wing of ♂ from north-west Asia Minor (Sogukpinar, Izmit, Ankara, Abant Gölü, Bolu, and Beycuma) 78.6 (76-81) (n=8), of ♀ 76.7 (73-79) (n=11), bill 15.9 (14.9-17.3) (n=5) (CSR, with additional data on wing from Kummerlöwe & Niethammer 1934-35); wing of ♂ from Priene (near the Menderes delta) 77.3 (74-81) (n=12), of ♀ 76.0 (74-78) (n=4) (Weigold 1913-14). Breeds also in the Levant, Cyprus, and Egypt.]

(E)   ***C. c. loudoni*** Zarudny, 1906, Gilan and Qazvin (northern Iran). [Upperparts and patch at sides of breast darker drab-brown than in *niediecki*, not as cinnamon as in *balcanica*, more umber-brown; patch at sides of breast larger, often without yellow at border. Wing of ♂ from northern Iran and south-west Azerbaijan 80.9 (78-84) (n=7), of ♀ 77.8 (75-80) (n=4), bill of ♂ 15.6 (14.3-16.3) (n=8), of ♀ 15.0 (14.2-16.1) (n=3) (A. J. van Loon); wing of ♂ from Iranian Azerbaijan 81.7 (80-85) (n=9), of ♀ 79.1 (76-83) (n=8) (Vaurie 1949). Breeds in Transcaucasia and northern Iran. This subspecies is often called *brevirostris* Zarudny, 1899, but the latter name is preoccupied by *brevirostris* (Moore), 1856, a name in use for a subspecies of *C. flavirostris*, and it cannot be used when *flavirostris* and *carduelis* are united in a single genus *Carduelis*.]

Subspecies recognized in Turkey: for convenience, birds from Thrace are included in *balcanica* and breeding birds from at least the western half of Asia Minor in *niedieki*. The differences between the various subspecies are not as easy to detect as suggested in the subspecies diagnosis above, as the influence of bleaching and wear on the plumage is enormous: virtually all subspecies gradually change from more cinnamon in fresh plumage to greyish in worn plumage; moreover, the bill length, which is often mentioned as a character for identifying subspecies, tends to be shorter in winter and longer in summer in all populations of the species. Wing of ♂ of *balcanica* from Thrace 78.0 (74-80) (n=4) (Rokitansky & Schifter 1971); for *niedieki* from western Asia Minor, see above. No breeding birds were examined from east of a line running from Beycuma through Ankara to Gaziantep; as winter birds from south-central Turkey (Amik area, Iskenderun, Haruniye, Tarsus, and Mersin), as well as single winter birds from Eregli and Erzurum, all seem referable to *loudoni*, this latter subspecies may occur as a breeding bird in eastern Turkey. Wing of ♂ of these wintering *loudoni* 78.8 (76-81.5) (n=6), of ♀ 75.9 (75-78) (n=5), bill of both sexes 14.6 (13.3-15.5) (n=10) (CSR).

CARDUELIS SPINUS

Eurasian Siskin   Karabas Iskete

Habitat Subalpine forest of spruce and fir, generally at 1500-2000m, but at 750m near Dursunbey (Western Anatolia) and at 1200m near Abant Gölü (western Black Sea Coastland).

Distribution See map 126. Locally common on mountain slopes in the Black Sea Coastlands, and in small numbers in the northern part of Western Anatolia (where perhaps not breeding annually). Recorded singing up to late May in the Belgrat Ormani (Thrace), near Yozgat (northern Central Plateau), and in the western Taurus, and may breed here.

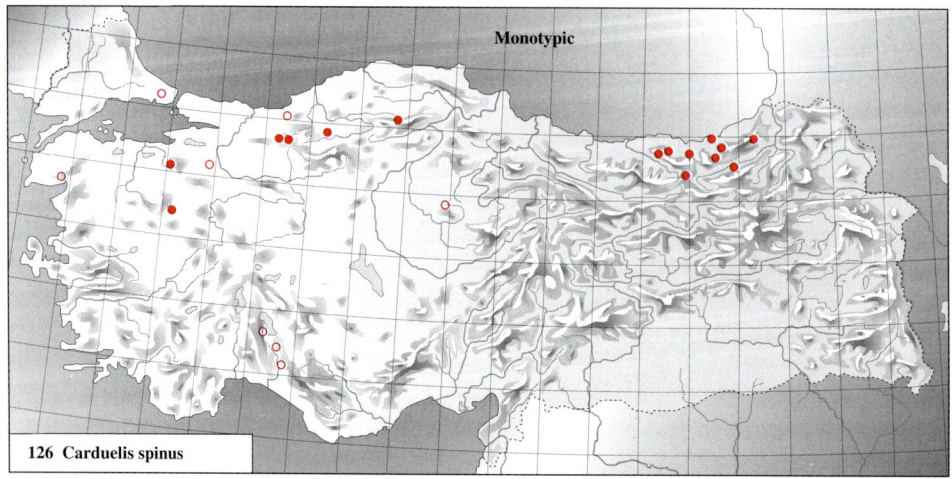

126 Carduelis spinus

Geographical Variation None. *C. spinus* (Linnaeus), 1758, was originally described from Sweden. Wing of ♂ from Turkey (not necessarily all breeding birds) 71.1 (68-73) (n=8), of ♀ 71 (n=1) (Kummerlöwe & Niethammer 1934-35, Rössner 1935); wing of ♂ from elsewhere in the species' range 73.0 (69-76) (n=52), of ♀ 70.7 (67-74) (n=31) (A. J. van Loon).

CARDUELIS CANNABINA
Linnet   Ketenkusu

Habitat Rocky slopes with sparse vegetation and scattered scrub or small trees, from open boulder-strewn maquis near the coast to alpine meadows with scattered low scrub and stunted junipers above the tree line; also, open forest and forest glades. Mainly on mountain slopes and plateaux at 700-2700m, but at 1600-3100m in the east and locally down to the foothills near the coast.

Distribution See map 127. Widespread, in the west generally in small numbers, in the east often common. Absent from treeless steppe and desert, from bare mountain tops, and from wooded country.

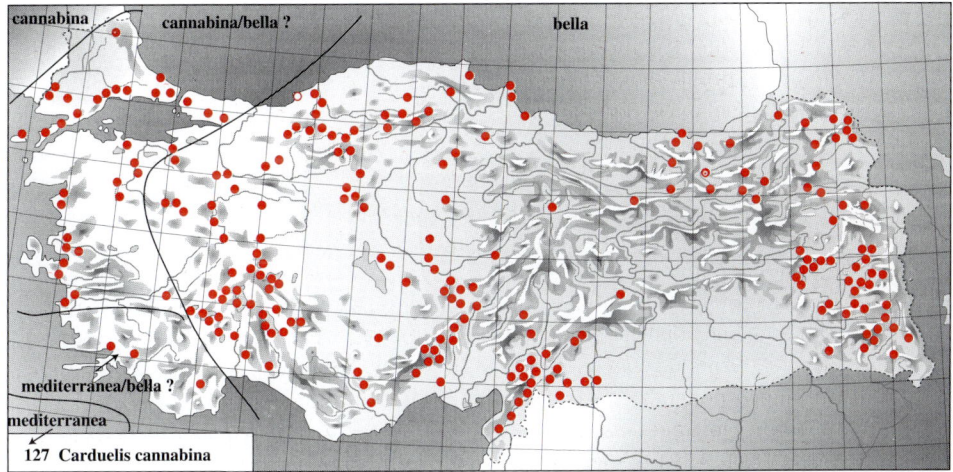

127 Carduelis cannabina

Geographical Variation

Subspecies described or recorded in the region:

(W)   ***C. c. cannabina*** (Linnaeus), 1758, Sweden. [Top and side of head of ♂ in spring medium ash-grey, mantle and scapulars chestnut-brown, lower rump and upper tail-coverts streaked black and white, breast carmine, flanks extensively rufous-cinnamon. Wing of ♂ from central Europe 80.8 (78-85) (n=75), of ♀ 78.7 (76-82) (n=42), bill of both sexes 13.0 (11.8-14.0) (n=53), bill depth 7.1 (6.7-7.4) (n=35); wing-tip (from tip of innermost primary to tip of longest primary in closed wing) generally over 25 mm (CSR). Breeds in central, northern, and north-eastern Europe, east to west Siberia and south to Romania, Bulgaria, and north-east Greece, but not in the Crimea and the Caucasus area.]

(W)   ***C. c. mediterranea*** (Tschusi), 1903, Kotor (Montenegro). [Like *cannabina*, but mantle and scapulars of ♂ in spring often slightly paler, more orange-brown. Rather small, bill rather thick. Wing of ♂ from the coast of former Yugoslavia (including the type specimen) 79.0 (77.5-80) (n=4), bill 12.5 (12.0-13.3) (n=4), bill depth 7.6 (7.0-8.3) (n=4). Birds from the

Mediterranean basin (Portugal, Spain, North Africa, southern Italy, western and southern mainland Greece, and many islands, from the Balearics to Crete and Karpathos) are usually included in *mediterranea*, but are distinctly smaller, and often show a more slender bill (except on the Balearic islands and in North Africa). These small-sized birds probably form a recognizable subspecies, for which the name *obscura* Von Jordans, 1923, is perhaps available (type specimen from Oporto, Portugal, not examined); on the other hand, *mediterranea* does not deserve recognition, birds from the type locality being variable intermediates between *cannabina* and *bella*. Wing of ♂ from Portugal and Spain 77.0 (74-79) (n=8), of ♀ 74.5 (72-77) (n=8), bill 12.5 (11.6-13.3) (n=14), bill depth at base 7.0 (6.6-7.4) (n=15); wing-tip in Iberia generally less than 24.5 mm (CSR). Wing of ♂ from Crete and Karpathos 76.7 (74-78) (n=4) (Niethammer 1942, Kinzelbach & Martens 1965), wing (both sexes) on Karpathos 75.0 (71-79) (n=55) (Kinzelbach & Martens 1965).]

(S/E)   ***C. c. bella*** (C. L. Brehm), 1845, Beirut (Lebanon). [Top and side of head of ♂ in spring pale ash-grey, mantle and scapulars cinnamon-brown or hazel, paler than in *mediterranea*, rump and upper tail-coverts mainly white, virtually unstreaked, rump sometimes with some rosy-red admixed; carmine of forehead and breast paler, more rosy-red, and less extensive; flanks paler and less extensive tawny-cinnamon. Large, bill rather thick. Wing of ♂ from Cyprus, the Levant, and west Iran 81.4 (78.5-84) (n=9) of ♀ 78.8 (77.5-79.5) (n=5), bill 13.3 (12.8-13.8) (n=13), bill depth 7.6 (7.2-8.2) (n=13); wing-tip as in *cannabina* (CSR). Breeds south to Rodhos, north to the Caucasus and Crimea, and east to western Turkmenistan.]

Subspecies recognized in Turkey: all breeding birds examined from Asia Minor belong to *bella*: wing of ♂ from Beysehir, Keciborlu, Burdur, Kütahya, Sogukpinar, Ulu Dag, Abant Gölü, Beycuma, Kastamonu, Çorum, Gevas, Van, and Yüksekova is 80.6 (78-84) (n=25), of ♀ 78.6 (76-80) (n=11), bill of both sexes 13.2 (12.0-14.2) (n=20), bill depth 7.7 (7.4-8.1) (n=19) (CSR, with some additional wing data from Kummerlöwe & Niethammer 1934-35 and Vauk 1973). Spring birds from Izmir and Priene in Western Anatolia are also large, and likely belong to *bella* (if not migrant *cannabina*): wing of ♂ 80.2 (78-82) (n=8), of ♀ 78.6 (77-80) (n=5) (Weigold 1913-14). *C. c. cannabina* is known to occur in winter in Turkey (e.g., examined from Erzurum in October). No birds from Thrace were seen; these may belong to *cannabina*, which occurs also in the nearby parts of Bulgaria, or they may be intermediate between *cannabina* and *bella*. Small-sized Mediterranean birds ('*mediterranea*') do not reach Turkey.

References Kinzelbach, R., & J. Martens (1965) Zur Kenntnis der Vögel von Karpathos (Südliche Ägäis). *Bonner zool. Beitr.* 16, 50-91.

CARDUELIS FLAVIROSTRIS
Twite   Sarigaga Ketenkusus

Habitat Sparsely vegetated stony slopes above the tree line, at *c*. 1000m above Burdur, at 1700-2400m in the eastern Taurus, and at 1900-3000m in the east.

Distribution See map 128; also, Kumerloeve (1967d). Largely confined to mountain slopes in the east of the country, west to Gümüsane and the Nemrut Dag, and probably widespread on all mountains over 2000m in this area. Very local in the eastern Taurus, from the Ala Dag and (probably) Hodul Dagi to the mountains above Aziziye (Pinarbasi). Recorded in early August on the Katrancik Dagi (above Burdur) in the western Taurus, and perhaps breeds in small numbers in the alpine zone here and elsewhere in the

Taurus, as well as in the mountains between Aziziye and Erzincan. More data are needed.
Geographical Variation
Subspecies described or recorded in the region:
(T)   ***C. f. brevirostris*** (Moore), 1856, Erzurum (eastern Turkey). [A fairly heavily streaked subspecies; extent and width of dark streaks on body as in *flavirostris* from Scandinavia, but ground-colour paler and more contrasting: on upperparts pale tawny-cinnamon (when plumage fresh) to grey-buff (when worn) (less tawny-brown than in *flavirostris*), on side of head, throat, and breast buff, on remainder of underparts extensively cream-buff (when fresh) to off-white (when worn). Wing of ♂ from Turkey (Eregli, Verçenik, Erzurum, near Eleskirt, and the Baskale area) 75.9 (74-78) (n=14), of ♀ 73.6 (72-75) (n=7), bill (both sexes) 9.4 (8.5-10.5) (n=16) (A. J. van Loon, CSR), or (using largely the same birds) wing of ♂ 75.4 (74-77) (n=10), of ♀ 73.4 (71-75) (n=7) (Kumerloeve 1967d). Breeds also in the Caucasus, Transcaucasia, and northern Iraq and Iran; all other subspecies occur far away, in Great Britain and northern Europe (*pipilans*, *bensonorum*, and *flavirostris*), as well as from the Transcaspian plains east to the mountain steppes and plains of central and eastern Asia (all other subspecies).]

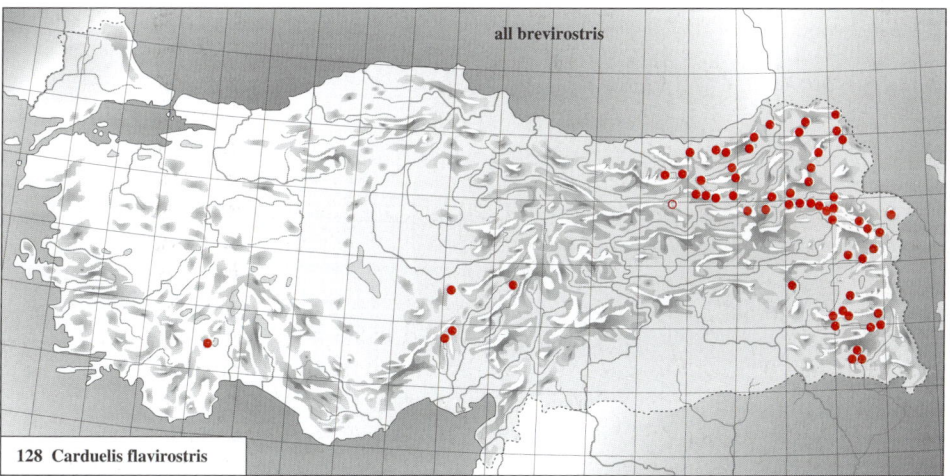

128 Carduelis flavirostris

Subspecies recognized in Turkey: all birds examined were *brevirostris*. Wing of ♂ from Çildir Gölü and Ispir 75.0 (73-77) (n=3) (Kumerloeve 1968). Wing of ♂ from the Caucasus 76.1 (74-79) (n=7), of ♀ 74.6 (73-76) (n=8), bill 9.3 (8.9-10.1) (n=14) (A. J. van Loon, CSR).
References Kumerloeve, H. (1967d) Contribution à la connaissance de *Carduelis (Acanthis) flavirostris brevirostris* (Bonaparte, 1855). *Alauda* 35, 118-124.

LOXIA CURVIROSTRA
Common Crossbill   Çaprazgaga
Habitat Open or dense coniferous forest, mainly of *Pinus, Picea,* and *Abies*, sometimes in mixed forest, and after the breeding season occasionally in deciduous forests, orchards, and gardens. Mainly at 800-2000m, on the Ulu Dag at 1400-1900m, locally at lower levels in the Black Sea Coastlands wherever suitable coniferous forests exist, and almost down to sea level near Kesan (Thrace).
Distribution See map 129. Locally fairly common in the Black Sea Coastlands, in the Marmara

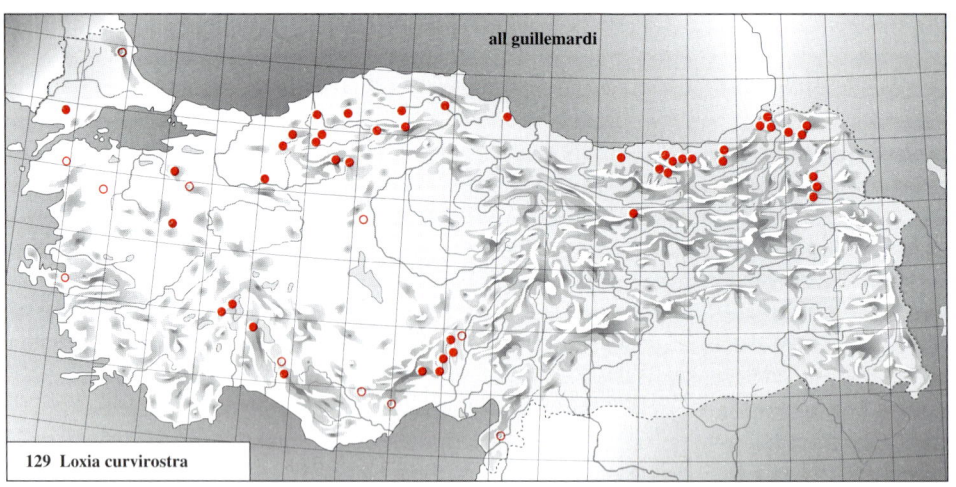

129 Loxia curvirostra

sub-region of Western Anatolia, and in the Taurus. Has bred near Kesan (Thrace). Breeding rather erratic (though apparently less so in the Mediterranean basin and Asia Minor than in northern and central Europe), and scattered observations of small flocks from late April to July outside the areas mentioned above may either refer to wandering birds or may point to (occasional) local breeding (e.g., recorded in Beynam forest near Ankara and in the Amanus mountains).

Geographical Variation

Subspecies described or recorded in the region:

- (W) **L. c. curvirostra** Linnaeus, 1758, Sweden. [Ground-colour of head and body brown-grey to medium ash-grey, feathers tipped rosy-scarlet in adult ♂, green, yellow, orange, rosy-scarlet, or a mixture of these in 1st year ♂, and yellow-green to olive-green in ♀, coloured feather tips largely hiding grey of feather bases. Wing of ♂ in central and northern Europe 98.3 (93-104) (n=87), of ♀ 95.2 (91-99) (n=54), bill of both sexes 22.7 (21.0-24.5) (n=70), bill depth at base 11.1 (10.5-11.8) (n=67), width of bill at base of lower mandible 10.6 (9.7-11.3) (n=69) (CSR). Breeds from western Europe east to eastern Asia, and in Europe south to northern Spain, central Italy, the northern Balkan countries, and northern Ukraine.]

- (T) **L. c. vasvarii** Keve, 1943, Bolu Dagh near Bolu (west part of the Black Sea Coastlands, Turkey). [Ground-colour of head and body darker brown-grey or ash-grey than in *curvirostra*, causing overlying colour to appear darker, and contrasting colour of feather-tips somewhat more restricted, with more grey of feather bases visible; adult ♂ appears darker carmine-red, adult ♀ darker green; 1st year ♂ usually red, as adult ♂, not usually orange as in 1st adult ♂ of *curvirostra*. Wing of ♂ from Thrace, Abant Gölü, Bolu Dagh (including the type specimen), and north-east Turkey 100.7 (97-105.5, once 94) (n=14), of ♀ 95.2 (93.5-98) (n=3), bill 22.1 (20.7-23.7) (n=16), bill depth 11.5 (10.7-12.7) (n=16), bill width 10.8 (9.8-11.8) (n=17) (CSR). Data from the literature (but note that some authors measure wing shorter than is now usually done): on the Ulu Dag, wing of ♂ 93-100 (n=4), of ♀ 94-96 (n=4) (Jordans & Steinbacher 1948); in the Ilgaz Daglari, wing of ♂ 98 (n=1), of ♀ 95 (n=1) (Kummerlöwe & Niethammer 1934-35).]

- (S) **L. c. guillemardi** Madarász, 1903, Cyprus. [Plumage as in *vasvarii*. Wing of ♂ from Cyprus 99.0 (95-102) (n=12), of ♀ 96.5 (92-99) (n=6), bill 23.9 (22.3-26.3) (n=16), bill depth 11.8 (11.4-12.5) (n=15), bill width 11.2 (10.6-11.9) (n=17) (CSR).]

Finches

(E) ***L. c. caucasica*** Buturlin, 1907, Caucasus and Transcaucasia. [Red of adult ♂ deeper and duller than in *curvirostra*, contrasting colour of feather tips more restricted, bill heavier; wing of ♂ from the Caucasus 92-100 (n=10), bill depth of ♂ 12.0-12.4 (n=10) (Buturlin 1907). Only four winter birds from the northern Caucasus were examined, and these were inseparable in colour and size from *curvirostra*; they were either migrants, or the northern Caucasus is inhabited by *curvirostra*, with *caucasica* restricted to the southern Caucasus and Transcaucasia; wing of ♂ of these four 97.7 (96-99.5) (n=3), of ♀ 95.5 (n=1), bill 22.4 (21.5-23.5) (n=4), bill depth 11.2 (10.4-11.8) (n=4), bill width 10.5 (10.3-10.8) (n=4) (CSR). *L. c. mariae* Dementiev, 1932, described from the Crimea and said to be characterized by a deep dark red plumage in ♂ (Hartert 1921-22), is a synonym of *caucasica* (Stepanyan 1990). Wing of ♂ from the Crimea 98.0 (93-102) (n=11), bill depth of ♂ 11.8 (10.5-13.0) (n=11) (Dementiev & Gladkov 1951-54); wing of ♂ 100, of ♀ 97.5 (n=1 in each), bill of both 22.6, 24.2, bill depth 11.9, 12.1, bill width 10.6, 11.2 (CSR).]

Subspecies recognized in Turkey: the subspecies *vasvarii*, *guillemardi*, and *caucasica*, for which the diagnoses are given above, are all closely similar in colour and are best united into a single subspecies, which should be named after the first taxon described, *guillemardi*. All Turkish populations, as well as those from Transcaucasia, the Caucasus (but see above), the Crimea, Cyprus, Greece, and, according to Matvejev (1976), north probably to Bulgaria and to the south of former Yugoslavia, thus form a single subspecies. The characters of *guillemardi* are not too different from those of *curvirostra*, but when series of specimens are compared the plumage differences given above under *vasvarii* are readily visible, and the slightly heavier bill-base of *guillemardi* separates most birds from those of *curvirostra* when both bill depth at base and width of lower mandible at base are used in an analysis.

References Buturlin, S. A. (1907) Neue Formen aus dem Kaukasus. *Orn. Monatsber.* 15, 8-9.

RHODOPECHYS SANGUINEA

Crimson-winged Finch   **Kizilsakrak**

Habitat Rough slopes with grasses, clumps of *Astragalus* or *Artemisia*, scattered scrub, stones, and ridges, as well as rocky valleys with *Berberis* bushes, or flat open country with low xerophytic scrub. At 1100-1500m in the Ilgaz area (Black Sea Coastlands), at c. 1000m

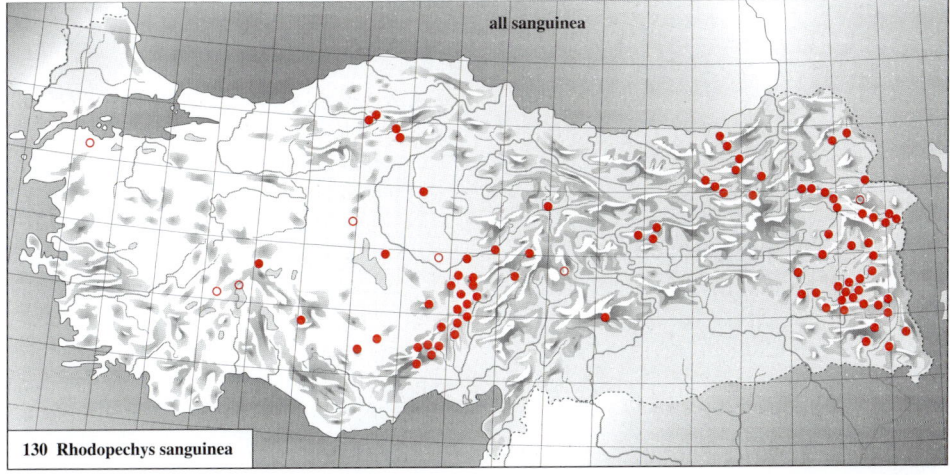

130 *Rhodopechys sanguinea*

and over at the fringes of the Central Plateau, at 1600-2900 (-3300)m in the Taurus, at 1500-2500m in the Munzur Daglari, and at 1500-3200m in the east.

Distribution See map 130; also, Kumerloeve (1966c). Rather common on mountain slopes in the east, west to the Munzur Daglari. First found on the Central Plateau by Lehmann & Mertens (1969) and Lehmann (1971), and considered to be rare in central Turkey then, but now known to be widespread (though only locally common) in the central and eastern Taurus and its foothills, with scattered breeding season records west to above Çay in the Sultan Daglari and north to the Ilgaz Daglari.

Geographical Variation

Subspecies described or recorded in the region:

(T)   *R. s. sanguinea* (Gould), 1838, Erzurum (eastern Turkey). [A large boldly coloured finch, occurring over the entire Asiatic range of the species, replaced by an isolated subspecies, *aliena*, in north-west Africa, which is closely similar in size to *sanguinea* but rather different in colour. Wing of ♂ from Turkey 107.6 (103-111.5) (n=8), of ♀ 100.3 (100-101) (n=3) (CSR).]

Subspecies recognized in Turkey: as to be expected, all Turkish birds examined belong to *sanguinea*. Wing of ♂ from Syria, Lebanon, and Israel 107.0 (103-109) (n=11), of ♀ 103.7 (101-106) (n=6), of ♂ from Armenia and northern Iran 105.6 (102.5-107) (n=7), of ♀ 101, 105.5 (n=2) (CSR). Wing of ♂ from north-west Iran 105.9 (105-109) (n=7), of ♀ 104 (n=1) (Vaurie 1949, Érard & Etchécopar 1970). No geographical variation in size or colour is apparent within the Asiatic range of the species.

References Lehmann, H., & R. Mertens (1969) The Red-winged Bullfinch (*Rhodopechys sanguinea*) as a breeding bird in Central Anatolia. *Ool. Rec.* 43, 1-16. Lehmann, H. (1971) Der Rotflügelgimpel (*Rhodopechys sanguinea*) auf dem Hochplateau Zentral-Anatoliens. *Jahrb. Naturwiss. Ver. Wuppertal* 24, 101-120.

RHODOSPIZA OBSOLETA

Desert Finch   **Aksakrak**

Habitat Plateaux and hillsides at the edge of the (semi-) desert, sparsely covered with tall bushes, occasional trees, or hedgerows, often close to or amidst cultivated fields or gardens, in Turkey mainly at 200-1000m.

Distribution See map 131; also, Kumerloeve (1966c). Restricted to the plateau area of the South-East, close to the known breeding sites in Syria. Breeding records 'east of Tuz Gölü' and in the Soganli Dagi ('south-east of Kayseri'), cited in Kumerloeve (1970b), require confirmation, as these sites are now known to be inhabited by *Rhodopechys sanguinea* and a misidentification may have occurred; likewise, breeding season records in the Nemrut crater (above Tatvan) (Dufourry 1990), on Kuh Dagi (above Erçek) (Gallner 1976), and near Dogubayazit (Frost & Hornbuckle 1992) are in need of verification, as in these cases confusion with *Bucanetes githagineus* (above Tatvan) and *B. mongolicus* (near Erçek and Dogubayazit) cannot be entirely ruled out.

Geographical Variation None. *R. obsoleta* Lichtenstein, 1823, was originally described from the Kara-ata spring near Bukhara (south-west Uzbekistan). *R. o. kaschgarica* Portenko, 1962, described from Lob-Nor, Sinkiang, China, and *R. o. chuancheica* Portenko, 1962, described from Gui-Dui oasis (Khuankhe area, Muni-Ula mountains, Kansu, China), are synonyms. Apparently only once collected in Turkey: a female near Urfa on 18 April, with wing 81 (Weigold 1912-13). Wing of ♂ from the Levant 86.7 (85-89) (n=4), of ♂ from Turkmenistan and western Uzbekistan 87.8 (86-90) (n=7), of ♂ from eastern Uzbekistan

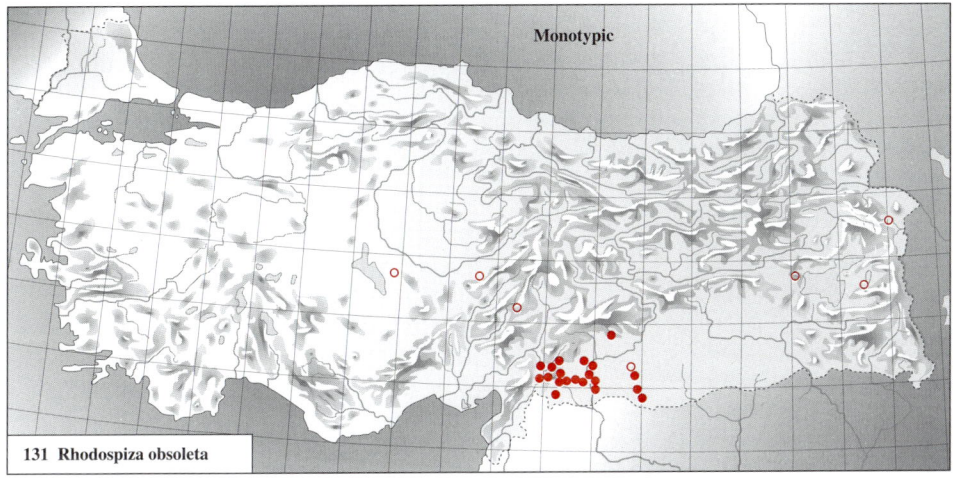

131 Rhodospiza obsoleta

to Kazakhstan 88.0 (85.5-91) (n=8), of ♂ from the Tarim basin (China) 89.3 (86.5-91.5) (n=7) (CSR). Wing of ♂ from southern Iran 86.2 (85-87) (n=5), of ♀ 81 (n=1) (Érard & Etchécopar 1970), wing of ♂ from Iran and Afghanistan 89.1 (87.5-92) (n=6), of ♀ 84.8 (83-87) (n=14) (Vaurie 1949). The differences in size are too small to warrant recognition of subspecies, and the differences in colour are largely due to influence of bleaching and wear.

BUCANETES MONGOLICUS
Mongolian Finch

Habitat Stony plateaux and slopes with boulders and sparse low vegetation, at c. 2100-2900m; outside Turkey (in areas where *B. githagineus* absent), also at lower altitudes on sparsely vegetated stony ridges and moraines in dry steppe and semi-desert.

Distribution See map 132; also Barthel *et al.* (1992). Restricted to mountain slopes in the extreme east, close to Nakhichevan where both this species and *B. githagineus* are known to occur (Panov & Bulatova 1972).

Geographical Variation Some, but no subspecies have been formally described and further information is required. *B. mongolicus* (Swinhoe), 1870, was originally described from the Nan-k'ou pass (north-west of Beijing, eastern China). Apparently never collected in Turkey. Wing of ♂ from western Turkmenistan, north-east Iran, and Bahrain 88.8 (87-92) (n=8), of ♀ 86.9 (85-88) (n=5), bill of both sexes 12.1 (11.8-12.5) (n=13); these birds are rather pale and small, in spring showing cold drab-grey upperparts, pale pink face, and light grey side of head and sides of breast (CSR). Birds from Turkey are probably similar to these. Birds from the Tien Shan and Tarbagatay mountains in west-central Asia differ in size and colour: wing of ♂ 91.9 (89-96) (n=14), of ♀ 88.6 (86-92) (n=10), bill 12.5 (11.8-13.6) (n=24); upperparts in spring darker drab-brown, face darker pink, side of head and sides of breast drab-brown (CSR). No subspecies are named here, however, as only a few non-comparable autumn birds from the Chinese type-locality of the species were examined, and, moreover, birds seen from Afghanistan, Pakistan, and Ladakh combined characters of birds from Iran-Turkmenistan and of birds from the Tien Shan and Tarbagatay: wing of ♂ from Afghanistan, Pakistan, and Ladakh 93.2 (90-97) (n=11), of ♀ 89.5 (87-93) (n=7), bill of both sexes 12.3 (11.5-13.2) (n=18), colour more or less intermediate (CSR).

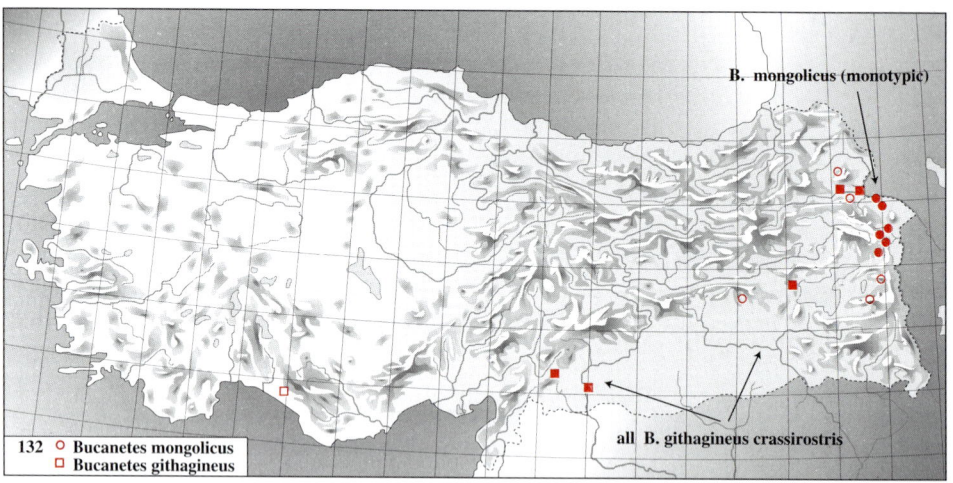

132
○ Bucanetes mongolicus
□ Bucanetes githagineus

all B. githagineus crassirostris

B. mongolicus (monotypic)

References Panov, E. N., & N. Sh. Bulatova (1972) [On the common habits and interrelations of *Bucanetes githagineus* (Lichtenstein) and *Bucanetes mongolicus* (Swinhoe) in Transcaucasia.] *Byull. Mosk. Obshch. Ispyt Prir. Otd. biol.* 77(4), 86-94. Barthel, P. H., W. Hanoldt, K. Hubatsch, H.-M. Koch, V. Konrad, & R. Lannert (1992) Der Mongolengimpel *Bucanetes mongolicus* in der Westpaläarktis. *Limicola* 6, 265-286.

BUCANETES GITHAGINEUS
Trumpeter Finch   Çol Sakragi

Habitat Arid stony loam steppe, undulating stony plateaux, desolate ravines, and stony desert fringes, sparsely covered with clumps of low vegetation, preferably with access to some water nearby. In the overlap zone with *B. mongolicus*, occurs generally lower down and on less stony terrain than that species (see, e.g., Vaurie 1949 for Iran).

Distribution See map 132; also, Krieger (1988). Known to occur in the Yesilce area (Büyük Araptar, north-west of Gaziantep), in the crater of the Nemrut Dag above Tatvan, and in the Aras valley near the Armenian border, at the latter site close to breeding localities in Armenia and Nakhichevan (Panov & Bulatova 1972). Seen in the breeding season between Manavgat and Alanya (Southern Coastlands) (Krieger 1988) and near Birecik (Bakker & Steenge 1990), but no proof of breeding.

Geographical Variation
Subspecies described or recorded in the region:

(S/E) **B. g. crassirostris** (Blyth), 1847, Afghanistan. [Upperparts paler and more extensively drab-grey than in other subspecies, less pink; bill heavier at base. Wing of ♂ from Israel and the Syrian desert 88.8 (86.5-91.5) (n=7), of ♀ 78.5, 84 (n=2), bill (both sexes) 12.4 (11.4-12.9) (n=9), bill depth at base 9.2 (8.8-9.9) (n=8); wing of ♂ from the Sinai peninsula 86.4 (84-89) (n=9), of ♀ 84.3 (83.5-85) (n=3), bill 13.4 (12.4-14.1) (n=8), bill depth 9.2 (8.9-9.9) (n=8) (CSR). Breeds from Sinai and Arabia east to Pakistan and Uzbekistan. The other subspecies are all confined to northern Africa.]

Subspecies recognized in Turkey: apparently never collected, but unlikely to be anything else other than *crassirostris*, as this is the only subspecies which breeds nearby in Syria, Iran, and Armenia.

References Krieger, H. (1988) The Trumpeter Finch, *Bucanetes githagineus*, in Turkey. *Zool. Middle East* 2, 43-45.

CARPODACUS ERYTHRINUS

Common Rosefinch   Karmen Sakragi

Habitat Tangled deciduous bushes and hedge rows amidst damp meadowland and gardens, as well as thickets along streams, roads, and forest edges, more rarely in open forest of *Pinus brutia* or in stunted conifers above the tree line. Mainly on coastal plains, in valley bottoms, and on lower slopes, from sea level in the Black Sea Coastlands up to c. 2200m in the interior, but up to 2700m in the extreme South-East

Distribution See map 133. Common in the Black Sea Coastlands, west to the Bolu area, in a zone up to 90 km from the coast, as well as in the eastern part of the country, south to the Yüksekova area; abundant in the extreme north-east. Range apparently expanding, found breeding on the Ulu Dag since about 1960 (Géroudet 1963, Kumerloeve 1966c), and has bred Sultansazligi near Kayseri (Kasparek 1985). Singing birds in May in the south of Turkey, as far west as Güllük (Brock 1989) and Köycegiz (Kumerloeve 1975c, Versluys 1992) may point to a further extension of the breeding range.

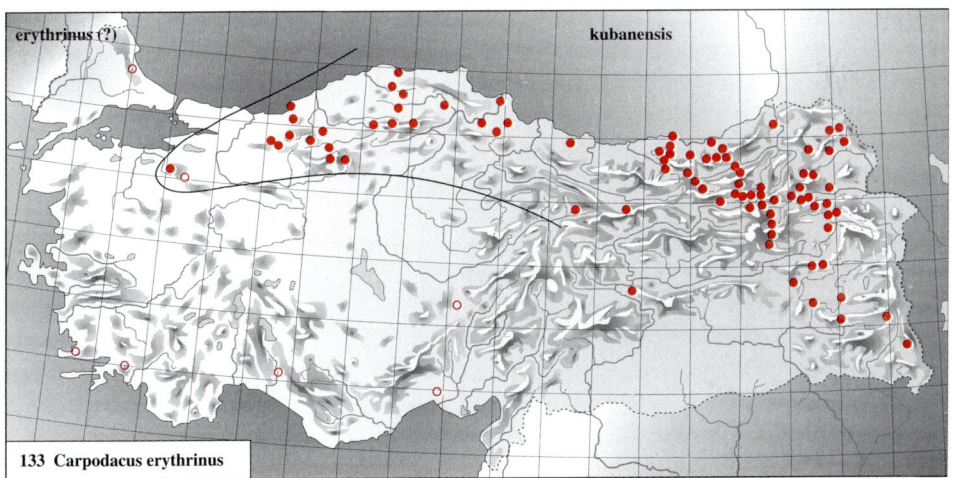

133 *Carpodacus erythrinus*

Geographical Variation

Subspecies described or recorded in the region:

(W)   ***C. e. erythrinus*** (Pallas), 1770, Volga and Samara (= Bol'shaya Kinelyu) Rivers (near Samara = Kuybyshev, south-east European Russia). [Head and breast of adult ♂ mainly scarlet-red, remainder of upperparts brown with slight red suffusion; flanks and upper belly pink-white, contrasting with the red of the breast; rather small. Wing of red adult ♂ from central and eastern Europe east to central Asia 84.6 (82-88) (n=21), of ♀ 82.7 (80-85) (n=13), bill of both sexes 14.2 (13.1-15.3) (n=44) (CSR). Breeds in Europe south to the north of former Yugoslavia, Romania, Ukraine, and the lower Volga, but is expanding westwards and southwards.]

(E)   ***C. e. kubanensis*** Laubmann, 1915, Karaul Kisha, Kuban area (north-west Caucasus, southern European Russia). [Head and breast of adult ♂ more rosy-red than in *erythrinus*, remainder of upperparts brown with more extensive red suffusion; flanks and upper belly darker pink-red, less contrasting with the red of the breast; slightly larger. No constant difference from *erythrinus* in brown 1st-year ♂ and in ♀. Wing of red adult ♂ from the northern Caucasus 86.6 (84-91) (n=18), of ♀ 83, 83.5 (n=2), bill 14.5 (13.8-

15.3) (n=16) (CSR); wing of ♂ from the Caucasus 86.1 (84-89.5) (n=17), of ♀ 84.5 (81.5-87.5) (n=5) (Hesse 1915); wing of red adult ♂ from Iran 87.0 (82-92) (n=22), of ♀ 85 (n=1) (Stresemann 1928, Paludan 1940, Vaurie 1949, Schüz 1959). Breeds from the Caucasus south to northern Iran and western Turkmenistan.]

Subspecies recognized in Turkey: all breeding birds examined from Asia Minor are *kubanensis* according to plumage; wing of ♂ 85.8 (84.5-87) (n=5), bill 13.6 (13.2-13.8) (n=3) (Kummerlöwe & Niethammer 1934-35, CSR). Birds singing in spring in Thrace may either belong to *erythrinus* or *kubanensis*.

References Hesse, E. (1915) Bemerkungen über *Carpodacus erythrinus* und seine Formen. *Orn. Monatsber.* 23, 112-118.

CARPODACUS RUBICILLA
Great Rosefinch   Kaya Sakragi

Habitat Breeds in the Caucasus below the snow line at *c.* 3000-3500m, feeding lower down in the (sub-) alpine belt from *c.* 2500m upwards; winters at *c.* 1900-2000m (Loskot *in press*).

Distribution No map. 2-3 pairs first found on 9 June 1910 and present up to July at Kyrk Deirmen, near Erzurum, by McGregor (1917); recorded in the same area in the '1940s' by Kosswig (Kumerloeve 1961a). As it was found singing, it may have bred locally, but there was no proof. A large red finch seen flying into a gorge near Ani (east of Kars in the extreme East) by J. Hornbuckle may have been this species (Frost & Hornbuckle 1992). Otherwise, in our region only known to occur in the Caucasus.

Geographical Variation
Subspecies described or recorded in the region:

(E) ***C. r. rubicilla*** Güldenstädt, 1775, Caucasus Mountains. [Head and upperparts of adult ♂ dark blood-red on dark grey-brown ground-colour; head and throat marked with small white spots; upperparts of ♀ and 1st-year ♂ dark grey-brown, underparts with broad dark grey-brown streaking on cream ground-colour. In other races of the species, all occurring extralimitally in central Asia, the colour of the ♀ and 1st-year ♂, as well as the ground-colour of the adult ♂, are paler drab-grey, the red of the adult ♂ is more scarlet, and the white spots on the head and throat of the adult ♂ are larger. Wing of red adult ♂ from the northern Caucasus 119.1 (116-122) (n=26), of ♀ 114.2 (111-118) (n=9), bill of both sexes 20.2 (18.9-22.0) (n=29) (CSR).]

Subspecies recognized in Turkey: apparently never collected. Unlikely to be anything else other than *rubicilla*, as this subspecies breeds nearby in the Caucasus area, and the other races breed far away in central Asia.

References Loskot, V. M. (*in press*) [Distribution and life history of the Caucasus Great Rosefinch *Carpodacus rubicilla rubicilla* (Güld.)] *Trudy Zool. Inst. Akad. Nauk SSSR* 231, 43-116.

PYRRHULA PYRRHULA
Eurasian Bullfinch   Sakrak

Habitat Dense mature coniferous or mixed forest at *c.* 1000-1900m.

Distribution See map 134. Restricted to the moister well-wooded parts of the Black Sea Coastlands, south to the Çubuk Baraj area (Vermeulen 1987), as well as locally in the northern part of Western Anatolia. Not uncommon but difficult to find. In winter, recorded in the Istranca Daglari in Thrace, in the Beynam forest on the Central Plateau, and in the Upper Aras area, and the species may breed here; the few birds observed in winter in the Taurus are likely to be wanderers from further north.

Finches

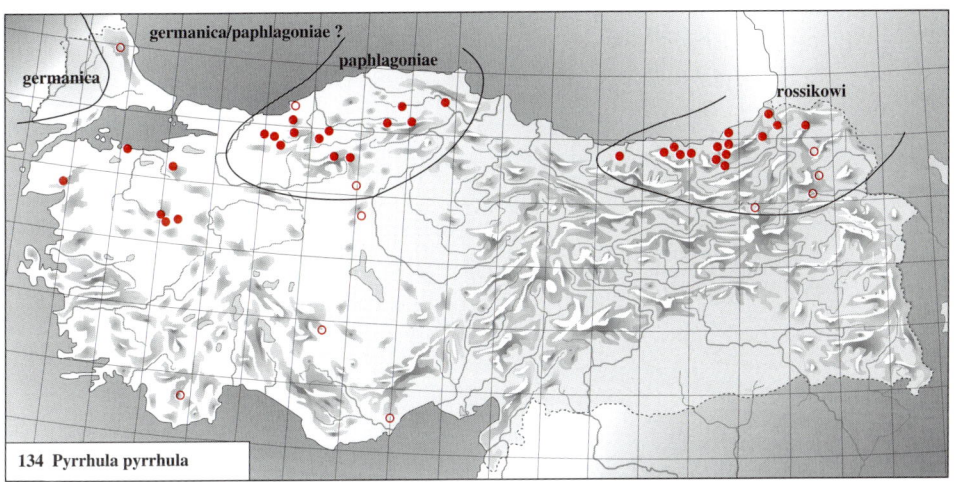

134 Pyrrhula pyrrhula

Geographical Variation

Subspecies described or recorded in the region:

(W) **P. p. pyrrhula** (Linnaeus), 1758, Sweden. [Mantle and scapulars of ♂ medium blue-grey; side of head and underparts down to belly rosy-pink. Large; wing of ♂ from Scandinavia 93.8 (90-97) (n=21), of ♀ 91.8 (87-95) (n=13), average bill depth at base for both sexes 9.6 (n=25), average width of lower mandible at base 10.3 (n=26) (CSR); though the sides of the upper mandible are strongly bulging, the cutting edges, as seen from above, are straight, the outline of the bill from above being a regular triangle. Breeds also in northern European Russia and in north-west and north-central Asia, south to central European Russia, Belorussia, and the Carpathian mountains.]

(W) **P. p. germanica** C. L. Brehm, 1831, Renthendorf (Thüringen, Germany). [Mantle and scapulars very slightly darker grey than in *pyrrhula* and side of head and underparts slightly deeper rosy-red. Slightly smaller; wing of ♂ from Bulgaria 89.7 (87-92) (n=13), of ♀ 86.5 (86-87) (n=4), average bill depth 9.5 (n=8), average width 9.9 (n=9) (Jordans 1940, Niethammer 1950, CSR). Bill shape as in *pyrrhula*. Breeds in a narrow zone from eastern Denmark, easternmost Germany, and Poland (except the north-east) to the Alps, Bulgaria, and northern Greece. This is one of the intermediate forms between the large and pale *pyrrhula* from northern and eastern Europe and the small and dark *europoea*, the latter occurring from north-west Germany and the Netherlands south to western France. The characters of *germanica* are closer to those of *pyrrhula* than to *europoea*. Wing of ♂ *europoea* from the Netherlands 82.8 (79-87) (n=50), of ♀ 81.4 (78-85) (n=44), average bill depth 8.9 (n=18), average bill width 9.3 (n=20) (A. J. van Loon, CSR). Another intermediate, *coccinea*, breeds in a zone between *europoea* and *germanica*, from western Denmark south to northern Switzerland as well as in Italy and in the coastal zone of former Yugoslavia; wing of ♂ from south-west Germany, north-east France, and northern Switzerland 85.4 (82-89) (n=44), of ♀ 84.7 (81-88, occcasionally up to 92) (n=16), average bill depth 8.7 (n=7), average width 9.6 (n=7) (Bacmeister & Kleinschmidt 1918-20, Stresemann 1919, CSR). This form is close to *europoea* in colour and size.]

(T) **P. p. paphlagoniae** Roselaar. See chapter V. Similar to *rossikowi* (see below), but smaller. Wing of ♂ from Karadere (including the type) as well as from the Abant Gölü, Bolu Dagh, and Ilgaz Daglari 88.0 (84-91.5) (n=6), of ♀ 86.8 (85-88) (n=3) (Jordans and Steinbacher 1948, CSR). In colour, *paphlagoniae* is close to *coccinea*, but the mantle and scapulars of the ♂ are slightly paler, more ash-grey, less blue-grey than in *pyrrhula*, *germanica*, and *coccinea*, and the side of the head and underparts are slightly deeper flame-red, less cerise-red or pink-red; the upperparts of the ♀ are greyer, less washed with drab-brown than in *coccinea*, *germanica*, and *pyrrhula*. The measurements and structural characters are near those of *germanica*, but the bill is more swollen at the base, with convex cutting edges when seen from above, and the depth at the base is about equal to the bill width (average depth 9.6 (n=7), average width 9.8 (n=8) (CSR).]

(T) **P. p. rossikowi** Derjugin & Bianchi, 1900, Çoruh (=Artvin, eastern part of the Black Sea Coastlands, Turkey). [Similar to *paphlagoniae* in structural characters and colour (see above), but size larger, near that of *pyrrhula*; differs from *pyrrhula* in bill structure and in the deeper sealing wax-red underparts of the male. Wing of ♂ from north-east Turkey (Erzurum, Sumela above Trabzon, Cayeli, and Sarikamis) 91.2 (90.5-92) (n=3), of ♀ 89, 90 (n=2) (CSR); wing of ♂ from the Caucasus and Transcaucasia 91.4 (90-95) (n=11), of ♀ 88.9 (87-91) (n=8) (Buturlin 1906, Vaurie 1949, A. J. van Loon, CSR), average bill depth 10.3 (n=4), average width 10.2 (n=4) (CSR). Birds from Iranian Azarbaijan probably belong to this subspecies or are intermediate between *rossikowi* and *caspica*; wing of ♂ from this area 90.4 (88-91) (n=8), of ♀ 88.5 (87.5-89) (n=3) (Vaurie 1949). *P. p. caspica* is a small subspecies, restricted to a few localities in northern Iran; it is similar in size and structure to *paphlagoniae*, but it is slightly darker grey above and deeper red below; wing of ♂ 86.7 (84-88) (n=4) (Stresemann 1928, CSR).

Subspecies recognized in Turkey: birds from the western Black Sea Coastlands are *paphlagoniae*, those of the north-east *rossikowi* (see above). The subspecies of the birds breeding in north-west Anatolia and (if any breed) in Thrace is not known; they may belong to *paphlagoniae* (especially those of Western Anatolia) or to *germanica* (especially those of Thrace); the latter subspecies is known to breed nearby in northern Greece and Bulgaria. A winter bird examined from the Istranca Daglari in northern Thrace was *pyrrhula* (male, wing 93: CSR), but this subspecies is not known to breed in Turkey (it may do so when one includes *germanica* in *pyrrhula*).

References Bacmeister, W., & O. Kleinschmidt (1918-20) Zur Ornithologie von Nordost-Frankreich. *J. Orn.* 66, 245-284; 68, 1-32, 97-123. Stresemann, E. (1919) Über die europäischen Gimpel. *Beitr. Zoogeogr. paläarktischen Region* 1, 25-56.

COCCOTHRAUSTES COCCOTHRAUSTES
Hawfinch   Kocabas

Habitat Dense or open stands of mature mixed and deciduous forest as well as parkland, at 0-2000m.

Distribution See map 135. Widespread in Thrace and the western part of the Black Sea Coastlands, but generally not numerous; very local in the north of Western Anatolia. Probably also breeds in very small numbers further east along the coast of the Black Sea, as 5 birds were seen on 19 July at Sumela (above Trabzon) and 2 birds were recorded at Pülümur on 3 August. Stated by Danford to breed in the Taurus mountains, but his observations were in spring and the species was not found in summer by later visitors. However, it may breed occasionally in the area, as it was seen in the mountains south-west of Aksaray on 3 July 1990.

Geographical Variation
Subspecies described or recorded in the region:
(W)   **C. c. coccothraustes** (Linnaeus), 1758, Italy. [Mantle and scapulars umber-brown to dark

Finches

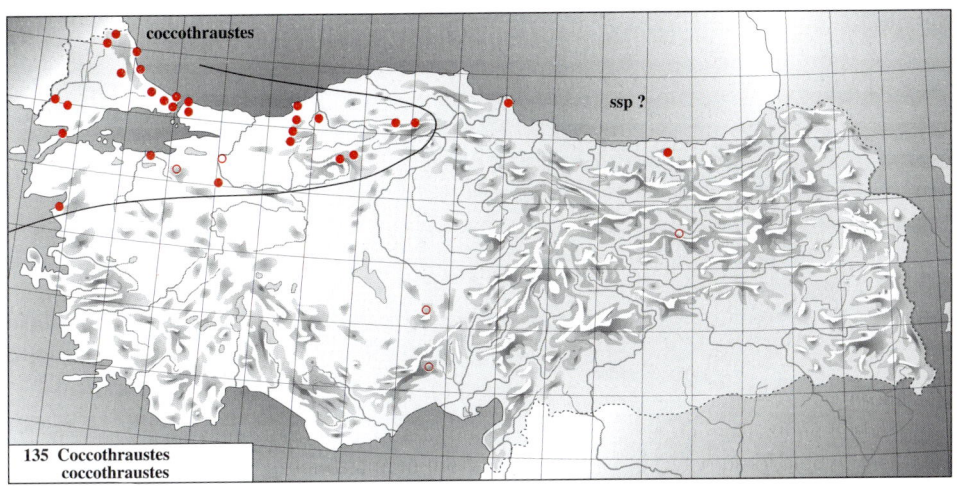

135 Coccothraustes coccothraustes

rufous sepia-brown (♂) or dark cinnamon-brown (♀), underparts largely light vinous drab-grey (♂) or drab-buff (♀). Wing of both sexes from western and central Europe 102.8 (97-110) (n=83) (A. J. van Loon, CSR), bill depth at base of ♂ 16.5 (15.8-17.3) (n=20), of ♀ 15.3 (14.5-15.9) (n=10) (CSR); wing of birds from the Balkan countries and Greece 102.2 (99-107) (n=17), bill depth of ♂ 16.4 (15.9-16.7) (n=5), of ♀ 14.8 (14.5-15.2) (n=4) (CSR, including wing data from Stresemann 1920).]

(E) **C. c. nigricans** Buturlin, 1908, Tbilisi (western Transcaucasia, Georgia). [Mantle and scapulars of ♂ slightly darker dull umber-brown, underparts pinker, less brown; ♀ whiter below (Vaurie 1949). Wing in the Caucasus area 100.3 (96-106) (n=54) (Dementiev & Gladkov 1951-54), in northern Iran 102.6 (95-110) (n=38) (Witherby 1910, Stresemann 1928, Vaurie 1949, Vaurie 1959, Diesselhorst 1962, CSR); bill depth of ♂ at base 16.6 (15.5-17.4) (n=3), of ♀ 14.3, 16.1 (n=2) (CSR). A very poor race, probably not separable from *coccothraustes* on plumage, as the differences cited are mainly those between fresh and old specimens, or are caused by individual variation (Witherby 1910, Hartert 1921-22, Stresemann 1928, Bird 1937, Diesselhorst 1962, Érard & Etchécopar 1970, CSR); however, the subspecies is considered to be valid by, e.g., Vaurie (1949, 1959), Matvejev & Vasic (1973), Matvejev (1976), and Stepanyan (1990).]

Subspecies recognized in Turkey: not known. Matvejev (1976) considers birds from eastern and southern Balkan countries to be about halfway along a cline running from *coccothraustes* to *nigricans*, and birds from Turkey may also be on this cline. As the population in north-western Turkey seems not to be in contact with that of the Caucasus, being separated by a gap in north-east Turkey, the occurrence of such a cline seems doubtful. Moreover, the separation of the population of the Caucasus, Transcaucasia, and northern Iran as *nigricans* is also doubtful, and as the few Turkish birds examined are not separable from *coccothraustes*, Turkish populations are probably better considered to belong to this subspecies. Juveniles examined from Devrek in the western part of the Black Sea Coastlands are inseparable from juveniles of *coccothraustes* (*contra* Maas Geesteranus 1959). Wing of both sexes from Turkey 100.4 (97-104) (n=9) (Rössner 1935, Kumerloeve 1961a, CSR).

EMBERIZA CITRINELLA

Yellowhammer **Sarı Kirazkuşu**

Habitat Bushes and isolated trees alongside arable fields and pastures, scrub at fringes of ditches and gardens, forest edges, and low growth in open forest.

Distribution See map 136. Quite a number of single singing males have been recorded from late April to June in various years, especially in the north-west, but also east to the Trabzon area, and the species probably breeds in at least Thrace, as a continuation of the breeding range in Bulgaria.

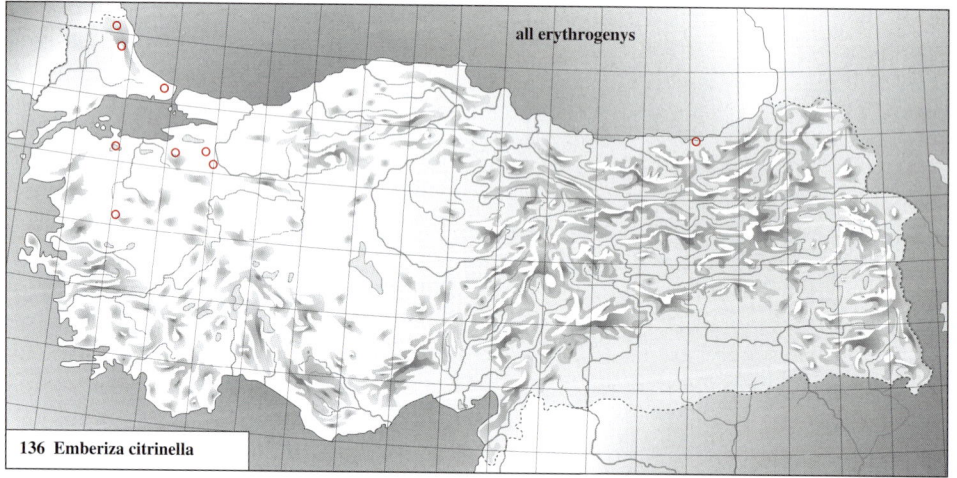

136 *Emberiza citrinella*

Geographical Variation

Subspecies described or recorded in the region:

(W) *E. c. citrinella* Linnaeus, 1758, Sweden. [A rather dark subspecies; greenish olive-brown on mantle and scapulars, greenish-olive on the upper breast; a restricted number of rufous streaks on the lower breast and flanks. Wing of ♂ in western, central, and northern Europe 88.3 (80-96) (n=162), of ♀ 83.6 (79-90) (n=74) (G. O. Keyl).]

(E) *E. c. erythrogenys* C. L. Brehm, 1855, Sarepta, near the lower Volga river (south European Russia). [Paler; more sandy-brown on mantle and scapulars; lower breast and flanks extensively washed and streaked rufous. Wing of ♂ from Romania 92.6 (88-95) (n=8), of ♀ 87.7 (85-91) (n=5) (CSR). See also Simeonov & Doichev (1973).]

Subspecies recognized in Turkey: apparently never collected in summer, but wintering birds belong to *erythrogenys* (Bird 1937, Kumerloeve 1961a), and as this subspecies also breeds as near as eastern Bulgaria and Romania, birds breeding in Turkey (if any regularly do) are also likely to belong to *erythrogenys*.

References Simeonov, S., & R. Doichev (1973) [Studies on the taxonomy of the Yellowhammer (*Emberiza citrinella*) in Bulgaria.] *Ann. Univ. Sofia, Fac. Biol.* I, 65, 163-171.

EMBERIZA CIRLUS

Cirl Bunting **Çit Kirazkuşu**

Habitat Hilly country with plenty of scrub, hedgerows, copses, or isolated trees on moist fertile soil, at borders of pastures and fields, or in olive groves, gardens, or overgrown river banks, mainly at 0-500m but up to 1200m in the Black Sea Coastlands.

Buntings

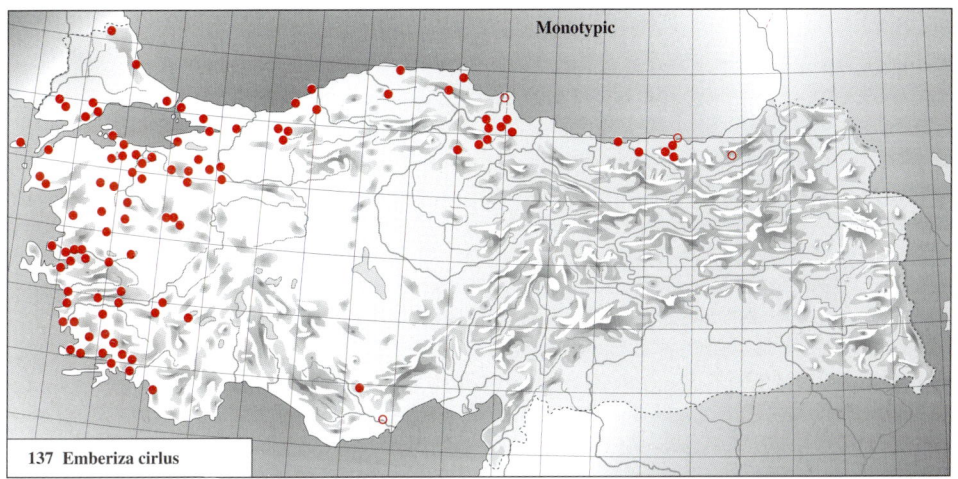

137 Emberiza cirlus

Distribution See map 137; also, Kumerloeve (1961a). Generally common in suitable habitat in Thrace, in the western part of the Black Sea Coastlands, and in Western Anatolia, in the latter area becoming scarce away from the coast and south of Mugla. In the Black Sea Coastlands, widespread east to the Samsun area, more local and uncommon further east, east to above Ispir (Broome 1989). Scattered records from late April to July at the southern foot of the Taurus (Ganso & Spitzer 1967, Groh 1968, Amcoff et al. 1986) may point to occasional breeding in this area.

Geographical Variation Slight, if any. *E. cirlus* Linnaeus, 1766, was originally described from southern Europe. Birds from Corsica and Sardinia are sometimes separated as *nigrostriata*, a subspecies said to be more distinctly streaked on the upperparts or flanks, but no constant difference is apparent when large series of various populations are compared. Hence, the species is considered to be monotypic. Wing of ♂ from south-west Europe 80.6 (79-83) (n=20), of ♂ from central and south-east Europe 82.7 (80-85) (n=16), of ♀ from both regions 77.6 (75-80) (n=16) (CSR); wing of ♂ from Turkey 81.2 (79-83) (n=4), of ♀ 78, 80 (n=2) (Jordans & Steinbacher 1948, Rokitansky & Schifter 1971, CSR).

EMBERIZA CIA
Rock Bunting   Kaya Kirazkusu

Habitat Rocky gorges and sunny outcrops in open forest and woodland, or boulder-covered slopes, fairly densely vegetated with ferns, bushes, and small trees (e.g., pine, oak, or juniper), especially on limestone. Mainly at (0-) 500-1300m in the western third of the country, but up to 2300m on the Ulu Dag; at 500-1700m in the Black Sea Coastlands, at 600-2600m in the Taurus, and at 1200-2600m in the East and South-East.

Distribution See map 138; also, Kumerloeve (1961a). Widespread and locally common in the Black Sea Coastlands, the Taurus Mountains, and the East, more local in Thrace, Western Anatolia, and the mountains of the South-East. Virtually absent from the Central Plateau, from all coastal zones, and from the plains of the South-East.

Geographical Variation
Subspecies described or recorded in the region:
(W)   ***E. c. cia*** Linnaeus, 1766, Lower Austria. [Upperparts dark rufous-cinnamon with rather broad black streaks; tips of median upper wing-coverts white; rather small. Wing of ♂ from central Europe south-east to northern Greece 84.1 (80-87) (n=18), of ♀ 77.0 (74-80) (n=15) (CSR).]

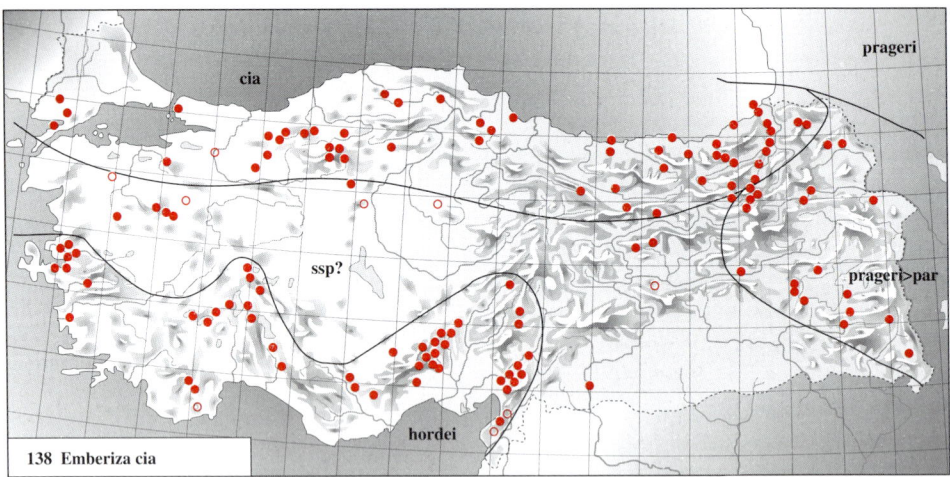

138 Emberiza cia

(W) **E. c. hordei** C. L. Brehm, 1831, 'south-east Europe'. [Like *cia*, but ground-colour of upperparts paler, pink-cinnamon when fresh, more pinkish sandy-grey when worn; width of dark streaks as in *cia*; tips of median coverts rufous-pink to pink-white. Size as in *cia*.]

(E) **E. c. prageri** Laubmann, 1915, Psebay (north-west Caucasus, south European Russia). [Colour as *hordei*, but size larger. Wing of ♂ from the northern Caucasus 88.0 (85-90) (n=5) (Vaurie 1956).]

(E) **E. c. par** Hartert, 1904, Gaudan (south of Ashkhabad, Turkmenistan). [Ground-colour of upperparts as in *hordei* and *prageri* or very slightly paler, but dark streaks narrower; underparts on average paler, but much individual variation due to influence of age, sex, bleaching, and wear; tips of median coverts rufous. Large, wing of ♂ in north-east Iran and Turkmenistan 91.0 (88-94) (n=4) (CSR), in Gorgan (northern Iran) 89.0 (85-92) (n=10) (Vaurie 1956), and in central and south-west Iran (south of Elburz) 88.0 (86-90) (n=8), of ♀ from the last area 79, 81 (n=2) (Witherby 1910, Diesselhorst 1962, Érard & Etchécopar 1970, CSR). Birds similar to *par* in colour but smaller in size occur from Afghanistan through the Tien Shan to the Altai mountains in central Asia; these are separable as *serebrowskii*, and have wing of ♂ 85.5 (82-89) (n=8), of ♀ 80.0 (78-82) (n=7) (CSR).]

Subspecies recognized in Turkey: bleaching and wear have a strong influence on the ground-colour of the body and on the colour of the median coverts, and the subspecies are hard to separate on these characters without adequate series of birds available for comparison. Wing length and width of the dark streaks on the upperparts are easier for identifying birds. All birds from northern Turkey are dark, rather broadly streaked above, and small, similar to or close to *cia*. Wing of ♂ of this population of *cia*, examined from the Ulu Dag east to Borçka, 83.2 (81-85) (n=6), of ♀ 78, 79 (n=2) (CSR, with some additional information from Kummerlöwe & Niethammer 1934-35 and Rokitansky & Schifter 1971). In the south, small birds with paler ground-colour but with rather broad streaking above occur, distributed from Izmir through the Taurus to the Anti-Taurus, and these birds are indistinguishable from *hordei* from southern Greece and southern Italy. Wing of ♂ of *hordei* from south-west Turkey 83.8 (81-87) (n=8), of ♀ 78.9 (77-81) (n=9) (CSR). Birds from south-east Turkey (Erzurum, Tatvan) are larger, wing of ♂ 86, 88.5 (n=2), of ♀ 80 (n=1), and similar to birds from north-west Iran (north of the Elburz); they are *prageri* or intermediates between *prageri* and *par*.

Buntings

EMBERIZA CINERACEA

Cinereous Bunting   Gri Kirazkusu

Habitat Grassy slopes and glades with scattered scrub and some isolated trees, gravelly gorges partly covered with patches of herbaceous plants, or dry rocky outcrops covered with scattered boulders and low vegetation as well as some junipers, mainly in semi-arid situations, at c. 100-500m in the west of the country, at c. 800-1500m in the centre, and up to at least c. 1800m in the east.

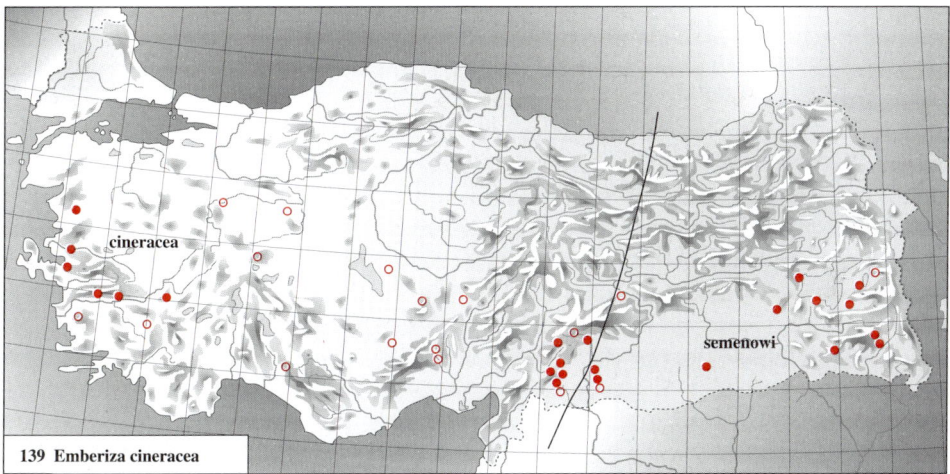

139 Emberiza cineracea

Distribution See map 139; also, Kumerloeve (1961a), Chappuis et al. (1973), and Knijff (1991). Widely scattered in small numbers over the entire southern half of the country, generally away from the coastal zone. Locally not uncommon, e.g. in the Yesilce area (north-west of Gaziantep). Most records from the edge of the Central Plateau and from the main Taurus range refer probably to migrants (see, e.g., Kummerlöwe & Niethammer 1934a, Kumerloeve 1961a, Bezzel 1964, Warncke 1964, Ganso & Spitzer 1967, Kasparek 1985), though sparse local breeding cannot be excluded.

Geographical Variation

Subspecies described or recorded in the region:

(T)   **E. c. cineracea** C. L. Brehm, 1855, Izmir (Western Anatolia, Turkey). [Cap of ♂ greenish-yellow, chest and flanks grey, belly whitish, under tail-coverts uniform off-white; upperparts of ♀ drab-brown. Wing of ♂ from Izmir and Aydin (Western Anatolia) 91.2 (87-96) (n=54), of ♀ 86.6 (84-89) (n=11) (CSR).]

(E)   **E. c. semenowi** Zarudny, 1904, 'Jebel Tnüe' (at the upper Karun River west of Izeh, south-west Iran). [Cap of ♂ yellow, underparts bright sulphur-yellow (tinged olive-grey on sides of breast and flanks), under tail-coverts grey-white with grey centres and dark shaft streaks; upperparts of ♀ greyish-olive. Wing of ♂ from south-west Iran (Paludan 1938, CSR) and from migrants and birds on winter quarters (CSR) 93.1 (88-98.5) (n=12), of ♀ 87.5 (85-90) (n=10). Wing of ♂ from Baykan (south-east Turkey) 92.0 (90-94) (n=4) (Chappuis et al. 1973).]

Subspecies recognized in Turkey: E. c. cineracea breeds from Western Anatolia east to the Gaziantep area (Kumerloeve 1961a, Broome 1989), but birds from Haydan and from between Van and Çatak (Chappuis et al. 1973), those from above Tatvan (Broome 1989), and even those as far west as Halfeti (Hüni 1982) are *semenowi*, and the gap between both subspecies, between Gaziantep and Halfeti, thus appears to be narrow. However, both subspecies may grade into each other, as a number of migrants in Elat (Israel) appear to be intermediate (H. Shirihai), and even among many birds examined from the Izmir area (the type locality of *cineracea*) a few show a slight yellow tinge on the belly, thus tending somewhat to *semenowi*. The nearest relatives of *E. cineracea* are *E. hortulana*, *E. caesia*, and *E. buchanani* (Chappuis *et al.* 1973, Vuilleumier 1977, Martens 1979); in structure (e.g., wing/tail ratio) and moult, *E. cineracea* is very close to *E. buchanani*, while *E. hortulana* is close to *E. caesia*. The latter two are both somewhat different from the former two, which is justification to deviate from the species sequence of the Voous order. Many specimens of this rare species were collected in the Izmir area in the last century by Krüper and Schrader; to those cited by Chappuis *et al.* (1973), a series of 36 skins in the Leiden Museum can be added.

References Martens, J. (1979) Gesang und Verwandtschaft des Steinortolan (*Emberiza buchanani*). *Natur und Museum* 109, 337-343. Chappuis, C., H. Heim de Balsac, & J. Vielliard (1973) Distribution, reproduction, manifestations vocales et affinités du Bruant cendré, *Emberiza cineracea*. *Bonner zool. Beitr.* 24, 302-316. Knijff, P. de (1991) Little-known West Palearctic birds: Cinereous Bunting. *Birding World* 4, 384-391.

EMBERIZA BUCHANANI
Grey-necked Bunting   Tas Kirazkusu

Habitat Arid rocky slopes, barren steppe on mountain tops, or stony fields near patches of snow, covered with a varying amount of scattered low vegetation and scrub (e.g., *Artemisia*, *Astragalus*) at (1850-) 2200-3000m, overlapping with *E. hortulana* at 1850-2600m, but the latter also occurs lower down; habitats of both are roughly similar, but *E. buchanani* generally occurs on more barren and arid sites than *E. hortulana*. At some sites, the ranges of *hortulana* and *buchanani* overlap completely (see, e.g., Kumerloeve 1969h), at others, they exclude each other (see, e.g., Martens 1979)

Distribution See map 140; also, Kumerloeve (1969h). Restricted to the Van region and the Kurdish Alps, where locally common or even abundant, west to the Nemrut Dag above Tatvan. Perhaps also in the mountains bordering the upper Aras valley.

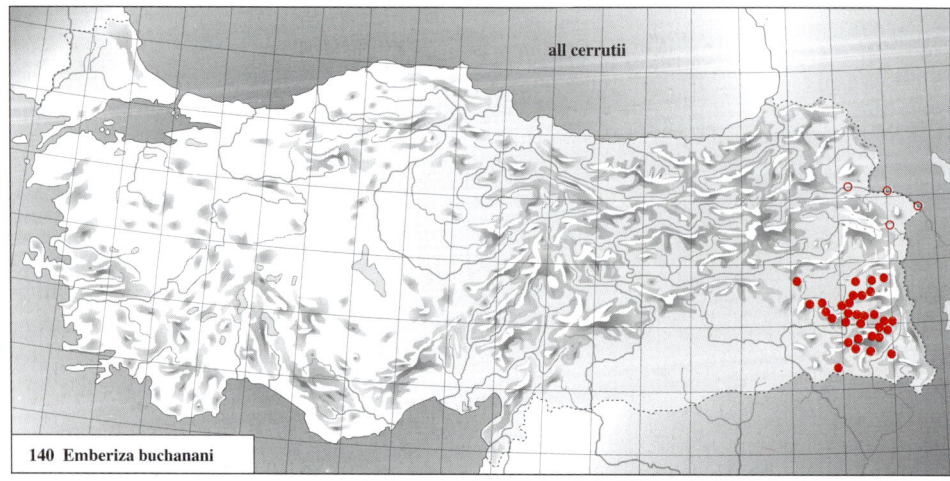

140 *Emberiza buchanani*

Geographical Variation
Subspecies described or recorded in the region:
(E) ***E. b. cerrutii*** De Filippi, 1863, Sadarak (Nakhichevan, at Turkish border) and Sainkalé (=Shahin Dezh, north-west Iran). [A pale subspecies, more sandy-buff on mantle and scapulars than the races of central Asia, less rusty or chestnut on upperparts and belly. Wing of ♂ in Iran 86.9 (86-89.5, one each 82 and 82.5, but these perhaps wrongly sexed) (n=11), of ♀ 80.5 (76.5-83.5) (n=9) (Paludan 1938, 1940, Érard & Etchécopar 1970, CSR). Breeds Transcaucasia, as well as northern Iran east to Gorgan.]

Subspecies recognized in Turkey: all birds examined are inseparable from *cerrutii*. Wing of ♂ from Turkey 89.4 (88-92) (n=9) (CSR), of ♀ 85 (n=1) (Kumerloeve 1969a), thus perhaps somewhat larger than in Iran, but the difference is probably largely due to a different measuring technique used by the authors cited above, as Iranian birds measured by the author have an average wing of 89.6 (♂, n=4) and 82.0 (♀, n=3).

References Kumerloeve, H. (1969h) Der Steinortolan, *Emberiza buchanani*, als türkischer Brutvogel. *J. Orn.* 110, 110-111.

EMBERIZA HORTULANA
Ortolan Bunting   Kirazkusu

Habitat Grassy slopes with rocky outcrops, scattered trees or scrub, and irregular patches of herbaceous plants; at borders of cornfields, pastures, wasteland, thickets, or copses of oak, pine, fir, and juniper, or in open forest. At (100-) 250-800m in Western Anatolia and Thrace, 100-1700m in the Black Sea Coastlands, 900-1800m in the western Taurus and the Elmali area, at 400-2100 (-2500)m in the central and eastern Taurus, at 600-1500m in the Amanus mountains, and at (1000-) 1500-3300m in the mountains of the East and South-East. See also Kumerloeve (1989).

Distribution See map 141; also, Kumerloeve (1961a, 1962c). Widespread and sometimes locally common, especially in the mountains. Largely absent from south-west Anatolia, the Central Plateau, the plateau of the South-East, and from well-wooded sites, as well as barren mountain peaks devoid of vegetation.

Geographical Variation None. *E. hortulana* Linnaeus, 1758, was originally described from Sweden. Some variation in colour exists, e.g. in greyer or greener cap, narrower or wider

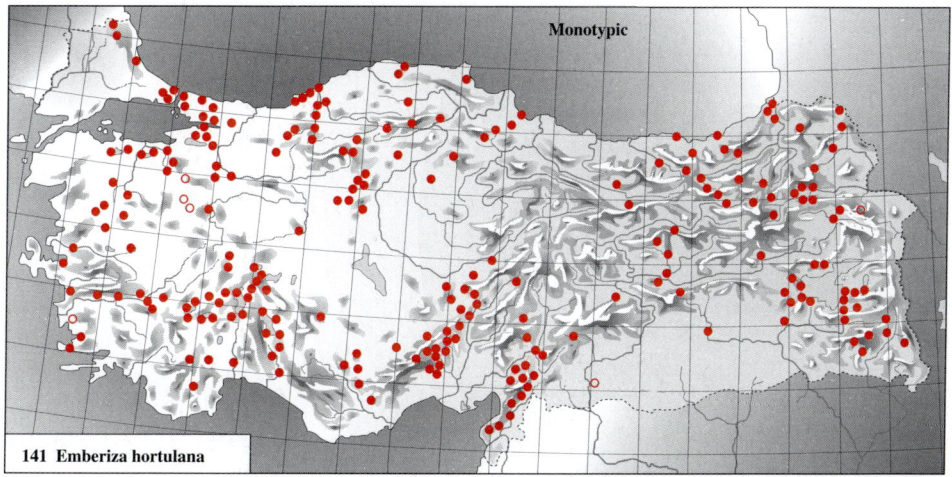

141 *Emberiza hortulana*

dark streaking on mantle and scapulars, paler or darker yellow throat (occasionally, the throat is even white), and in colour of underparts (especially under tail-coverts), the latter grading from light yellowish-buff to deep rufous-cinnamon. However, these differences are either individual or are due to influence of bleaching and wear, not due to geographical variation. Throughout the species' range, birds in June show a greyer head, a paler yellow throat, and darker rufous-cinnamon underparts than birds in May. Wing of breeding ♂ from western Asia Minor (Hacin Dagi, Beysehir, Sindirgi, Sogukpinar, Beynam forest, and Beycuma) 89.2 (83-92, once, in a bird from Beynam, 97) (n=11), of ♀ 84.0 (83.5-84.5) (n=3), of breeding ♂ from the Van and Hakkari areas 93.4 (92-96) (n=7), of ♀ 87.5, 88 (n=2) (CSR); wing of migrant ♂ from western and southern Turkey (Weigold 1912-13, Kummerlöwe & Niethammer 1934-35, Rokitansky & Schifter 1971, Vauk 1973, CSR) and of probable migrants collected late summer near Rize and Verçenik in the north-east 87.4 (82-92) (n=12), of ♀ 83.5 (78-86.5) (n=11) (CSR). Wing of ♂ from central Europe 88.2 (86-89.5) (n=10), of ♀ 83.7 (82.5-85.5) (5); of ♂ from Greece 92.7 (91-95.5) (n=7), of ♀ 90 (n=1); of ♂ from Iran, Caucasus, Turkmenistan, and southern European Russia 90.2 (86.5-92) (n=7) (CSR). Thus, variation in wing ength is markedly large, but without apparent trends; as adults average larger than 1st-year birds, part of the variation found may be due to a different proportion of each age group in various samples; also, birds breeding at higher altitudes are perhaps larger than those of the lowlands.

References Kumerloeve, H. (1962c) Zur Brutverbreitung der beiden Ortolan-Arten *Emberiza hortulana* L. und *Emberiza caesia* Cretzschmar in Kleinasien. *Bonner zool. Beitr.* 13, 327-332. Kumerloeve, H. (1989) Le Bruant ortolan *Emberiza hortulana* nicheur à haute altitude en Anatolie. *Oiseau* 59, 179.

EMBERIZA CAESIA
Cretzschmar's Bunting   Kizil Kirazkusu
Habitat Similar to that of *E. hortulana*, but generally lower down where the ranges of both species overlap, and often on more open and more sparsely vegetated stony slopes (see, e.g., Niethammer 1943, Mauersberger 1960, Kumerloeve 1961a, 1962c, and Reid 1979). Mainly occurs at 100-600m, but up to *c.* 900m in the Taurus and locally to *c.* 1200-1300m between Burdur and Elmali.

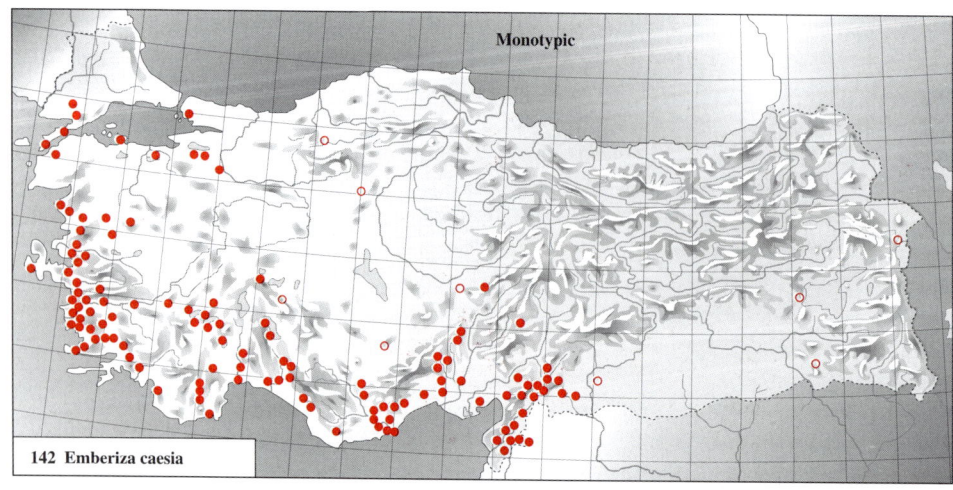

142 Emberiza caesia

Buntings

Distribution See map 142; also, Kumerloeve (1961a, 1962c). Restricted to Western Anatolia, the coastal zone of the Sea of Marmara, and the Southern Coastlands, in the latter area east to Gaziantep and north to the foothills of the Ala Dag and to Çukurbag. Records of singing birds further north in May-July (e.g., in Gerede, Çubuk Baraj, near Ürgüp, and near Eber Gölü and Aksehir) as well as those near Halfeti, Sirnak, and Tatvan (if the latter is properly identified; see, e.g., *Bull. Orn. Soc. Middle East* (1981) 6, 9) may point to occasional breeding outside the normal range.

Geographical Variation None. *E. caesia* Cretzschmar, 1826, was originally described from a migrant collected on Kurgos Island (in the Nile River near Berber, northern Sudan). Wing of ♂ from Greece 86.0 (84.5-87) (n=6), of ♀ 82.1 (81-83) (n=4), of ♂ from Cyprus 85.2 (84-86.5) (n=6), of ♀ 80.5 (79-81.5) (n=3), of ♂ from Syria and Egypt 85.4 (83.5-88) (n=4), of ♂ from Turkey 85.0 (82-88.5) (n=32), of ♀ 81.0 (78-83, once 87) (n=17) (CSR; wing of Turkish birds includes some data from Weigold 1913-14). Cyprus birds differ slightly in colour from those of Izmir, but these Cyprus birds were collected earlier in the spring, and when birds of similar collecting dates are compared no geographical variation in colour is apparent.

References Mauersberger, G. (1960) *Emberiza caesia* Cretzschmar, in: E. Stresemann & L. A. Portenko (eds) Atlas der Verbreitung Palaearktischer Vögel 1. Berlin. Reid, J. C. (1979) The Ortolan and Cretzschmar's Buntings: an ornithological enigma (Aves, Emberizidae). *Bonner zool. Beitr.* 30, 357-366.

EMBERIZA SCHOENICLUS

Reed Bunting **Batak Kirazkusu**

Habitat Large reedbeds in marshes and swamps in lowlands and on plateaux.

Distribution See map 143. Recorded as common only in the Meriç delta, at the Eber Gölü, and in the marshes near Bulanik and Van, apparently rare in most other areas. Sites shown in black have at least one record during June or July in any year, pointing to breeding, though the species is sometimes not found in other years. Around Amik Gölü, recorded breeding by Aharoni (1930, in Kumerloeve 1963a), but apparently absent in May 1953 and 1962 (Kumerloeve 1963a), though present in May and late August 1956 (Hollom 1959, Kumerloeve 1963a), and many seen on 31 May 1975 (*Bull. Orn. Soc. Turkey*

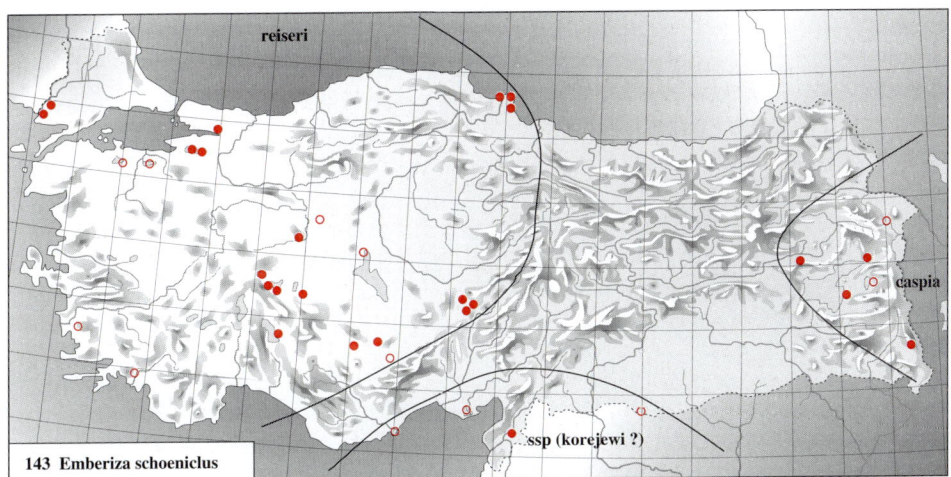

143 *Emberiza schoeniclus*

(1976) 13, 2-5); similar conflicting data between years are available for some other localities. Either the species does not breed annually at all sites or it occurs in very small numbers and is easily overlooked.

Geographical Variation

Subspecies described or recorded in the region:

(W) *E. s. schoeniclus* Linnaeus, 1758, Sweden. [A small dark subspecies with a slender bill; culmen almost straight. Wing of ♂ from southern Sweden, western Germany, the Netherlands, and Britain 80.4 (76-84) (n=50), of ♀ 74.8 (71-79) (n=22), bill of both sexes 12.1 (11.2-13.3) (n= 47), depth of bill at base 5.2 (4.6-5.8) (n=42) (CSR).]

(W) *E. s. stresemanni* Steinbacher, 1930, Overbász (Vojvodina, northern Serbia). [Colour like *schoeniclus* or slightly paler, bill thicker at base, culmen slightly arched. Wing of ♂ from eastern Austria, Hungary, and the north of former Yugoslavia 82.5 (78-87) (n=30), of ♀ 75.1 (72.5-78) (n=10), bill 12.7 (11.8-13.6) (n=40), bill depth 6.0 (5.6-6.4) (n=39) (W. R. R. de Batz, CSR).]

(W) *E. s. tschusii* Reiser & Almasy, 1898, Dunavat (between Tulcea and Sfintu Gheorghe, Dobrogea, Romania). [Larger and paler, bill fairly thick, culmen moderately arched; ground-colour of upperparts paler, more buff when fresh, more cream when worn, dark streaks on average narrower; rump paler; flanks less streaked; lesser upper wing-coverts tawny instead of chestnut. Wing of ♂ from north-east Romania, southern Ukraine, Crimea, and the Sarpa area of southern European Russia 85.1 (81.5-87.5) (n=16), of ♀ 80.7 (78-84) (n=3), bill 12.6 (11.8-13.5) (n=18), bill depth 7.0 (6.4-7.5) (n=18) (W. R. R. de Batz, CSR).]

(W) *E. s. othmari* Hartert, 1910, Sultanlar marshes, between Varna and Provadiya, C-E Bulgaria. [Size as in *tschusii*, bill larger than in *stresemanni* but smaller than in *reiseri*. Colour dark, as in *E.s. schoeniclus, stresemanni* or *reiseri*, darker than *tschusii*. Wing of ♂ from northern and eastern Bulagaria 81.1 (77-86) (n-4), bill 13.0 (12,3-13.8) (n=4), bill depth 7.0 (6.5-8.1) (n=4) (CSR).]

(W) *E. s. reiseri* Hartert, 1904, Lamia (Thessalia, Greece). [Large; bill heavy with strongly arched culmen; colour dark, as in *schoeniclus* or *stresemanni*, or flanks even more heavily streaked. Wing of ♂ from Greece and the south of former Yugoslavia 87.2 (82-94) (n=26), of ♀ 79.9 (78-83) (n=4), bill 14.1 (13.4-14.7) (n=7), bill depth 8.4 (7.9-9.0) (n=7) (CSR; includes wing data from Stresemann 1920, Makatsch 1950, and Vaurie 1958); wing of ♂ from south-east Albania mainly 87-92, of ♀ 78-81, bill 13.0-14.5, bill depth mainly 8.0-9.0 (Ticehurst & Whistler 1932). Restricted to a few sites in Greece, the south of former Yugoslavia, and (formerly) south-east Albania.]

(E) *E. s. caspia* Ménétries, 1832, 'near Bechebermak close to Caspian Sea' (near Baku, eastern Azerbaijan). [Large, rather pale, and thick-billed; colour similar to *tschusii* above, but bill longer and culmen more strongly arched; bill slightly less heavy at base than in *reiseri*; colour of plumage less pale and bill less heavy than in the subspecies *pyrrhuloides* (below). Wing of ♂ *caspia* from Enzeli (=Bandar-e Pahlavi, along the southern shore of the Caspian Sea, Iran) 85, 86 (n=2), of ♀ 80.7 (77-83.5) (n=3), bill length 13.7 (13.0-14.4) (n=5), bill depth 7.3 (7.0-7.5) (n=5) (W. R. R. de Batz, CSR).]

(E) *E. s. korejewi* (Zarudny), 1907, Seistan and Persian Baluchestan (south-east Iran). [Like *caspia*, but slightly paler, upperparts buffer, less rufous; not as pale as *pyrrhuloides*. Bill slightly thicker than in *caspia*, almost as heavy as in *reiseri*, but not as thick as in *pyrrhuloides* and longer than in that subspecies, less stubby. Wing of ♂ from Seistan 88, of ♀ 80.5, bill of both 13.4, 14.1, bill depth 7.7, 8.7 (CSR). Outside the Seistan basin, this race apparently occurs (or has occurred) along the Hari Rud (at the border of north-east Iran and Afghanistan), as well as locally in the Zagros mountains of south-west Iran, in the Al Jazirah area of eastern Syria and at Tell el Abyad in northern Syria.]

(E) *E. s. pyrrhuloides* (Pallas), 1811, Astrakhan (Volga delta, south European Russia). [The largest and palest subspecies; ground-colour of upperparts cream-buff to grey-white, black streaks relatively narrow; underparts cream to white, flanks unstreaked; bill very thick, strongly arched, as in *reiseri* but shorter, appearing more stubby. Wing of ♂ from the Volga and Ural river deltas 88.9 (85-91) (n=7), bill 13.4 (13.2-13.6) (n=5), bill depth 8.0 (7.6-8.5) (n=6); wing of ♂ from Uzbekistan and Kazakhstan 89.8 (86-93) (n=11), of ♀ 83.1 (81-87) (n=6), bill 13.0 (12.0-13.7) (n=17), bill depth 8.2 (7.7-9.2) (n=17) (W. R. R. de Batz, CSR).]

Subspecies recognized in Turkey: all birds recorded breeding in Turkey belong to a thick-billed subspecies, e.g., those at Eber Gölü (Kumerloeve 1964b), Balikdami (Porter 1969), Kizilirmak delta (Renkhoff 1972), Beysehir Gölü (Broome 1989), as well as at Bataklik Gölü, near Van, and near Dogubayazit. Birds from the Greek side of the Meriç delta are pure *reiseri*, and undoubtedly the ones from the Turkish side in Thrace also belong to this subspecies. The size and colour of the birds examined from Asia Minor fit in a cline running from *reiseri* (from Greece) to *caspia* (from the southern shore of the Caspian Sea); birds from Eber Gölü are very close to *reiseri* and therefore included in this subspecies, those of Van Gölü are slightly paler and less thick-billed and are here united with *caspia*. Wing of ♂ of *reiseri* from Eber Gölü 87.5 (85-91.5) (n=4), bill 14.0 (13.8-14.1) (n=4), bill depth 8.1 (7.9-8.4) (n=4); wing of ♂ of *caspia* from Van Gölü 85.4 (83.5-88) (n=5), bill 13.8 (12.8-14.4) (n=5), bill depth 7.8 (7.3-8.1) (n=5) (CSR). The subspecies of the birds breeding from the Göksu delta east to Amik Gölü is not known, as apparently no breeding birds have been collected. Possibly, the pale thick-billed *korejewi* is involved, as in nearby Syria, or perhaps the darker *caspia* or the even darker *caspia-reiseri* intergrades. Birds from Amik Gölü were assigned to the pale thick-billed subspecies *pyrrhuloides* by Aharoni (in Kumerloeve 1963a), but Aharoni was either not aware of the existence of *korejewi* or his birds were collected in winter, when *pyrrhuloides* is known to occur in the Middle East. Small-billed birds, probably *E. s. schoeniclus* and the paler-coloured *ukrainae*, occur on migration and in winter, with some lingering into May; also, the pale intermediate-billed *tschusii* occurs in winter (e.g., collected at Amik Gölü), while another intermediate-billed subspecies, the dark *othmari* from northern and eastern Bulgaria and south-east Romania, may occur in northern Thrace.

EMBERIZA MELANOCEPHALA
Black-headed Bunting   Karabas Kirazkusu

Habitat Cultivated fields or wasteland, fairly densely covered with scattered bushes, hedges, patches of herbaceous plants, or small trees; often near streams or irrigation ditches; in gardens, orchards, vineyards, or in open forest. Occurs at all levels so long as cultivation is present, mainly in lowlands, on plateaux, and on lower slopes of wide valleys, from 0-1200 (-2000)m in the west and the centre of the country and to 2100 (-2900)m in the east.

Distribution See map 144. Widespread and common to abundant in many places, but sparse or absent from the coastal zone of the Black Sea Coastlands.

Geographical Variation None. *E. melanocephala* Scopoli, 1769, was originally described from Carniola (=Krain, central and eastern Slovenija). Wing of ♂ from Italy and from former

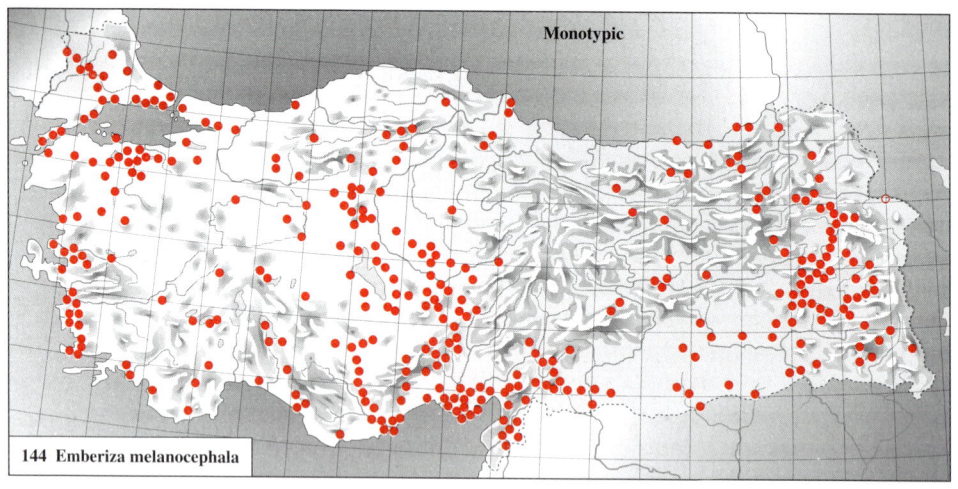

144 Emberiza melanocephala

Yugoslavia 98.2 (92.5-103) (n=7), of ♀ 90.1 (88-93.5) (n=8); of ♂ from Romania and from the plains of south European Russia 95.3 (92-100) (n=8); of ♂ from Greece 95.0 (92-97) (n=7), of ♀ 91, 91 (n=2) (CSR); of ♂ from Thrace 94.0 (92-95) (n=4), of ♀ 89 (n=1) (Rokitansky & Schifter 1971); of ♂ from Western Anatolia 96.0 (93-100.5) (n=10), of ♀ 90.9 (89.5-92.5) (n=4) (CSR); of ♂ from the Central Plateau 95.5 (92.5-98.5) (n=7) (Kummerlöwe & Niethammer 1934-35, CSR); of ♂ from the Taurus area 94.2 (89-98) (n=16), of ♀ 89.8 (85-92.5) (n=10) (Jordans & Steinbacher 1948, Rokitansky & Schifter 1971, Vauk 1973, CSR); of ♀ from south-east Turkey 89, 91 (n=2); of ♂ from Cyprus and the Levant 94.4 (91-97) (n=10), of ♀ 86.5 (n=1) (CSR). Thus, average size perhaps decreases slightly from north-west to south-east.

MILIARIA CALANDRA

Corn Bunting   Tarla Kirazkusu

Habitat Plains and valleys with fertile and not-too-dry cultivated fields, stony slopes, or steppe country, alternating with scattered shrubs, thorny hedges, and isolated trees, generally more open than the habitat favoured by *Emberiza melanocephala* which is also common in cultivation; sometimes in swamps of *Carex*. At 0-1300 (-1600)m in the west and centre, mainly at 1000-2000 (-2300)m in the mountain valleys of the eastern third of the country.

Distribution See map 145. Widespread and often common or even abundant, especially in the centre and east of Turkey; numbers are more moderate in Western Anatolia, where scarcer still or even locally absent in the south. Rather local in the coastal zone of the Black Sea Coastlands and apparently rare in the area round Kars in the north-east.

Geographical Variation

Subspecies described or recorded in the region:

(W)   **M. c. calandra** (Linnaeus), 1758, Sweden. [Ground-colour of upperparts and sides of head slightly darker brown-grey, of underparts creamy-yellow; dark streaks of underparts slightly broader and more extensive; throat spotted. Wing of ♂ in central and southern Europe (Denmark to Portugal and Greece) 102.6 (98-107) (n=58), of ♀ 93.8 (90-98) (n=27) (CSR). Breeds North Africa and Europe, south to the Levant, Transcaucasia, and northern Iran; variation slight and clinal, obscured by influence of bleaching, abrasion and individual variation; populations of *calandra* gradually merge into those of the next subspecies towards the east (CSR).]

Buntings

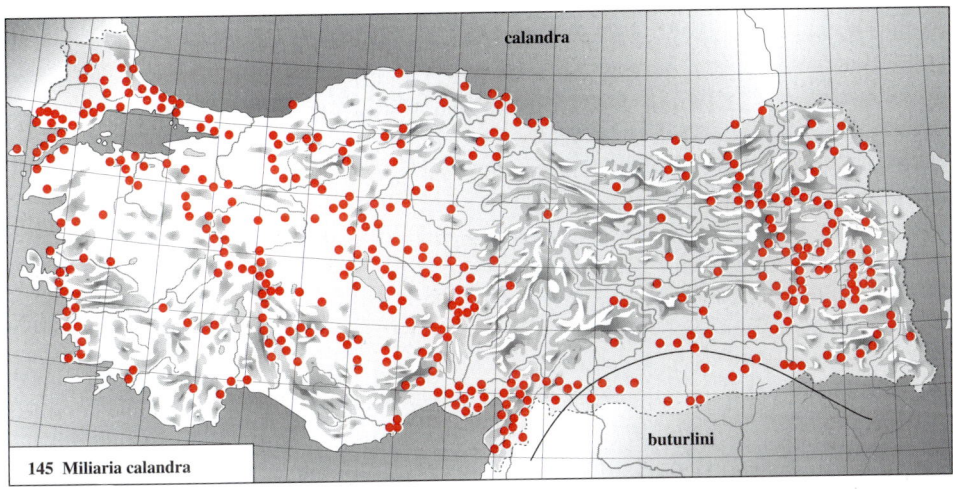

145 Miliaria calandra

(E)   ***M. c. buturlini*** (Hermann Johansen), 1907, mouth of the Kastek river (near Tokmak, at the border of south-east Kazakhstan). [Ground-colour of upperparts and sides of head paler, more olive-grey when plumage fresh, more sandy-grey when worn; ground-colour of underparts whitish; dark streaks on underparts narrower and less extensive; throat unspotted; fringes of flight feathers whiter. Wing of ♂ in central Asia 100.9 (97-104) (n=9), of ♀ 92.8 (91-95) (n=5) (M. Platteeuw, CSR), of ♂ in Iran (where some birds throughout the country are *buturlini*, but others, especially in the north and west, are inseparable from *calandra*) 100.8 (98-104.5) (n=12), of ♀ 92.5 (90-95) (n=10) (Stresemann 1928, Paludan 1938, 1940, Schüz 1959, Diesselhorst 1962, Érard & Etchécopar 1970, CSR). Various other subspecies have been described (e.g., in Portenko 1962), more or less intermediate between *calandra* and *buturlini*, but the difference between these extremes is too slight to warrant recognition of intermediate forms.]

Subspecies recognized in Turkey: most birds examined are inseparable from *calandra*, though some tend to become paler in worn plumage than is usual in central and northern Europe. Birds from Adana to Gaziantep are *calandra*, those from Ceylanpinar in the South-East *buturlini*, as are birds from northern Syria (Kumerloeve 1970a). Wing of ♂ *calandra* from Turkey 102.3 (98-106) (n=21), of ♀ 93.2 (90-96) (n=19) (CSR, including some data from Kummerlöwe & Niethammer 1934-35, Kumerloeve 1963a, and Rokitansky & Schifter 1971). Wing of ♂ of *buturlini* from Ceylanpinar 100.5-105 (n=2), of ♀ 94.5 (n=1) (Kumerloeve 1970a). Abberant birds are occasionally recorded (Lees-Smith & Madge 1982).

References Lees-Smith, D. T., & S. C. Madge (1982) Aberrant Corn Bunting suggesting 'lost' male plumage. *Bull. Orn. Soc. Middle East* 9, 5-6.

XI   LIST OF NON-PASSERINE BIRDS BREEDING IN TURKEY

In contrast to the passerine birds listed above, this survey of the non-passerine birds of Turkey is largely tentative. It gives the names of all subspecies known to breed in Turkey or presumed to occur as a breeding bird. Although I examined a great number of non-passerine bird skins from Turkey (see, e.g., the headings and taxonomic sections of Roselaar in Cramp & Simmons 1977, 1980 and 1983, and Cramp 1985), not all those available were checked as critically as the passerine skins, especially the species which are thought to show little geographical variation. The ranges outlined for the subspecies below are generally those found after personal examination of specimens, the results of which are shown in the species headings I made in Cramp & Simmons and Cramp (op. cit.); additional evidence was obtained from the literature (mainly from Kummerlöwe & Niethammer 1934-35, Jordans & Steinbacher 1948, Kumerloeve 1961a, 1968, 1969a, Vaurie 1965, Rokitansky & Schifter 1971, and Stepanyan 1990). The species included are generally those mentioned as breeding birds or probable breeding birds by Kasparek (1992), to which the reader should refer for details on the status of each bird and for its general distribution within Turkey. As in the passerine list, the English names are derived from British Birds (1993) or Beaman (1994), the Turkish names are from Kiziroglu (1989).

Tachybaptes ruficollis   Little Grebe   Bahri
   T. r. ruficollis: north-west Africa, Europe, western Turkey, and Israel.
   T. r. capensis: Africa (except north-west), eastern Turkey (probably), the Caucasus area and northern Iran east to central Asia.
Podiceps cristatus   Great Crested Grebe   Tepeli Batagan
   P. c. cristatus: North Africa, Europe, and Asia.
Podiceps grisegena   Red-necked Grebe   Kirmizi Boyun Batagan
   P. g. grisegena: Europe and western Asia.
Podiceps nigricollis   Black-necked Grebe   Karaboyun Batagan
   P. n. nigricollis: Eurasia.
Calonectris diomedea   Cory's Shearwater   Sarigaga Yelkovan
   C. d. diomedea: the Mediterranean basin, possibly including Turkey.
Phalacrocorax carbo   Great Cormorant   Karabatak
   P. c. sinensis: Europe (except for the coasts of the northern Atlantic Ocean) and Asia.
Phalacrocorax aristotelis   Shag   Tepeli Karabatak
   P. a. desmarestii: coasts of the Mediterranean basin and the Black Sea.
Phalacrocorax pygmaeus   Pygmy Cormorant   Cüce Karabatak
   Monotypic.
Anhinga rufa   African Darter   Yilanboyun Kusu
   A. f. chantrei: Iraq and (formerly) southern Turkey and Israel.
Pelecanus onocrotalus   Great White Pelican   Beyaz Pelikan
   Monotypic.
Pelecanus crispus   Dalmatian Pelican   Tepeli Pelikan
   Monotypic.
Botaurus stellaris   Great Bittern   Balaban
   B. s. stellaris: North Africa and Eurasia.
Ixobrychus minutus   Little Bittern   Cüce Balaban
   I. m. minutus: North Africa, Europe, and western Asia.

Nycticorax nycticorax　　Black-crowned Night Heron　　Gece Balikçili
　　N. n. nycticorax: Europe, Africa, and Asia.
Ardeola ralloides　　Squacco Heron　　Alaca Balikçil
　　Monotypic.
Bubulcus ibis　　Cattle Egret　　Öküz Balikçili
　　B. i. ibis: Africa, southern Europe, and some localities from the Middle East east to the shores of the Caspian Sea; also, established in North and South America.
Egretta garzetta　　Little Egret　　Küçük Akbalikçil
　　E. g. garzetta: Africa, southern Europe, and southern Asia.
Egretta alba　　Great White Egret　　Büyük Akbalikçil
　　E. a. alba: central and southern Europe, east to the Middle East and central Asia.
Ardea cinerea　　Grey Heron　　Gri Balikçil
　　A. c. cinerea: Africa (except Mauritania), Europe, and northern and south-western Asia.
Ardea purpurea　　Purple Heron　　Erguvani Balikçil
　　A. p. purpurea: Africa (except the Cape Verde Islands and Madagascar), Europe, and from the Middle East to Kazakhstan.
Ciconia nigra　　Black Stork　　Karaleylek
　　Monotypic.
Ciconia ciconia　　White Stork　　Akleylek
　　C. c. ciconia: Africa, Europe, and the Middle East.
Plegadis falcinellus　　Glossy Ibis　　Çeltikçi
　　P. f. falcinellus: North Africa, Europe, and from the Middle East to central Asia; also, south-east U. S. A. and the Greater Antilles.
Geronticus eremita　　Northern Bald Ibis　　Kelaynak
　　Monotypic.
Platalea leucorodia　　Eurasian Spoonbill　　Kasikçil
　　P. l. leucorodia: Eurasia.
Phoenicopterus ruber　　Greater Flamingo　　Flamingo
　　P. r. roseus: Africa, southern Europe, and western Asia.
Cygnus olor　　Mute Swan　　Kugu
　　Monotypic.
Anser anser　　Greylag Goose　　Bozkaz
　　A. a. rubrirostris: south-eastern and eastern Europe, as well as northern Asia.
Tadorna ferruginea　　Ruddy Shelduck　　Pasrenkli Angit
　　Monotypic.
Tadorna tadorna　　Common Shelduck　　Suna
　　Monotypic.
Anas strepera　　Gadwall　　Külrengi Ördek
　　A. s. strepera: Eurasia and North America.
Anas crecca　　Common Teal　　Krik Ördek
　　A. c. crecca: Eurasia.
Anas platyrhynchos　　Mallard　　Yesilbas Ördek
　　A. p. platyrhynchos: Eurasia and western North America.
Anas acuta　　Northern Pintail　　Kilkuyruk
　　A. a. acuta: Eurasia and North America.
Anas querquedula　　Garganey　　Bagirtlak
　　Monotypic.

Anas clypeata   Northern Shoveler   Kasikgaga
   Monotypic.
Marmaronetta angustirostris   Marbled Duck   Dargaga
   Monotypic.
Netta rufina   Red-crested Pochard   Pas Rengi Ördek
   Monotypic.
Aythya ferina   Common Pochard   Elmabas
   Monotypic.
Aythya nyroca   Ferruginous Duck   Akgöz
   Monotypic.
Aythya fuligula   Tufted Duck   Tepeli Ördek
   Monotypic.
Melanitta fusca   Velvet Scoter   Kadife Ördek
   M. f. fusca: northern Europe, north-west Siberia, Transcaucasia, and (probably this subspecies) eastern Turkey.
Bucephala clangula   Common Goldeneye   Altingöz
   B. c. clangula: northern Eurasia and (probably this subspecies) occasionally Turkey.
Oxyura leucocephala   White-headed Duck   Akbas
   Monotypic.
Pernis apivorus   European Honey Buzzard   Aricil
   Monotypic.
Milvus migrans   Black Kite   Karaçaylak
   M. m. migrans: Europe, North Africa, and the Middle East, including Turkey.
Milvus milvus   Red Kite   Kizilçaylak
   M. m. milvus: North Africa, Europe, Transcaucasia, and possibly Turkey.
Haliaeetus albicilla   White-tailed Eagle   Akkuyruk
   Monotypic.
Gypaetus barbatus   Lammergeier   Sakkali Akbaba
   G. b. aureus: Eurasia, including the Middle East.
Neophron percnopterus   Egyptian Vulture   Beyaz Akbaba
   N. p. percnopterus: Africa, Europe east to central Asia, and the Middle East.
Gyps fulvus   Griffon Vulture   Kizil Akbaba
   G. f. fulvus: North Africa, Europe east to central Asia, and the Middle East.
Aegypius monachus   Eurasian Black Vulture   Kara Akbaba
   Monotypic.
Circaetus gallicus   Short-toed Eagle   Yilan Kartali
   C. g. gallicus: North Africa, Europe, the Middle East, and west Siberia.
Circus aeruginosus   Marsh Harrier   Saz Delicesi
   C. a. aeruginosus: Europe east to central Asia and the Middle East.
Circus pygargus   Montagu's Harrier   Çayir Dogani
   Monotypic.
Accipiter gentilis   Northern Goshawk   Çakir Kusu
   A. g. marginatus: southern Europe through the southern Balkan countries and Greece to the Middle East and the Caucasus.
Accipiter nisus   Eurasian Sparrowhawk   Dogu Atmacasi
   A. n. nisus: Europe and the Middle East.
Accipiter brevipes   Levant Sparrowhawk   Kisa Parmak Atmaca
   Monotypic.

Buteo buteo   Common Buzzard   Sahin
> *B. b. buteo*: central and parts of northern and southern Europe, east to the Balkan countries and western Turkey.
> *B. b. menetriesi*: central and eastern Turkey, east to the Caucasus and Iran.

Buteo rufinus   Long-legged Buzzard   Kizil Sahin
> *B. r. rufinus*: Europe, east to central Asia and the Middle East.

Aquila pomarina   Lesser Spotted Eagle   Küçük Bagirgan Kartal
> *A. p. pomarina*: central and south-east Europe and the Middle East.

Aquila nipalensis   Steppe Eagle   Bozkir Kartali
> *A. n. orientalis*: eastern Europe, east to central Asia, and (probably this subspecies) occasionally in Turkey.

Aquila heliaca   Imperial Eagle   Sah Kartal
> Monotypic.

Aquila chrysaetos   Golden Eagle   Kaya Kartali
> *A. c. chrysaetos*: Europe (except south-west) to central Asia, including the Balkan countries and (apparently this subspecies) Thrace (European Turkey).
> *A. c. homeyeri*: south-west Europe, North Africa, and from the Middle East to the Caucasus, including Asia Minor.

Hieraaetus pennatus   Booted Eagle   Küçük Kartal
> Monotypic.

Hieraaetus fasciatus   Bonelli's Eagle   Atmaca Kartali
> *H. f. fasciatus*: North Africa and Europe east to southern and eastern Asia.

Pandion haliaetus   Osprey   Balikkartali
> *P. h. haliaetus*: North Africa and Europe east to eastern Asia and the Middle East.

Falco naumanni   Lesser Kestrel   Kizil Kerkenez
> Monotypic.

Falco tinnunculus   Common Kestrel   Kerkenez
> *F. t. tinnunculus*: north-west Africa, Europe, northern Asia, and the Middle East.

Falco subbuteo   Eurasian Hobby   Delice Dogan
> *F. s. subbuteo*: North Africa, Europe, northern Asia, and the Middle East.

Falco eleonorae   Eleonora's Falcon   Karadogan
> Monotypic.

Falco biarmicus   Lanner Falcon   Biyikli Dogan
> *F. b. feldeggii*: Italy, the Balkan countries, Greece, and Turkey.

Falco cherrug   Saker Falcon   Uludogan
> *F. c. cherrug*: central and south-east Europe east to the steppes of central Asia, and (formerly) Thrace and the Bosporus area.
> *F. c. milvipes*: Turkmenistan east to the mountains of central Asia, apparently Iran, and (apparently this subspecies) Asia Minor.

Falco peregrinus   Peregrine Falcon   Gezginci Dogan
> *F. p. brookei*: the Mediterranean basin east to Turkey, the Caucasus, and the Crimea.

Tetrao tetrix   Black Grouse   Kara Horozu
> *T. t. tetrix*: Europe (except Britain and south-east European Russia) to central Siberia; perhaps this subspecies formerly in the Bosporus area.

Tetrao mlokosiewiczi   Caucasian Black Grouse   Kafkas Horozu
> Monotypic.

Tetraogallus caspius   Caspian Snowcock   Urkeklik

   T. c. tauricus: the Taurus, eastern Turkey, and the neighbouring part of Armenia. *T. c. semenowtianschanskii*, breeding in the Zagros mountains of south-west Iran, may be the subspecies occurring in the Amadiyah area of northern Iraq and the Hakkari area of Turkey; however, the race is doubtfully separable from *tauricus* (Roselaar in Cramp & Simmons 1980).

Alectoris chukar   Chukar Partridge   Kinali Keklik

   A. c. kleini: north-east Greece, south-east Bulgaria, northern Turkey, and the Caucasus.

   A. c. cypriotes: Crete, the Aegean islands, southern Turkey east to the Amanus mountains, and Cyprus.

   A. c. kurdestanica: south-east Turkey, Transcaucasia, northern Iraq, and northern Iran. See Roselaar (in Cramp & Simmons 1980) for influence of *A. c. sinaica* on the birds of south-east Turkey; *sinaica* occurs from the Sinai and the Levant to central Syria.

Ammoperdix griseogularis   See-see Partridge   Çöl Kekligi

   Monotypic.

Francolinus francolinus   Black Francolin   Turaç

   F. f. francolinus: Cyprus and southern Turkey (except Amik Gölü area) east to Armenia and northern Iran.

   F. f. billypayni: the Amik Gölü area, now apparently extinct.

Perdix perdix   Grey Partridge   Çil Keklik

   P. p. perdix: northern and central Europe south to Balkan countries, Greece, and Thrace (European Turkey).

   P. p. canescens: Asia Minor, the Caucasus area, and northern Iran.

Coturnix coturnix   Common Quail   Bildircin

   C. c. coturnix: mainland North Africa and Europe east to central Asia and northern India.

Phasianus colchicus   Common Pheasant   Sülün

   P. c. colchicus: western Transcaucasia. One would expect this subspecies in northern Turkey if the species occurs naturally. However, most records apparently refer to other subspecies and, if *colchicus* bred in the past, introduction of more exotic subspecies into Turkey may have swamped the *colchicus* genes.

Rallus aquaticus   Water Rail   Suyelvesi

   R. a. aquaticus: North Africa and Europe east to south-west Siberia and the Middle East.

Porzana porzana   Spotted Crake   Benikli Suyelvesi

   Monotypic. Probably breeds in Turkey.

Porzana parva   Little Crake   Küçük Benekli Suyelvesi

   Monotypic. Probably breeds in Turkey.

Porzana pusilla   Baillon's Crake   Benekli Çüce Suyelvesi

   P. p. pusilla: south European Russia east to Japan, south to Iran and China.

   P. p. intermedia: North Africa and central and southern Europe, east to Romania and Bulgaria. It is not known what race breeds in Turkey, if any regularly does so.

Crex crex   Corn Crake   Bildircin Kilavuzu

   Monotypic. Probably breeds in Turkey.

Gallinula chloropus   Moorhen   Yesilayak Sutavugu

   G. c. chloropus: North Africa and Eurasia.

Porphyrio porphyrio   Purple Swamp-hen   Saz Horozu

   P. p. caspius: southern Turkey, Iran, and the shores of the Caspian Sea.

Fulica atra   Common Coot   Sakarmeki
 F. a. atra: North Africa and Eurasia.
Grus grus   Common Crane   Turna
 G. g. lilfordi: Asia Minor, Transcaucasia, and northern Asia.
Anthropoides virgo   Demoiselle Crane   Telli Turna
 Monotypic.
Tetrax tetrax   Little Bustard   Mezgeldek
 Monotypic.
Chlamydotis undulata   Houbara Bustard   Yakali Toykusu
 C. u. macqueenii: the Sinai peninsula, the Levant, south-east Turkey (formerly), and from Iran east to Mongolia.
Otis tarda   Great Bustard   Büyük Toykusu
 O. t. tarda: north-west Africa, Europe, the Middle East, and west and central Asia.
Haematopus ostralegus   Oystercatcher   Denizsaksagani
 H. o. longipes: south-eastern and eastern Europe, east to the Middle East and central Asia.
Himantopus himantopus   Black-winged Stilt   Uzunbacak
 H. h. himantopus: Africa, as well as Eurasia east to Mongolia and Sri Lanka.
Recurvirostra avosetta   Avocet   Kiliçgaga
 Monotypic.
Burhinus oedicnemus   Stone-curlew   Kocagöz
 B. o. oedicnemus: Europe (except Greece), the Caucasus, Transcaucasia, north-east Turkey, and north-west Iran.
 B. o. saharae: North Africa, Greece, Cyprus, the Levant, and western, central, and southern Turkey.
 B. o. harterti: south-east Turkey, and from south-west Iran and south-east European Russia east to the plains of Kazakhstan.
Cursorius cursor   Cream-coloured Courser   Kosar Kus
 C. c. cursor: North Africa east to the Middle East, Arabia, and possibly south-east Turkey.
Glareola pratincola   Collared Pratincole   Bataklik Kirlangici
 G. p. pratincola: North Africa and southern Europe east to Kazakhstan.
Charadrius dubius Little Ringed Plover   Koyleyli Küçük Yagmurkusu
 C. d. curonicus: North Africa, Europe, the Middle East, and northern Asia.
Charadrius alexandrinus   Kentish Plover   Kesik Koyleyli Küçük Yagmurkusu
 C. a. alexandrinus: North Africa, Europe, the Middle East, and western and central Asia.
Charadrius leschenaultii   Greater Sand Plover   Çöl Yagmurkusu
 C. l. columbinus: central Asia Minor and Syria east through Transcaucasia and Iran to western Afghanistan.
Hoplopterus spinosus   Spur-winged Lapwing   Mahmuzlu Kizkusu
 Monotypic.
Hoplopterus indicus   Red-wattled Lapwing
 H. i. aigneri: south-east Turkey and Iraq east to Pakistan.
Chettusia leucura   White-tailed Lapwing   Akkuyruklu Kizkusu
 Monotypic.
Vanellus vanellus   Northern Lapwing   Kizkusu
 Monotypic.
Gallinago gallinago   Common Snipe   Bekazin
 G. g. gallinago: Europe (except Iceland to Orkney), northern Asia, and possibly Turkey.

Scolopax rusticola   Eurasian Woodcock   Çulluk
Monotypic. Occurs Armenia and thus may possibly breed in north-east Turkey.

Tringa totanus   Common Redshank   Kizilbacak
T. t. britannica: England, and from the Netherlands and France throughout central Europe, possibly east to the Caspian Sea, and perhaps this subspecies or an unknown one in Turkey (but certainly not *T. t. totanus*, which is restricted to northern Europe). See also Kumerloeve (1969a).

Actitis hypoleucos   Common Sandpiper   Akkarin Yesilbacak
Monotypic.

Larus melanocephalus   Mediterranean Gull   Karakafa Marti
Monotypic.

Larus ridibundus   Black-headed Gull   Gülen Marti
Monotypic.

Larus genei   Slender-billed Gull   Incegaga Marti
Monotypic.

Larus audouinii   Audouin's Gull   Pembegaga Marti
Monotypic.

Larus cachinnans   Yellow-legged Gull   Gümüsi Marti
L. c. michahellis: the west coasts of France, Iberia and Morocco, and throughout the Mediterranean basin east to the coasts of western and southern Turkey.
L. c. cachinnans: the coasts of the Black Sea (including those of northern Turkey), and inland east to Kazakhstan.

Larus armenicus   Armenian Gull   Van Gölü Martisi
Monotypic (restricted to inland eastern Turkey, Armenia, and north-west Iran).

Gelochelidon nilotica   Gull-billed Tern   Gülen Sumru
G. n. nilotica: North Africa, Europe, the Middle East, and western and central Asia.

Sterna caspia   Caspian Tern   Hazar Martisi
Monotypic.

Sterna hirundo   Common Tern   Adi Denizkirlangici
S. h. hirundo: western and northern Africa, Europe, the Middle East, western Asia, and North and Central America.

Sterna albifrons   Little Tern   Akalin Denizkirlangici
S. a. albifrons: North Africa, Europe, the Middle East, and western and central Asia to India.

Chlidonias hybridus   Whiskered Tern   Akbiyik Denizkirlangici
C. h. hybridus: North Africa and Eurasia.

Chlidonias niger   Black Tern   Siya Denizkirlangici
C. n. niger: Eurasia.

Chlidonias leucopterus   White-winged Black Tern   Palamut Kusu
Monotypic.

Pterocles orientalis   Black-bellied Sandgrouse   Karakarin Steptavugu
P. o. orientalis: Iberia, North Africa, Turkey, and western Iran.

Pterocles alchata   Pin-tailed Sandgrouse   Kilkuyruk Steptavugu
P. a. caudacuta: North Africa, the Middle East, and south-west Asia.

Columba livia   Rock Dove   Kaya Güvercini
C. l. livia: Europe east to the Balkan countries, Greece, and possibly Thrace (European Turkey). Birds from western, central, and northern Asia Minor as well as from the Caucasus are distinctly larger than *livia*, but scarcely paler; they are usually referred to as *gaddi*,

but this subspecies, though large indeed, is distinctly paler. Following Roselaar in Cramp (1985), most populations from Asia Minor and the Caucasus are provisionally included in *livia*, but they may form a separable subspecies.

C. l. palaestinae: Syria to Arabia; may occur from Gaziantep to Cizre in south-east Turkey.

C. l. gaddi: south-eastern Turkey (in the Van-Hakkari area), Transcaucasia, and from Iran east to Uzbekistan.

Columba oenas   Stock Dove   Mavi Güvercin

C. c. oenas: North Africa, Europe, the Middle East, and western and north-central Asia.

Columba palumbus   Wood Pigeon   Tahtali Güvercin

C. p. palumbus: North Africa, Europe, western Siberia, and the Middle East.

C. p. iranica: Iran and south-west Turkmenistan, grading into *palumbus* in south-east Turkey.

Streptopelia decaocto   Collared Dove   Kumru

S. d. decaocto: Europe and from the Middle East east to India.

Streptopelia turtur   Turtle Dove   Üveyik

S. t. turtur: Europe (except the Balearic Islands), western Siberia, the Middle East, the northern Levant, Cyprus, Turkey, and the Caucasus area.

S. t. arenicola: the Balearic Islands, north-west Africa, the southern Levant, and from Iraq east to central Asia, grading into *turtur* in south-east Turkey.

Streptopelia senegalensis   Laughing Dove   Küçük Kumru

S. s. phoenicophila: north-west Africa and (apparently introduced) Turkey and Syria.

Clamator glandarius   Great Spotted Cuckoo   Tepeli Gugukkusu

C. g. glandarius: northern Africa, southern Europe, and the Middle East.

Cuculus canorus   Common Cuckoo   Gugukkusu

C. c. canorus: central, eastern, and northern Europe, the Middle East, and northern Asia.

Tyto alba   Barn Owl   Peçeli Baykus

T. t. alba: Great Britain, France, and through southern Europe east to northern Greece and north-west Turkey.

T. a. erlangeri: North Africa, southern Greece, Cyprus, southern Turkey, and from the Levant and Iran to Arabia.

Otus brucei   Striated Scops Owl

O. b. obsoletus: northern Syria, south-east Turkey, northern Iraq, and through Turkmenistan and northern Afghanistan to the lowlands of Uzbekistan.

Otus scops   European Scops Owl   Ishak Kusu

O. s. scops: from France and Sardinia throughout central and southern Europe to the lower Volga river and northern Greece, as well as in northern Turkey and Transcaucasia.

O. s. cycladum: southern Greece, southern Turkey east to the Çukurova area, and the Levant.

O. s. turanicus: Iraq, Iran, and through southern Transcaspia to Afghanistan, and probably this subspecies in south-east Turkey.

Bubo bubo   Eurasian Eagle Owl   Puhu

B. b. interpositus: western and northern shores of the Black Sea, Turkey, and from the Levant to north-west Iran and the Caucasus area.

B. b. nikolskii: Iraq and Iran (except the north-west) to western Pakistan; perhaps grading into *interpositus* in south-east Turkey.

Ketupa zeylonensis   Brown Fish Owl   Balikçi Puhu

K. z. semenowi: southern Turkey and the Levant east to western and northern Pakistan.

Athene noctua   Little Owl   Kukumav Kusu

A. n. indigena: the southern Balkan countries and Greece through Turkey and the northern coastal Levant to Transcaucasia, north to southern European Russia and south-west Siberia.

A. n. lilith: Cyprus, inland parts of the Levant, and south-east Turkey, grading into *bactriana* in Iraq, the latter subspecies occurring from Iran to Kazakhstan.

Strix aluco   Tawny Owl   Alaca Baykus
   S. a. sylvatica: Great Britain, France, and southern Europe east to the western half of Asia Minor.
   S. a. willkonskii: north-east Turkey, the Caucasus area, and from northern Iran east to western Turkmenistan.
   S. a. sanctinicolai: northern Iraq (Dihok area), south-west Iran, and possibly in south-east Turkey.

Asio otus   Long-eared Owl   Kulakli Ormanbaykusu
   A. o. otus: North Africa and Eurasia.

Asio flammeus   Short-eared Owl   Bataklik Baykusu
   A. f. flammeus: Eurasia and North America.

Aegolius funereus   Tengmalm's Owl   Çipkayakli Baykus
   A. f. funereus: central and northern Europe and (probably this subspecies) north-west Turkey.
   A. f. caucasicus: the Caucasus area and (probably this subspecies) north-east Turkey.

Caprimulgus europaeus   European Nightjar   Çobanaldatan
   C. e. meridionalis: North Africa, southern Europe, Turkey, Cyprus, the Caucasus area, and north-west Iran.

Apus apus   Common Swift   Ebabil
   A. a. apus: North Africa, Europe, the Middle East, and western and north-central Asia.
   A. a. pekinensis: Iran to south-central Asia, apparently grading into *apus* in south-east Turkey.

Apus pallidus   Pallid Swift   Gri Ebabil
   A. p. brehmorum: coastal North Africa, southern Europe, north-west Turkey, Cyprus, and (perhaps this subspecies) southern Turkey.
   A. p. pallidus: Sahara, Egypt, from the Levant east to Pakistan, and (perhaps this subspecies) eastern Turkey.

Apus melba   Alpine Swift   Akkarin Ebabil
   A. m. melba: northern Morocco, southern Europe, Turkey, the Caucasus area, and north-west Iran.
   A. m. tuneti: North Africa (except northern Morocco), Lebanon, Syria, and from south-west Iran east to western Pakistan, perhaps grading into *melba* in south-east Turkey.

Apus affinis   Little Swift   Akkuyruksokumlu Ebabil
   A. a. galilejensis: northern Africa and from the Middle East east to Uzbekistan and western Pakistan.

Halcyon smyrnensis   Smyrna Kingfisher   Izmir Yaliçapkini
   H. s. smyrnensis: from Asia Minor to north-west India.

Alcedo atthis   Common Kingfisher   Yaliçapkini
   A. a. atthis: north-west Africa, southern Europe, Turkey, and from the Levant east to central Asia and Pakistan.

Ceryle rudis   Pied Kingfisher   Gri Yaliçapkini
   C. r. syriaca (see chapter V): the Middle East from Turkey to south-west Iran and south to Israel and Jordan.

Merops persicus   Blue-cheeked Bee-eater   Maviyanak Arikusu
  M. p. persicus: northern Egypt, and from the Middle East east to Kazakhstan and north-west India.

Merops apiaster   European Bee-eater   Arikusu
  Monotypic.

Coracias garrulus   European Roller   Kuzgun
  C. g. garrulus: North Africa, Europe, south-west Siberia, Turkey, the Levant, and north-west Iran.
  C. g. semenowi: from Iraq and south-west Iran east to central Asia, perhaps grading into *garrulus* in south-east Turkey and Syria.

Upupa epops   Eurasian Hoopoe   Hüthüt
  U. e. epops: north-west Africa, Europe, the Middle East, and western and central Asia.

Jynx torquilla   Eurasian Wryneck   Boyunçeviren
  J. t. torquilla: northern and central Europe east to Bulgaria and Caucasus and (perhaps this subspecies) northern Turkey.

Picus canus   Grey-headed Woodpecker   Gri Agaçkakan
  P. c. canus: Europe, east to western and central Asia and north-west Turkey.

Picus viridis   Green Woodpecker   Yesil Agaçkakan
  P. v. karelini: the Balkan countries, Greece, Turkey, the Caucasus area, northern Iran, and south-west Turkmenistan.

Dryocopus martius   Black Woodpecker   Kara Agaçkakan
  D. m. martius: Eurasia, except south-west China.

Dendrocopos major   Great Spotted Woodpecker   Büyük Alaca Agaçkakan
  D. m. candidus: southern Ukraine, the southern Balkan countries, and Greece, east to Thrace (European Turkey).
  D. m. paphlagoniae: north-west Anatolia, east through northern Asia Minor to at least Sebinkarahisar and Sarikamis; perhaps this subspecies in the Taurus mountains.
  D. m. tenuirostris: the Caucasus, Transcaucasia, and possibly in extreme north-east Turkey.

Dendrocopos syriacus   Syrian Woodpecker   Suriye Alaca Agaçkakan
  D. s. syriacus: central Europe, the Balkan countries, western and southern Turkey, the Levant, and south-west Iran.
  D. s. transcaucasicus: Transcaucasia and northern Iran, grading into *syriacus* in northern Turkey.

Dendrocopos medius   Middle Spotted Woodpecker   Albas Agaçkakan
  D. m. medius: Europe, south-east to Thrace (European Turkey).
  D. m. anatoliae: western and southern Asia Minor.
  D. m. caucasicus: northern Asia Minor, Transcaucasia, and the Caucasus.
  D. m. sanctijohannis: south-west Iran, grading into *anatoliae* in the Dihok area (northern Iraq) and perhaps in extreme south-east Turkey.

Dendrocopos leucotos   White-backed Woodpecker   Aksirt Agaçkakan
  D. l. lilfordi: southern Europe, Turkey, Transcaucasia, and the Caucasus.

Dendrocopos minor   Lesser Spotted Woodpecker   Küçük Agaçkakan
  D. m. danfordi: Greece and the western half of Turkey.
  D. m. colchicus: the Caucasus, Transcaucasia, and (probably this subspecies) north-east Turkey.
  D. m. morgani: south-west Iran and (probably this subspecies) extreme south-east Turkey.

XII BIBLIOGRAPHY

No single complete list of references is given in this book. For convenience, all full references to distribution and geographical variation of Turkish passerine birds are given in chapter III, while those on the same topics for neighbouring countries are given in chapter IV. It was considered to be more appropriate to have separate lists for these areas. Details of references pertaining to a single species or to a few species are given at the end of the species account in which they were used. To find the place where the complete reference is given, all authors and years for which papers have been used in this book are given below in alphabetical sequence. This list includes some complete references of general papers which are not specific to chapters III or IV or the species accounts.

Abs (1963) - see Galerida cristata
Aharoni (1931) - see Ramphocoris clotbey
Aharoni (1932) - see Ramphocoris clotbey
Albrecht, J. S. M. (1979) Atlas of breeding birds in Turkey. *Bull. Orn. Soc. Middle East* 2, 6-8.
Albrecht (1981) - see Ficedula parva
Albrecht (1986) - see chapter III
Albrecht (1987) - see Phylloscopus trochiloides
Amcoff *et al.* (1986) - see chapter III
Andrew *et al.* (1972) - see Sylvia mystacea
Ayvaz (1990) - see chapter III
Ayvaz (1991) - see chapter III
Bacmeister & Kleinschmidt (1918-20) - see Pyrrhula pyrrhula
Bakker & Steenge (1990) - see chapter III
Balance (1958) - see chapter III
Baris, Y. S. (1989) Turkey's bird habitats and ornithological importance. *Sandgrouse* 11, 42-51.
Baris *et al.* (1984) - see chapter III
Barthel *et al.* (1992) - see Bucanetes mongolicus
Bates (1935a) - see Pycnonotus xanthopygos
Bates (1935b) - see Oenanthe xanthoprymna
Bauer *et al.* (1969) - see chapter IV (Greece)
Baumgart (1971) - see Oenanthe pleschanka
Baumgart & Stephan (1987) - see chapter IV (Syria)
Beaman (1975) - see chapter III
Beaman (1978) - see chapter III
Beaman (1986) - see chapter III
Beaman, M. (1994) Palearctic Birds: A checklist of the Birds of Europe, North Africa and Asia north of the foothills of the Himalayas. Stonyhurst
Beaudoin (1976) - see chapter III
Beers (1982) - see chapter III
Beretzk *et al.* (1969) - see Carduelis chloris
Berg & Bosman (1988) - see Acrocephalus agricola
Berk & Letschert (1988) - see chapter III
Berk *et al.* (1993) - see chapter III
Bezzel (1964) - see chapter III
Bijlsma & De Roder (1986) - see chapter III
Bird (1937) - see chapter III

Bibliography

Blondel (1967) - see Phoenicurus phoenicurus
Braun (1908) - see chapter III
British Birds (1993) The 'British Birds' List of English Names of Western Palearctic Birds. Blunham.
Brock (1989) - see chapter III
Broome (1989) - see chapter III
Bruin (1988) - see chapter III
Bub & Herroelen (1981) - see Eremophila alpestris
Buturlin (1906) - see chapter IV (USSR)
Buturlin (1907) - see Loxia curvirostra
Catuneanu (1975) - see Parus lugubris
Chappuis et al. (1973) - see Emberiza cineracea
Christensen (1974) - see Oenanthe pleschanka
Clancey (1975) - see Acrocephalus palustris
Clancey (1987) - see Anthus trivialis
Clancey (1989) - see Hippolais icterina
Colin (1982) - see chapter III
Colin (1989) - see chapter III
Colin (1990) - see chapter III
Cramp (1971) - see Passer moabiticus
Cramp, S. (ed) (1985) Handbook of the Birds of Europe, the Middle East, and North Africa 4. Oxford.
Cramp, S. (ed) (1988) Handbook of the Birds of Europe, the Middle East, and North Africa 5. Oxford.
Cramp, S. (ed) (1992) Handbook of the Birds of Europe, the Middle East, and North Africa 6. Oxford.
Cramp, S., & C. Perrins (eds) (1993) Handbook of the Birds of Europe, the Middle East, and North Africa 7. Oxford.
Cramp, S., & C. Perrins (eds) (1994a) Handbook of the Birds of Europe, the Middle East, and North Africa 8. Oxford.
Cramp, S., & C. Perrins (eds) (1994b) Handbook of the Birds of Europe, the Middle East, and North Africa 9. Oxford.
Cramp, S., & K. E. L. Simmons (eds) (1977) Handbook of the Birds of Europe, the Middle East, and North Africa 1. Oxford.
Cramp, S., & K. E. L. Simmons (eds) (1980) Handbook of the Birds of Europe, the Middle East, and North Africa 2. Oxford.
Cramp, S., & K. E. L. Simmons (eds) (1983) Handbook of the Birds of Europe, the Middle East, and North Africa 3. Oxford.
Crosland (1989) - see chapter III
Ctyroky (1987) - see chapter IV (Iraq)
Curio (1959) - see Ficedula semitorquata
Danford (1877-78) - see chapter III
Danford (1880) - see chapter III
Delacour & Vaurie (1950) - see Parus major
Dementiev & Gladkov (1951-54) - see chapter IV (USSR)
Desfayes & Praz (1978) - see chapter IV (Iran)
Diesselhorst (1962) - see chapter IV (Iran)
Dijksen & Kasparek (1985) - see chapter III
Dijksen & Kasparek (1988) - see chapter III
Dorèl (1991) - see chapter III
Dott (1967) - see Oenanthe pleschanka
Dufourny (1990) - see chapter III

Songbirds of Turkey

Dunajewski (1934) - see Sitta europaea
Dunajewski (1938) - see Sylvia borin
Eck (1975a) - see Luscinia megarhynchos
Eck (1975b) - see Luscinia megarhynchos
Eck (1980) - see Parus lugubris
Eck (1990) - see Luscinia megarhynchos
Eggers & Lemke (1964) - see chapter III
Engelmoer, M., C. S. Roselaar, G. C. Boere, & E. Nieboer (1983) Post-mortem changes in measurements of some waders. *Ringing & Migration* 4, 245-248.
Etchécopar (1959) - see chapter III
Eyckerman *et al.* (1976) - see Sylvia melanocephala
Eyckerman *et al.* (1992) - see chapter III
Érard & Etchécopar (1968) - see chapter III
Érard & Etchécopar (1970) - see chapter IV (Iran)
Erol, O. (1982-83) Die naturräumliche Gliederung der Türkei. Tübinger Atlas der Vorderen Orients, Karte A VII 2 & Beihefte A 13, 1-245. Wiesbaden.
Ertan, A., A. Kiliç, & M. Kasparek (1989) Türkiye'nin önemli kus alanlari. Istanbul.
Fischer & Fischer (1976) - see Regulus ignicapillus
Flint & Stewart (1992) - see chapter IV (Cyprus)
Forster & Numminen (1990) - see chapter III
Franckx (1982) - see chapter III
Frankis (1991) - see Sitta krueperi
Frost & Hornbuckle (1992) - see chapter III
Gallner (1976) - see chapter III
Ganso & Spitzer (1967) - see chapter III
Gaston (1968) - see chapter III
Géroudet (1963) - see Serinus pusillus
Glimmerveen & Hols (1986) - see Oenanthe moesta
Glutz von Blotzheim, U. N., & K. M. Bauer (1980) Handbuch der Vögel Mitteleuropas 9. Wiesbaden.
Gooders (1988) - see chapter III
Grant (1975) - see Sitta europaea
Greenway & Vaurie (1958) - see Cinclus cinclus
Grimmett, R. F. A., & T. A. Jones (1989) Important bird areas in Europe. ICBP Technical Publications no. 9.
Groh (1968) - see chapter III
Grote (1937) - see Oenanthe isabellina
Grote (1939) - see Oenanthe pleschanka
Grote (1943) - see Irania gutturalis
Haffer (1977) - see Oenanthe pleschanka
Harrap (1985) - see chapter III
Harrison (1955) - see Sitta europaea
Harrison (1959) - see chapter IV (Iraq)
Harrison & Pateff (1937) - see chapter IV (Bulgaria)
Harrison *et al.* (1972) - see Sylvia mystacea
Harrison *et al.* (1973) - see Sylvia mystacea
Hartert (1903-10) - see chapter IV (general works)
Hartert (1921-22) - see chapter IV (general works)
Hartert & Steinbacher (1932-38) - see chapter IV (general works)
Härms (1925) - see Oenanthe xanthoprymna

Heinzel, H., R. Fitter, & J. Parslow (1972) The birds of Britain and Europe, with North Africa and the Middle East. London.
Helb *et al.* (1982) - see Phylloscopus bonelli
Helbig (1984) - see chapter III
Hennipman *et al.* (1961) - see chapter III
Herrn (1966) - see chapter III
Hesse (1915) - see Carpodacus erythrinus
Hollom (1955) - see chapter III
Hollom (1959) - see chapter III
Hollom, P. A. D., R. F. Porter, S. Christensen, & I. Willis (1988) Birds of the Middle East and North Africa. Calton.
Husband & Kasparek (1984) - see chapter III
Hustings & Van Dijk (1994) - see chapter III
Hüni (1982) - see chapter III
Ivanov (1941) - see Oenanthe xanthoprymna
Jakobsen (1986) - see Corvus ruficollis
Johansen (1944-57) - see chapter IV (USSR)
Jonsson, L. (1993) Birds of Europe, with North Africa and the Middle East. London.
Jordans (1940) - see chapter IV (Bulgaria)
Jordans (1970) - see Parus major
Jordans & Steinbacher (1948) - see chapter III
Kasparek (1985) - see chapter III
Kasparek (1986) - see Oenanthe xanthoprymna
Kasparek (1987) - see chapter III
Kasparek (1988) - see chapter III
Kasparek (1989) - see Corvus frugilegus
Kasparek (1990a) - see chapter III
Kasparek (1990b) - see Sylvia atricapilla
Kasparek, M. (1991) Towards a Turkish Atlas. *Bull. Ornith. Soc. Middle East* 26, 8-12.
Kasparek, M. (1992) Die Vögel der Türkei. Eine Übersicht. Heidelberg.
Kasparek & Van der Ven (1983) - see chapter III
Kazakov (1973) - see Sylvia mystacea
Keve (1971) - see chapter III
Keve (1973) - see Garrulus glandarius
Keve (1978) - see Passer montanus
Keve-Kleiner (1943) - see Garrulus glandarius
Keve & Kohl (1978) - see Passer montanus
Keve & Rokitansky (1966) - see Lanius minor
Kiliç & Kasparek (1987) - see chapter III
Kiliç & Kasparek (1989) - see chapter III
Kinzelbach & Martens (1965) - see Carduelis cannabina
Kiraç (1993) - see chapter III
Kirwan (1990) - see chapter III
Kirwan (1993) - see chapter III
Kitson *et al.* (1983) - see Phylloscopus bonelli
Kiziroglu (1983) - see Parus ater
Kiziroglu, I. (1989) Türkiye Kuslari. Ankara.
Kiziroglu & Kiziroglu (1987) - see chapter III

Songbirds of Turkey

Kleiner (1939a) - see Garrulus glandarius
Kleiner (1939b) - see Garrulus glandarius
Kleiner (1939c) - see Pica pica
Kleiner (1939d) - see Corvus monedula
Knijff (1991) - see Emberiza cineracea
Konrad (1985) - see Sylvia mystacea
Krieger (1988) - see Bucanetes githagineus
Kumerloeve (1957a) - see chapter III
Kumerloeve (1957b) - see Pycnonotus xanthopygos
Kumerloeve (1957c) - see Turdus viscivorus
Kumerloeve (1958a) - see Turdus merula
Kumerloeve (1958b) - see Hippolais languida
Kumerloeve (1958c) - see Muscicapa striata
Kumerloeve (1958d) - see Panurus biarmicus
Kumerloeve (1958e) - see Sitta krueperi
Kumerloeve (1959) - see Prinia gracilis
Kumerloeve (1960) - see Certhia familiaris
Kumerloeve (1961a) - see chapter III
Kumerloeve (1961b) - see Lanius nubicus
Kumerloeve (1962a) - see chapter IV (Lebanon)
Kumerloeve (1962b) - see Sylvia borin
Kumerloeve (1962c) - see Emberiza hortulana
Kumerloeve ('1962') - see Kumerloeve (1964a)
Kumerloeve (1963a) - see chapter III
Kumerloeve (1963b) - see Calandrella cheleensis
Kumerloeve (1964a) - see chapter III
Kumerloeve (1964b) - see chapter III
Kumerloeve (1964c) - see Irania gutturalis
Kumerloeve (1964d) - see Regulus regulus
Kumerloeve (1965) - see Passer moabiticus
Kumerloeve (1966a) - see chapter III
Kumerloeve (1966b) - see Irania gutturalis
Kumerloeve (1966c) - see Serinus pusillus
Kumerloeve (1966-67) - see chapter III
Kumerloeve (1967a) - see Calandrella brachydactyla
Kumerloeve (1967b) - see Phylloscopus lorenzii
Kumerloeve (1967c) - see Corvus frugilegus
Kumerloeve (1967d) - see Carduelis flavirostris
Kumerloeve ('1967') - see Kumerloeve (1968)
Kumerloeve (1967-69) - see chapter IV (Syria)
Kumerloeve (1968) - see chapter III
Kumerloeve (1969a) - see chapter III
Kumerloeve (1969b) - see Calandrella brachydactyla
Kumerloeve (1969c) - see Saxicola rubetra
Kumerloeve (1969d) - see Oenanthe pleschanka
Kumerloeve (1969e) - see Hippolais languida
Kumerloeve (1969f) - see Panurus biarmicus
Kumerloeve (1969g) - see Passer moabiticus

Kumerloeve (1969h) - see Emberiza buchanani
Kumerloeve (1970a) - see chapter III
Kumerloeve (1970b) - see chapter III
Kumerloeve (1970c) - see Phylloscopus sibilatrix
Kumerloeve (1971) - see Alauda arvensis
Kumerloeve, H. (1975a) The history of ornithology in Turkey - a review. *Bird Rep. Orn. Soc. Turkey* 3, 289-319.
Kumerloeve (1975b) - see Oenanthe isabellina
Kumerloeve (1975c) - see Carpospiza brachydactyla
Kumerloeve (1978) - see Passer moabiticus
Kumerloeve, H. (1984) A chronological review of birds first described from Turkey with their taxonomic status in 1984. *Sandgrouse* 6, 62-68.
Kumerloeve, H. (1986) Bibliographie der Säugetiere und Vögel der Türkei (rezente Fauna). *Bonner zool. Monogr.* 21, 1-132.
Kumerloeve (1989) - see Emberiza hortulana
Kumerloeve *et al.* (1984) - see Oenanthe xanthoprymna
Kummerlöwe & Niethammer (1934a) - see chapter III
Kummerlöwe & Niethammer (1934b) - see chapter III
Kummerlöwe & Niethammer (1934-35) - see chapter III
Lees-Smith & Madge (1982) - see Miliaria calandra
Lehmann (1971) - see Rhodopechys sanguinea
Lehmann & Mertens (1969) - see Rhodopechys sanguinea
Loskot (1981) - see Luscinia megarhynchos
Loskot (1988) - see Prunella ocularis
Louette *et al.* (1977) - see chapter III
Lynes (1930) - see Cisticola juncidis
Maas Geesteranus (1959) - see chapter III
Mackworth-Praed & Grant (1951) - see Oenanthe oenanthe
Makatsch (1950) - see chapter IV (Greece)
Marien (1951) - see Prunella modularis
Marova & Leonovich (1993) - see Phylloscopus lorenzii
Martens (1979) - see Emberiza cineracea
Martens (1982) - see Phylloscopus lorenzii
Martens & Hänel (1981) - see Phylloscopus lorenzii
Martins (1989) - see chapter III
Mascara (1991) - see chapter III
Matvejev (1976) - see chapter IV (Yugoslavia)
Matvejev & Vasic (1973) - see chapter IV (Yugoslavia)
Mauersberger (1960) - see Emberiza caesia
Mauersberger (1971) - see Prunella modularis
Mayr, E. (1963) Animal species and evolution. Cambridge (Mass.).
Mayr & Stresemann (1950) - see Oenanthe hispanica
McGregor (1917) - see chapter III
Meinertzhagen (1924) - see Sturnus vulgaris
Meinertzhagen (1935) - see chapter III
Mertens (1971) - see chapter III
Mild (1993) - see Ficedula semitorquata
Mild (1994a) - see Ficedula semitorquata

Songbirds of Turkey

Mild (1994b) -. see Ficedula semitorquata
Molineux (1930-31) - see chapter IV (general works)
Moyle (1989) - see chapter III
Munteanu (1967) - see Sturnus vulgaris
Neufeldt & Wunderlich (1984) - see Sitta krueperi
Neumann (1915) - see Sturnus vulgaris
Neumann & Paludan (1937) - see Cinclus cinclus
Nicht (1961) - see chapter IV (USSR)
Niethammer (1942) - see chapter IV (Greece)
Niethammer (1943) - see chapter IV (Greece)
Niethammer (1950) - see chapter IV (Bulgaria)
Nijhoff & Swennen (1963) - see chapter III
Ogilvie (1954) - see chapter III
Oliver (1990) - see Oenanthe pleschanka
Paludan (1938) - see chapter IV (Iran)
Paludan (1940) - see chapter IV (Iran)
Paludan (1959) - see chapter IV (Afghanistan)
Panov (1986) - see Oenanthe pleschanka
Panov & Bulatova (1972) - see Bucanetes mongolicus
Panov & Ivanitskii (1975) - see Oenanthe pleschanka
Parrot (1905) - see Parus major
Pasquali (1990) - see chapter III
Pateff (1947) - see Sturnus vulgaris
Peklo (1987) - see Muscicapa striata
Petretti & Petretti (1980) - see chapter III
Piechocki & Bolod (1971) - see Oenanthe pleschanka
Piechocki et al. (1982) - see Oenanthe pleschanka
Portenko (1962) - see Anthus campestris
Porter (1969) - see chapter III
Porter, R. F., & M. Beaman (1977) The atlas of breeding birds of Turkey. *Bull. Orn. Soc. Turkey* 15, 4-5.
Ramsay (1914) - see chapter III
Reid (1979) - see Emberiza caesia
Renkhoff (1972) - see chapter III
Renkhoff (1973) - see chapter III
Roer (1962) - see Corvus frugilegus
Rokitasky & Schifter (1971) - see chapter III
Roselaar (1993) - see Carduelis chloris
Rössner (1935) - see chapter III
Salomonsen (1934) - see Oenanthe oenanthe
Sarudny & Härms (1923) - see Sitta tephronota
Schekkerman & Van Roomen (1993) - see chapter III
Schepers & Stuart (1991) - see chapter III
Schilperoort & Schilperoort-Huisman (1986) - see chapter III
Schmidtke & Utschick (1980) - see chapter III
Schubert (1979a) - see chapter III
Schubert (1979b) - see Hippolais languida
Schubert (1986) - see Regulus ignicapillus
Schüz (1957) - see chapter III

Bibliography

Schüz (1959) - see chapter IV (Iran)
Schweiger (1965) - see chapter III
Scott *et al.* (1975) - see chapter IV (Iran)
Selous (1900) - see chapter III
Shirihai (1986) - see Phylloscopus lorenzii
Shirihai & Colston (1992) - see Riparia riparia
Sick (1939) - see Certhia brachydactyla
Siegner (1983) - see Oenanthe pleschanka
Simeonov & Doichev (1973) - see Emberiza citrinella
Sluys (1983) - see Panurus biarmicus
Sluys & Van den Berg (1982) - see Oenanthe pleschanka
Smith (1960) - see chapter III
Smithe, F. B. (1974) Naturalist's color guide. New York.
Smithe, F. B. (1981) Naturalist's color guide 3. New York.
Snigirewski (1928) - see Sylvia nisoria
Snigirewski (1929) - see Sylvia curruca
Snow (1988) - see chapter III
Spitzer (1973) - see Panurus biarmicus
Stegmann (1928) - see Phoenicurus ochruros
Stegmann (1932) - see Calandrella rufescens
Stegmann (1934) - see Phylloscopus lorenzii
Stegmann & Stresemann (1935) - see Sturnus vulgaris
Steinbacher (1937) - see Oenanthe oenanthe
Steiner (1962) - see Phylloscopus trochiloides
Steiner (1970) - see Sylvia melanocephala
Stenhouse (1920) - see chapter III
Stepanyan (1964) - see Monticola saxatilis
Stepanyan (1967) - see Calandrella cheleensis
Stepanyan (1971) - see Oenanthe xanthoprymna
Stepanyan (1990) - see chapter IV (USSR)
Stepanyan & Matyukhin (1984) - see Acrocephalus agricola
Stresemann (1919) - see Pyrrhula pyrrhula
Stresemann (1920) - see chapter IV (Yugoslavia)
Stresemann (1926) - see Ficedula semitorquata
Stresemann (1928) - see chapter IV (Iran)
Stresemann (1962) - see Pycnonotus xanthopygos
Stresemann *et al.* (1937) - see Oenanthe pleschanka
Stresemann & Portenko (1960-) - see chapter IV (general works)
Stresemann & Schiebel (1925) - see Muscicapa striata
Sushkin (1933) - see Sturnus vulgaris
Sutton & Gray (1972) - see chapter III
Svensson, L. (1992) Identification guide to European passerines. Stockholm.
Ticehurst (1927) - see Oenanthe hispanica
Ticehurst (1931) - see Oenanthe oenanthe
Ticehurst *et al.* (1921-22) - see chapter IV (Iraq)
Ticehurst *et al.* (1926) - see chapter IV (Iraq)
Ticehurst & Whistler (1932) - see chapter IV (Albania)
Titcombe & Hatch (1989) - see chapter III

Songbirds of Turkey

Ullman (1992) - see Oenanthe pleschanka
Vader (1965) - see chapter III
Vauk (1973) - see chapter III
Vaurie (1949) - see chapter IV (general works)
Vaurie (1950) - see chapter IV (general works)
Vaurie (1951) - see chapter IV (general works)
Vaurie (1952) - see chapter IV (general works)
Vaurie (1953) - see chapter IV (general works)
Vaurie (1954) - see chapter IV (general works)
Vaurie (1955) - see chapter IV (general works)
Vaurie (1956) - see chapter IV (general works)
Vaurie (1956) - see chapter IV (general works)
Vaurie (1957) - see chapter IV (general works)
Vaurie (1958) - see chapter IV (general works)
Vaurie (1959) - see chapter IV (general works)
Vaurie, C. (1965) The Birds of the Palearctic Fauna, non-passeriformes. London.
Vaurie (1972) - see Ptyonoprogne rupestris
Vermeulen (1987) - see chapter III
Versluys (1992) - see chapter III
Vielliard (1968) - see chapter III
Vittery (1972) - see chapter III
Voous (1959) - see Turdus philomelos
Voous, K. H. (1977) List of recent Holarctic bird species, passerines. *Ibis* 119, 223-250, 376-406.
Voous & Van Marle (1953) - see Sitta europaea
Vuilleumier (1977) - see chapter IV (Iran)
Wadley (1951) - see chapter III
Warncke (1964-65) - see chapter III
Warncke (1966) - see chapter III
Warncke (1970) - see chapter III
Warncke (1972) - see chapter III
Watson (1961) - see Prunella modularis
Watson (1962a) - see Galerida cristata
Watson (1962b) - see Sylvia borin
Watson (1962c) - see Phylloscopus lorenzii
Weigold (1912-13) - see chapter III
Weigold (1913-14) - see chapter III
Williamson (1967) - see Phylloscopus bonelli
Williamson (1968) - see Sylvia curruca
Winden *et al.* (1989) - see chapter III
Witherby (1903) - see chapter IV (Iran)
Witherby (1907) - see chapter III
Witherby (1910) - see chapter IV (Iran)
Wolters (1968) - see Panurus biarmicus
Worobiev (1934) - see Hippolais languida

XIII Index of Turkish names

Agaç Incirkusu 51
Akgerdan 70
Akgerdan Ötlegen 116
Aksakal Ötlegen 110
Aksakrak 196
Ak Kuryuksallayan 56
Alaca Kuyrukkakan 79
Alkuyruk Kuyrukkakan 85
Alp Serçesi 64
Altintavukcuk 127
Anadolu Sivacisi 144
Ardiç Bülbülü 122

Bahçe Kizilkuyrugu 73
Bahçe Ötlegen 120
Bahçe Tirmasigi 152
Batak Kirazkusu 211
Bataklik Bastankara 136
Bataklik Saz Ardiçkusu 103
Bataklik Serçesi 176
Beyaz Mukallit 107
Biyikli Ardiçkusu 100
Biyikli Bastankara 133
Bogmakli Tarlakusus 31
Bozbogaz 62
Bozkir Toygari 34
Bugdaycil 70
Bülbül 68
Büyük Bastankara 143
Büyük Kaya Sivacisi 147
Büyük Saz Ardiçkusu 105

Çali Ötlegen 118
Çam Bastankara 139
Çaprazgaga 193
Çayir Taskusu 74
Çekirgekusu 156
Cif Caf 126
Çitkusu 60
Çit Ardiçkusu 101
Çit Kirazkusu 204
Çizgili Ötlegen 115
Çöl Kuyrukkakani 83
Çöl Sakragi 198
Çöl Toygari 30
Cüce Karga 166
Cüce Sinekkapan 131
Çulhakusu 153

Dag Kuyruksallayani 55
Dag Serçesi 178
Dag Söğütbülbülü 123
Dere Incirkusu 52
Duvar Tirmasigi 150

Ekinkargasi 167
Ev Kizilkuyrugu 71
Ev Serçesi 174

Florya 187
Fri Bülbül 57

Gri Kirazkusu 207
Gri Mukallit 106
Gri Sinekkapan 129

Ispinoz 184
Is Kirlangici 47

Kanarya 186
Karaalin Çekirgekusu 157
Karabas Iskete 190
Karabas Kirazkusu 213
Karabas Küçük Ötlegen 112
Karabas Ötlegen 121
Karaensse Kuyruksallayan 54
Karakarga 170
Karakulak Kuyrukkakan 81
Karasiri Kuyrukkakan 84
Karatavuk 91
Karmen Sakragi 199
Kar Ispinozu 183
Kayalik Serçesi 180
Kaya Ardici 87
Kaya Kirazkusu 205
Kaya Kirlangici 46
Kaya Kuyrukkakani 83
Kaya Sakragi 200
Kaya Sivacisi 148
Kestane Kargasi 160
Ketenkusu 191
Kirazkusu 209
Kir Incirkusu 49
Kizilalin Iskete 185
Kizilbasli Çekirgekusu 158
Kizilgaga Dagkarkasi 165
Kizilgerdan 66
Kizilsakrak 195
Kizil Çalibülbülü 65
Kizil Kirazkusu 210
Kizil Kirlangiç 48
Kocabas 202
Kolyeli Ardiç 89
Küçük Bogmakli Tarlakusu 33
Küçük Bozkir Toygari 36
Kulakli Tarlakusu 43
Kum Kirlangici 45
Kuyrukkakan 78

Leskargasi 168
Lorenz Bülbülø 124

Songbirds of Turkey

Mahsun Bastankara 137
Maskeli Çekirgekusu 159
Maskeli Ötlegen 114
Mavi Bastankara 141
Mavi Kayaardici 88

Ökseotu Ardici 93
Ölüdeniz Serçesi 177
Orman Söğütbülbülü 124
Orman Tirmasigi 151
Orman Toygari 41

Pembegöğüs Ötlegen 111
Pembesigircik 173
Pencere Kirlangici 48

Saka 189
Sakrak 200
Saksagan 162
Sariasma 155
Sarigaga Dagkarkasi 163
Sarigaga Ketenkusus 192
Sarigirtlak Serçe 179
Sari Incirkusu 53
Sari Kirazkusu 204
Sari Mukallit 109
Sarkici Ardiç 92

Savi'nin Dere Ardiçkusu 99
Saz Ardiçkusu 104
Setti Bülbülü 95
Sigircik 171
Sivacisi 145
Sürmeli Altintavukcuk 128
Sürmeli Çalilusu 97
Sürmeli Çitcerçesi 63
Su Karatavugu 58

Tarlakusu 42
Tarla Ardiçkusu 98
Tarla Kirazkusu 214
Taskusu 75
Tas Kirazkusu 208
Tas Serçesi 182
Tepeli Bastankara 139
Tepeli Toygar 39
Toprak Renkli Kuyrukkakan 77

Uzunkuyruk Bastankara 135

Yarimband Sinekkapan 132
Yelpazekuyruk 96

Zeytinlik 108

XIV Index of English and Scientific names

Accentor, Alpine 64
Accentor, Radde's 63
Acrocephalus agricola 102
Acrocephalus arundinaceus 105
Acrocephalus melanopogon 100
Acrocephalus palustris 103
Acrocephalus schoenobaenus 101
Acrocephalus scirpaceus 104
Aegithalos caudatus 135
Alauda arvensis 42
Alpine Accentor 64
Alpine Chough 163
Ammomanes deserti 30
Ammommanes cincturus 30
Anthus campestris 49
Anthus spinoletta 52
Anthus trivialis 51
Asian Short-toed Lark 37

Bar-tailed Desert Lark 30
Barn Swallow 47
Barred Warbler 115
Bearded Tit 133
Bimaculated Lark 33
Black-eared Wheatear 81
Black-headed Bunting 213
Blackbird 91
Blackcap 121
Black Redstart 71
Bluethroat 70
Blue Rock Thrush 88
Blue Tit 141
Bonelli's Warbler 123
Brown-necked Raven 169
Bucanetes githagineus 198
Bucanetes mongolicus 197
Bulbul, White-spectacled 57
Bullfinch, Eurasian 200
Bunting, Black-headed 213
Bunting, Cinereous 207
Bunting, Cirl 204
Bunting, Corn 214
Bunting, Cretzschmar's 210
Bunting, Grey-necked 208
Bunting, Ortolan 209
Bunting, Reed 211
Bunting, Rock 205

Chaffinch 184
Calandra Lark 31
Calandrella brachydactyla 34
Calandrella cheleensis 37
Calandrella rufescens 36
Carduelis cannabina 191
Carduelis carduelis 189
Carduelis chloris 187
Carduelis flavirostris 192

Carduelis spinus 190
Carpodacus erythrinus 199
Carpodacus rubicilla 200
Carpospiza brachydactyla 182
Carrion Crow 168
Caucasian Chiffchaff 124
Cercotrichas galactotes 65
Certhia brachydactyla 152
Certhia familiaris 151
Cetti's Warbler 95
Cettia cetti 95
Chiffchaff 126
Chiffchaff, Caucasian 124
Chough, Alpine 163
Chough, Red-billed 165
Cinclus cinclus 58
Cinereous Bunting 207
Cirl Bunting 204
Cisticola, Zitting 96
Cisticola juncidis 96
Citrine Wagtail 54
Coal Tit 139
Coccothraustes coccothraustes 202
Common Crossbill 193
Common Magpie 162
Common Raven 170
Common Redstart 73
Common Rosefinch 199
Common Starling 171
Common Stonechat 75
Common Whitethroat 118
Corn Bunting 214
Corvus corax 170
Corvus corone 168
Corvus frugilegus 167
Corvus monedula 166
Corvus ruficollis 169
Crag Martin 46
Crested Lark 39
Crested Tit 139
Cretzschmar's Bunting 210
Crimson-winged Finch 195
Crossbill, Common 193
Crow, Carrion 168

Dead Sea Sparrow 177
Delichon urbica 48
Desert Finch 196
Desert Lark 30
Desert Wheatear 83
Dipper, White-throated 58
Dunnock 62

Eastern Rock Nuthatch 147
Emberiza buchanani 208
Emberiza caesia 210
Emberiza cia 205

Songbirds of Turkey

Emberiza cineracea 207
Emberiza cirlus 204
Emberiza citrinella 204
Emberiza hortulana 209
Emberiza melanocephala 213
Emberiza schoeniclus 211
Eremophila alpestris 43
Erithacus rubecula 66
Eurasian Bullfinch 200
Eurasian Jackdaw 166
Eurasian Jay 160
Eurasian Nuthatch 145
Eurasian Siskin 190
Eurasian Skylark 42
Eurasian Treecreeper 151
European Goldfinch 189
European Greenfinch 187
European Robin 66
European Serin 186

Ficedula parva 131
Ficedula semitorquata 132
Finch, Crimson-winged 195
Finch, Desert 196
Finch, Mongolian 197
Finch, Trumpeter 198
Finsch's Wheatear 83
Firecrest 128
Flycatcher, Red-breasted 131
Flycatcher, Semi-collared 132
Flycatcher, Spottes 129
Fringilla coelebs 184

Galerida cristata 39
Garden Warbler 120
Garrulus glandarius 160
Goldcrest 127
Golden Oriole 155
Goldfinch, European 189
Graceful Warbler 97
Grasshopper Warbler 98
Great Reed Warbler 105
Great Rosefinch 200
Great Tit 143
Greenfinch, European 187
Greenish Warbler 122
Grey-necked Bunting 208
Grey Wagtail 55
Gymnoris xanthocollis 179

Hawfinch 202
Hippolais icterina 109
Hippolais languida 107
Hippolais olivetorum 108
Hippolais pallida 106
Hirundo daurica 48
Hirundo rustica 47
Horned Lark 43
House Martin 48

House Sparrow 174
Icterine Warbler 109
Irania gutturalis 70
Isabelline Wheatear 77

Jackdaw, Eurasian 166
Jay, Eurasian 160

Krüper's Nuthatch 144

Lanius collurio 156
Lanius minor 157
Lanius nubicus 159
Lanius senator 158
Lark, Asian Short-toed 37
Lark, Bar-tailed Desert 30
Lark, Bimaculated 33
Lark, Calandra 31
Lark, Crested 39
Lark, Desert 30
Lark, Horned 43
Lark, Lesser Short-toed 36
Lark, Short-toed 34
Lark, Wood 41
Lesser Grey Shrike 157
Lesser Short-toed Lark 36
Lesser Whitethroat 116
Linnet 191
Locustella luscinioides 99
Locustella naevia 98
Long-tailed Tit 135
Loxia curvirostra 193
Lullula arborea 41
Luscinia megarhynchos 68
Luscinia svecica 70

Magpie, Common 162
Marsch Warbler 103
Marsh Tit 136
Martin, Crag 46
Martin, House 48
Martin, Sand 45
Masked Shrike 159
Melanocorypha bimaculata 33
Melanocorypha calandra 31
Ménétries's Warbler 111
Miliaria calandra 214
Mistle Thrush 93
Mongolian Finch 197
Monticola saxatilis 87
Monticola solitarius 88
Montifringilla nivalis 183
Motacilla alba 56
Motacilla cinerea 55
Motacilla citreola 54
Motacilla flava 53
Mourning Wheatear 84
Moustached Warbler 100
Muscicapa striata 129

Index of English and Scientific names

Nightingale, Rufous 68
Northern Wheatear 78
Northern Wren 60
Nuthatch, Eastern Rock 147
Nuthatch, Eurasian 145
Nuthatch, Krüper's 144
Nuthatch, Western Rock 148

Oenanthe deserti 83
Oenanthe finschii 83
Oenanthe hispanica 81
Oenanthe isabellina 77
Oenanthe lugens 84
Oenanthe moesta 87
Oenanthe oenanthe 78
Oenanthe pleschanka 79
Oenanthe xanthoprymna 85
Olivaceous Warbler 106
Olive-tree Warbler 108
Oriole, Golden 155
Oriolus oriolus 155
Orphean Warbler 114
Ortolan Bunting 209
Ouzel, Ring 89

Paddyfield Warbler 102
Pale Rock Sparrow 182
Panurus biarmicus 133
Parus ater 139
Parus caeruleus 141
Parus cristatus 139
Parus lugubris 137
Parus major 143
Parus palustris 136
Passer domesticus 174
Passer hispaniolensis 176
Passer moabiticus 177
Passer montanus 178
Penduline Tit 153
Petronia petronia 180
Phoenicurus ochruros 71
Phoenicurus phoenicurus 73
Phylloscopus bonelli 123
Phylloscopus collybita 126
Phylloscopus lorenzi 124
Phylloscopus sibilatrix 124
Phylloscopus trochiloides 122
Pica pica 162
Pied Wagtail 56
Pied Wheatear 79
Pipit, Tawny 49
Pipit, Tree 51
Pipit, Water 52
Prinia gracilis 97
Prunella collaris 64
Prunella modularis 62
Prunella ocularis 63
Ptyonoprogne rupestris 46
Pycnonotus xanthopygos 57

Pyrrhocorax graculus 163
Pyrrhocorax pyrrhocorax 165
Pyrrhula pyrrhula 200

Radde's Accentor 63
Raven, Brown-necked 169
Raven, Common 170
Red-backed Shrike 156
Red-billed Chough 165
Red-breasted Flycatcher 131
Red-fronted Serin 185
Red-rumped Swallow 48
Red-rumped Wheatear 87
Red-tailed Wheatear 85
Redstart, Black 71
Redstart, Common 73
Reed Bunting 211
Reed Warbler 104
Regulus ignicapillus 128
Regulus regulus 127
Remiz pendulinus 153
Rhodopechys sanguinea 195
Rhodospiza obsoleta 196
Ring Ouzel 89
Riparia riparia 45
Robin, European 66
Robin, White-throated 70
Rock Bunting 205
Rock Sparrow 180
Rock Thrush 87
Rook 167
Rose-coloured Starling 173
Rosefinch, Common 199
Rosefinch, Great 200
Rufous-tailed Scrub-Robin 65
Rufous Nightingale 68
Rüppell's Warbler 114

Sand Martin 45
Sardinian Warbler 112
Savi's Warbler 99
Saxicola rubetra 74
Saxicola torquata 75
Scrub-Robin, Rufous-tailed 65
Sedge Warbler 101
Semi-collared Flycatcher 132
Serin, European 186
Serin, Red-fronted 185
Serinus pusillus 185
Serinus serinus 186
Short-toed Lark 34
Short-toed Treecreeper 152
Shrike, Lesser Grey 157
Shrike, Masked 159
Shrike, Red-backed 156
Shrike, Woodchat 158
Siskin, Eurasian 190
Sitta europaea 145
Sitta krueperi 144

239

Songbirds of Turkey

Sitta neumayer 148
Sitta tephronota 147
Skylark, Eurasian 42
Snowfinch 183
Sombre Tit 137
Song Thrush 92
Spanish Sparrow 176
Sparrow, Dead Sea 177
Sparrow, House 174
Sparrow, Pale Rock 182
Sparrow, Rock 180
Sparrow, Spanish 176
Sparrow, Tree 178
Sparrow, Yellow-throated 179
Spotted Flycatcher 129
Starling, Common 171
Starling, Rose-coloured 173
Stonechat, Common 75
Sturnus roseus 173
Sturnus vulgaris 171
Subalpine Warbler 110
Swallow, Barn 47
Swallow, Red-rumped 48
Sylvia atricapilla 121
Sylvia borin 120
Sylvia cantillans 110
Sylvia communis 118
Sylvia curruca 116
Sylvia hortensis 114
Sylvia melanocephala 112
Sylvia mystacea 111
Sylvia nisoria 115
Sylvia rueppelli 114

Tawny Pipit 49
Thrush, Blue Rock 88
Thrush, Mistle 93
Thrush, Rock 87
Thrush, Song 92
Tichodroma muraria 150
Tit, Bearded 133
Tit, Blue 141
Tit, Coal 139
Tit, Crested 139
Tit, Great 143
Tit, Long-tailed 135
Tit, Marsh 136
Tit, Penduline 153
Tit, Sombre 137
Treecreeper, Eurasian 151
Treecreeper, Short-toed 152
Tree Pipit 51
Tree Sparrow 178
Troglodytes troglodytes 60
Trumpeter Finch 198
Turdus merula 91
Turdus philomelos 92
Turdus torquata 89
Turdus viscivorus 93
Twite 192

Upcher's Warbler 107

Wagtail, Citrine 54
Wagtail, Grey 55
Wagtail, Pied 56
Wagtail, Yellow 53
Wallcreeper 150
Warbler, Barred 115
Warbler, Bonelli's 123
Warbler, Cetti's 95
Warbler, Garden 120
Warbler, Graceful 97
Warbler, Grasshopper 98
Warbler, Great Reed 105
Warbler, Greenish 122
Warbler, Icterine 109
Warbler, Marsch 103
Warbler, Ménétries's 111
Warbler, Moustached 100
Warbler, Olivaceous 106
Warbler, Olive-tree 108
Warbler, Orphean 114
Warbler, Paddyfield 102
Warbler, Reed 104
Warbler, Rüppell's 114
Warbler, Sardinian 112
Warbler, Savi's 99
Warbler, Sedge 101
Warbler, Subalpine 110
Warbler, Upcher's 107
Warbler, Wood 124
Water Pipit 52
Western Rock Nuthatch 148
Wheatear, Black-eared 81
Wheatear, Desert 83
Wheatear, Finsch's 83
Wheatear, Isabelline 77
Wheatear, Mourning 84
Wheatear, Northern 78
Wheatear, Pied 79
Wheatear, Red-rumped 87
Wheatear, Red-tailed 85
Whinchat 74
White-spectacled Bulbul 57
White-throated Dipper 58
White-throated Robin 70
Whitethroat, Common 118
Whitethroat, Lesser 116
Woodchat Shrike 158
Wood Lark 41
Wood Warbler 124
Wren, Northern 60

Yellow-throated Sparrow 179
Yellowhammer 204
Yellow Wagtail 53

Zitting Cisticola 96